EM Detection of Concealed Targets

EM Detection of Concealed Targets

DAVID J. DANIELS

WILEY

A JOHN WILEY & SONS, INC., PUBLICATION

Published by John Wiley & Sons, Inc Hoboken, New Jersey
Published simultaneously in Canada

Library of Congress Cataloging-in-Publication Data:

Daniels, D. J. (David J.)
EM detection of concealed targets / David J. Daniels.
p. cm.—(Wiley series in microwave and optical engineering)
ISBN 978-0-470-12169-6 (cloth/cd)
1. Radar. 2. Microwave detectors. 3. Imaging systems. 4. Target acquisition. I. Title.
TK6575.D25 2009
621.36′7–dc22

2009017251

10 9 8 7 6 5 4 3 2 1

Contents

Preface

The subject matter of this book is the radio-frequency (RF), millimetre-wave, microwave, and terahertz detection of concealed targets. It is intended for those interested in the development and application of such techniques and equipment. The application is of significant importance in view of the threat posed by suicide bombers and criminals in relation to weapons, explosives, and narcotics as well as human trafficking. In addition, the detection of targets buried in the ground is relevant, and the applications can range from the detection of buried land mines and improvised explosive devices to humans buried in avalanches or earthquakes to hostiles behind walls. The main aim of this book is to bring together in one volume the core information on a variety of electromagnetic techniques that are useful in detecting such concealed targets. While much of the information is available in a range of different publications, it is dispersed and therefore less accessible by virtue of the disparate nature of many of the sources. Baum [1] treated the topic with an emphasis on the identification of targets by means of the analysis of target resonances and scattering behavior with an emphasis on the lower RF frequencies. The author describes the detection of buried objects using ground-penetrating techniques in Daniels [2] and is grateful to the Institute of Engineering and Technology (IET) for permission to use material from this reference to describe both the physics of propagation in soils as well as antenna designs in this volume. Astanin and Kostylev [3] consider the subject from the view of ultra-wideband radar; however, their treatment is more applicable to operation at frequencies up to a few tens of gigahertz.

The targets of interest may be targets buried in the ground, concealed in building structures, hidden under clothing or within luggage, or may be stowaways in vehicles, as well as victims of avalanches or earthquakes. A variety of techniques may be used to examine potential targets or target areas. Well-established techniques such as ionizing radiation (X rays), vapor detection, optical techniques, and thermal imaging are readily available and in day-to-day use in airports and for maintaining security. However, they all have limitations such as range, penetration, specificity; so there is a role for RF and microwave/millimeter, and submillimeter-wave electromagnetic techniques for the detection of concealed targets can offer performance in the areas where conventional techniques struggle. Conversely, electromagnetic (EM) techniques have limitations that, for example, preclude their use for detecting objects concealed in metallic containers, submerged in saltwater, or in highly cluttered situations. The emphasis of the book is in developing an understanding of the factors that influence system performance and has been written from the perspective of the equipment designer to aid in understanding where such techniques can be applied successfully.

This book covers the following topics; radar systems and, in particular, ultra-wideband radar, ground-penetrating radar, millimeter, submillimeter, and terahertz imaging, and nonimaging systems, Doppler radar systems as well as radiometric systems, and nuclear quadrupole resonance systems. The key elements of the subject of electromagnetic techniques for the detection of concealed targets are considered, and in general terms the interrelationship between those topics in electromagnetism, system design, and signal processing that define system performance.

A further aim of this book is to provide an introduction for the newcomer to the field, as well as a useful source of further reading, information, and references for the current practitioner. By necessity, this book provides a snapshot of the field, and it is to be expected that further developments in hardware and signal-processing techniques will incrementally improve the performance and extent of applications. Note that the symbols i and j are both used for the complex operator.

The author has been involved with the design and application of the techniques associated with the detection of concealed targets and has developed equipment to meet many of the applications. Successful development and application is achieved from a sound understanding of the basic physics coupled with a realistic assessment of the technology. Satisfactory performance of the latter is achieved by close attention to the details of system design and engineering. The significant advances in device performance at microwave and millimeter wavelengths that have been achieved in the last decade has enormously benefited the development of sensors, and the designer has a wealth of choice in terms of component technology. This can only help to enable the end user to detect targets that were once invisible.

Acknowledgments

I am extremely grateful to many colleagues who provided much useful information, advice, and especially to those who assisted with reviewing the material in the preparatory stages. Professor John Smith from King's College, London, provided invaluable assistance with the chapter on nuclear quadrupole resonance, and my thanks are due to Dr. Tim Rayner from Rapiscan as well as Nigel Hunt from ERA Technology who kindly reviewed the material in that chapter. My thanks also go to Professor Steve Nightingale from ERA Technology who reviewed the material in the chapter on millimeter-wave systems and techniques and to Dr. Mike Phillipakis from ERA Technology who kindly reviewed the material on the chapter on antennas. My thanks also to Professor Hugh Griffiths of University College, London, who kindly the reviewed the manuscript and whose suggestions greatly aided the readability of the book.

I would also like to pay tribute to the scientific and engineering expertise and support of my colleagues at ERA Technology who have enabled many of the sensors and designs to be built and tested over the last decade. My special thanks to Nigel Hunt who designed and developed many of the NQR and radar subsystems, to Paul Curtis for the development of the MINEHOUND™ time-domain radar system for land-mine detection, to Dr. Jon Dittmer for his input on signal-processing techniques on GPR systems, and to Blair Graham for his contribution to trials in some very interesting parts of the world.

I am also most grateful to Lisa Reading of the IET for permission to use material previously published in my earlier book on ground-penetrating radar. This material has been revised and updated for the current book.

My thanks to Lucy Hitz at John Wiley for her patience while the manuscript was being written.

This book is dedicated to all the inventive engineers and scientists whose efforts make the concealed visible and help develop techniques and equipment

ETSI	European Telecommunications Standards Institute
EU	European Union
FAR	false alarm rate
FCC	Federal Communications Commission
FFT	fast Fourier transform
FID	free induction decay
FMCW	frequency-modulated continuous wave
FN	false negative
FP	false positive
FPGA	field-programmable gate array
GPR	Ground Penetrating Radar
GPS	Global Positioning System
GTD	geometric theory of diffraction
HMX	tetrahexamine tetranitramine
HPA	Health Protection Agency
HPBW	half-power beam width
HTS	high-temperature superconductor
ICNRP	International Commission on Non-Ionizing Radiation Protection
IED	Improvised explosive device
IEE	Institution of Electrical Engineers
IEEE	Institute of Electrical and Electronic Engineers
IET	Institute of Engineering and Technology
IF	intermediate frequency
ILDC	incremental length diffraction coefficient
INS	inertial navigation system
IR	infrared
IRA	impulse radiating antenna
ITU	International Telecommunication Union
LBRTM	Line-by-Line Radiative Transfer Model
LHCP	left-hand circular polarization
LIDAR	light detection and ranging
LNA	low-noise amplifier
LO	local oscillator

MoM	method of moments
MTI	moving target indication
NIR	non-ionizing radiation
NM	noise modulation
NMR	nuclear magnetic resonance
NQR	nuclear quadrupole resonance
NRPB	National Radiological Protection Board
NTIA	National Telecommunications and Information Administration
OSHA	Occupational Safety and Health Administration
PAPS	phase-alternated multiple-pulse sequence
PD	probability of detection
PETN	Pentaerythritol tetranitrate
PFA	probability of false alarm
PLL	phase lock loop
PPI	Plan position indicator
PRC	pseudo-random coded
PSL	pulsed spin locking
PTD	physical theory of diffraction
RAM	radar absorbing material
RCS	radar cross section
RDX	cyclotrimethylenetrinitramine
RF	radio frequency
RFI	radio-frequency interference
RHCP	right-hand circular polarization
ROC	receiver operating characteristic
RPD	Radiation Protection Division
SAR	Specific Absorption Rate
SFCW	stepped frequency continuous wave
SLSE	spin-locked spin echo
SNR	signal-to-noise ratio
TD	time domain
TEM	transverse electromagnetic
TN	true negative

TNT	trinitrotoluene
TP	true positive
TPR	total power radiometer
TR	transmit–receive
UV	ultraviolet
UWB	ultra-wideband
VCO	voltage-controlled oscillator
VSWR	voltage standing-wave ratio
WKB	Wentzel–Kramers–Brillouin

CHAPTER 1

Introduction

1.1 SCOPE AND OBJECTIVES

This book describes the principles and design of sensor systems using nonionizing radio-frequency (RF) radiation for the detection of concealed targets. Nonionizing radiation (NIR) is the term used to describe the parts of the electromagnetic spectrum covering optical radiation [ultraviolet (UV), visible, and infrared] and electromagnetic fields (EMFs) (power frequencies, microwaves, and radio frequencies). Equipment and techniques using either ionizing radiation from sources such as x-rays and γ-rays or vapor or gas detection are not covered but clearly have capability for particular applications.

This book covers the following topics: radar systems and, in particular, ultra-wideband (UWB) radar, ground-penetrating radar, Doppler radar, millimeter and submillimeter systems, as well as passive (radiometric) millimeter, submillimeter, and terahertz systems. Nuclear quadrupole resonance has been included because it is an RF technique, albeit at relatively long wavelengths, and its design has many aspects common to radar systems and it offers high specificity to explosives and narcotics. The key elements of the subject of electromagnetic techniques for the detection of concealed targets are considered as well as the interrelationship between physical principles, system design, and signal processing that define system performance.

In many cases the system diagrams are shown in terms of function, and this should not always be interpreted to mean hardware components. Often digital signal processing hardware can be used for signal processing or antenna beamforming functions. This book does not cover the signal and image processing methods that can be used to extract maximum information from the radar or radiometer system as this topic could form a future volume in its own right.

EM Detection of Concealed Targets, by David J. Daniels

1.2 STRUCTURE

This Chapter provides a background to the market for security sensors and then identifies each of the main areas of application. Several key issues such as the statistics of detection performance and radiological and licensing considerations are described. Each chapter contains a short summary of its key points. The physics of propagation is, of course, fundamental to a proper understanding of the use of electromagnetic fields for detecting concealed targets, and the characteristics of EMFs as well as propagation of energy in a dielectric and absorption in water, man-made, and natural materials are described in Chapter 2. This also includes consideration of aspects of human and animal physiology relevant to the propagation of electromagnetic (EM) waves as well as the EM characteristics of certain materials.

Most sensing techniques need antennas to transmit and receive the signals from concealed targets, and Chapter 3 provides an introduction to the types of antennas often used in UWB short-range radar systems. Nuclear quadrupole resonance (NQR), despite the "nuclear" in its title, is a radio-frequency technique that can be used to interrogate specific substances, including explosives and narcotics and is described in Chapter 4. Generic radar systems are described in Chapter 5, which provides an understanding of the various modulation techniques that can be employed and some of the issues critical to successful implementation. This chapter covers the principles of operation of Doppler radar, frequency-domain radar, and time-domain radar systems. Chapter 6 discusses passive or radiometric systems as applied to concealed target detection. It is followed in Chapter 7 by a review of the applications and the radar technology for the detection of concealed targets. This contains a review of developments in Doppler radar systems, through-wall radar systems, ultra-wideband, microwave, millimeter, submillimeter, and terahertz radar systems. The material is then summarized in Chapter 8. References cited throughout the book are found following Chapter 8.

1.3 MARKET NEEDS FOR SECURITY

Various estimates from sources such as Beckner and Shaheen, [4], Shaheen [5], Barami [6], as well as Frost and Sullivan [298], suggest that the annual global security market will be on the order of £1000 million by the end of the first decade of the twenty-first century and will grow at a rate of around 10%. The largest market will be the United States, but the demand from the rest of the world will eventually exceed that, with Europe and the wealthy nations of the Middle East and Asia being the main markets. The market can be segmented into three main groups: transportation, civil and government infrastructure, and military. Many of the threats that are now active cross these market boundaries and the difference between the threat from terrorists and insurgents are not always easy to define. In addition the segments cover two different

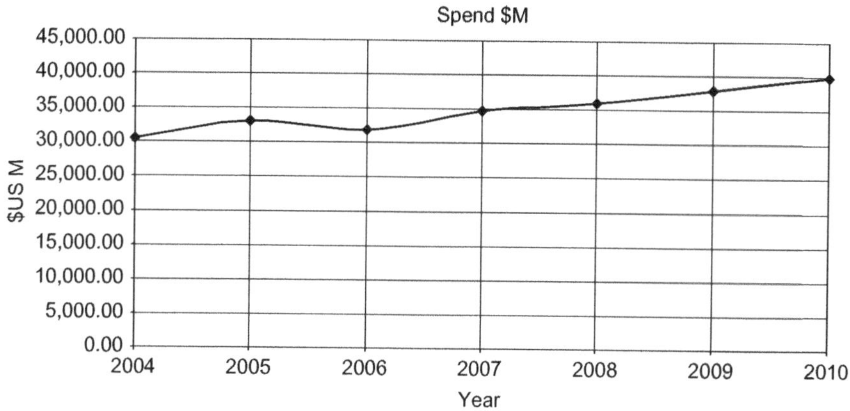

FIGURE 1.1 Estimated U.S. Department of Homeland Security: Spending forecasts 2004–2010.

customer bases: military and civilian. In the United States alone the estimated U.S. Department of Homeland Security spending forecasts for 2004–2010 is shown in Figure 1.1.

The products developed for these customer bases are, however, significantly different in that the civilian homeland security market has different requirements for the through-life costs associated with operation and support compared with the military requirements. The military has well-developed procedures for procurement, operation, and support and tends to expect significant equipment lifetimes and midlife upgrades. The civilian market views these aspects rather differently, and equipment replacement philosophy is more geared to shorter lifetimes due to product performance improvements. Manufacturers recognize that the civilian market does not have the same infrastructure as the military, and buyers expect products to be designed with this in mind. The civilian market is also more attuned to trading-off performance and cost in a way quite foreign to the military.

The market segments can be further subdivided represented as shown in Figure 1.2. Furthermore, each of these segments can be defined by the type of operation that is to be carried out, which then enables the type of equipment to be defined as shown in Figure 1.3.

The global market for homeland security, while enormous, is challenging because it is highly fragmented with decision makers having perspectives aligned to their own region, country, and segment. Competition from national suppliers is fierce, and technical edge, protection of market share, pricing, and national interest make for a keenly contested arena.

For example, in 2007, 61.3 million international passengers passed through London Heathrow (LHR), which handled 3.6% of the world's total international travelers and was at that date the largest international airport in the world. With 1702 million international passengers each year, the worldwide

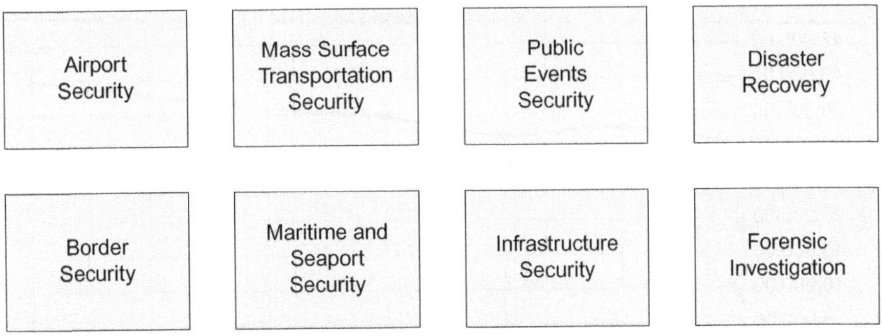

FIGURE 1.2 Market segments for security.

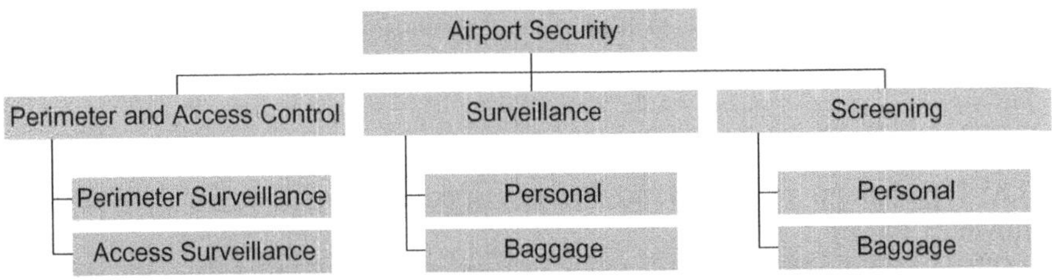

FIGURE 1.3 Types of operation for security processes.

requirement for surveillance and screening is clearly a major market. The market in one country is significant, and estimates of the UK homeland security market made by Frost and Sullivan [298], suggest this to be in excess of £5.40 billion across all eight threat domains by the year 2016 as shown in Figure 1.4.

The agencies that are responsible for security at ports of entry, security at airports, implementing customs regulations, and policing political, sporting, trade, and cultural events need to be able to detect concealed objects on suspects that are hidden in a variety of ways. These can range from:

Drugs inside body cavities
Explosives hidden on a person's clothes or apparel
Metallic and nonmetallic weapons, knives, guns, and suicide bombers
Illegal immigrants and stowaways

The operational requirements for rapid passenger throughput and market push for technology advancement as well as governments and the public need for risk reduction, combined with strong internal drivers such as government and corporate spending, indicate that the homeland security market will continue to grow over the next decade. The market can be characterized by a period of rapid technological growth as the development and deployment of a generation of technologies designed specifically for the market are introduced.

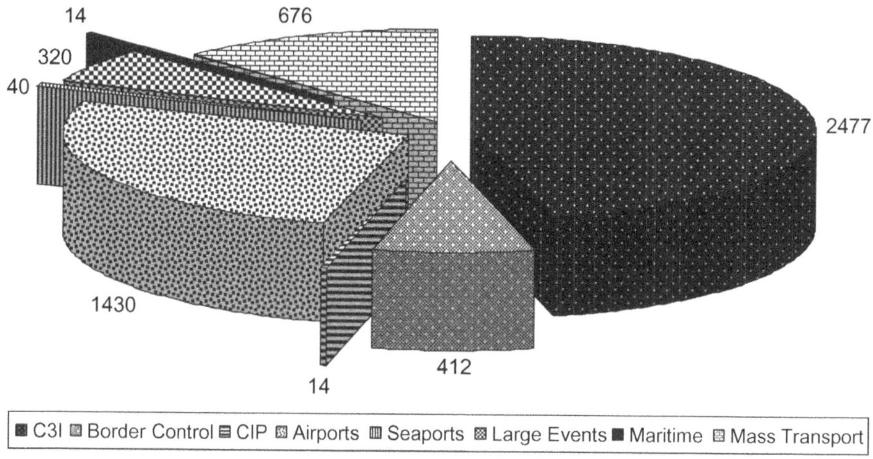

FIGURE 1.4 UK security market in £1,000,000 for, 2007–2016.

Researchers and developers are working swiftly to devise novel solutions and bring to market new generations of security technology. This environment should provide knowledgeable investors with good rewards provided there is a sound understanding of what can be achieved with physics and technology.

1.4 TARGETS INSIDE CONTAINERS

Explosives and weapons can be contained in hold baggage, suitcases, or any means of transportation. X-ray equipment is very successful in detecting anomalous material such as guns and concealed weapons. The main advantage of X-ray equipment is its capability to penetrate through conducting or metallic casing, which is not possible with EM techniques. However, X-ray techniques are not specific, and techniques such as NQR, which are highly specific, may be used to uniquely identify explosives and narcotics, provided the casing of the container is nonmetallic or has a sufficient amount of gapping to enable propagation of EM waves into the interior of the container. Provided the container casing is completely nonmetallic, then EM techniques can be used to image the interior of the case, and some reports suggest that terahertz systems can provide spectroscopic information on the suspect material.

1.5 BURIED LAND MINES

Ground-penetrating radar (GPR) is one of a number of technologies that has been extensively researched as a means of improving mine detection efficiency for both civil and military programs. The military programs are largely based on the requirement to maintain the pace of military operations and have different requirements in terms of speed and detection performance to civil or humanitarian programs. Land mines can be either buried or surface laid. They are emplaced by a variety of techniques, including being scattered on the surface by vehicles or helicopters. Thus, mines may be found in regular patterns or in irregular distributions. When environmental conditions caused by natural processes result in soil erosion and movement caused by rain over several seasons, the mines may be lifted and moved to new locations and can be covered or exposed. Mines are encountered in desert regions, mountains, jungles, as well as urban areas.

Land-mine detection equipment has to be designed to work in a wide range of physical environments, and the statement of operational requirements issued by end users will reflect this need. Detection equipment must be able to be operated in climatic conditions that range from arid desert, hillside scree, to overgrown jungle. Ambient operating temperatures can range from below −20 to 60°C. Rain, dust, humidity, and solar insolation, all must be considered in the design and operation of equipment. The transport conditions of equipment can be arduous, and these as well as human–machine interface issues are vitally important to the design of detectors.

In general, most pressure-sensitive mines are not designed to operate when buried deeply. The overburden ground material acts as a mechanical bridge and inhibits triggering of the detonator mechanism and also reduces the force of the explosion. This fact is often taken into account in the specification of performance for a mine detector, which typically has a depth range of 300 mm, although some land mines can be encountered at depths well beyond the range of current detection systems. Land-mine detection systems can be employed in several different roles: for close-in handheld detection, for vehicle-mounted standoff detection, or as a remote sensor mounted on low-flying fixed or rotary wing aircraft.

1.6 FORENSIC DETECTION OF BURIED BODIES

Forensic investigation using GPR has become a recognized method of forensic archeology because of some high-profile case histories. GPR techniques are one among a number of methods that can be used in forensic investigations and can greatly assist police investigations by pinpointing suspicious areas and thus saving unnecessary excavation. GPR provides the means to conduct rapid, nondestructive investigations that can alleviate the need to carry out extensive and expensive excavations. In skilled hands, GPR can be employed to search for and locate the types of target sought by forensic investigators. Unlike searching for utilities and services, where the radar signatures tend to be well defined, radar signatures of forensic targets are varied and, due to the wide variety of targets and scenarios, never the same. However, with experience an operator will begin to be accustomed to general patterns within the radar images. It must be emphasized that GPR cannot provide an image of the buried remains. Because of the variability of the sites, those without significant experience of GPR forensic surveying do not find it easy to carry out the interpretation of GPR B-scan images. For good GPR survey results it is also vital that the ground be undisturbed by either the passage of heavy vehicles (which tend to compress the soil and give rise to unwanted reflections) or by digging. Once digging has been carried out, the soil structure is disturbed and effective GPR surveying becomes extremely difficult. The golden rule is that the site should be in a virgin state for best GPR results.

Forensic investigations should be multidisciplinary, and the following aspects should be considered when dealing with GPR:

Availability of site plans of utilities
Availability of site information on burial sites of pet animals
Correct recording of scene and site evidence
Methods of removing surface vegetation and obstructing artefacts (note that GPR requires a reasonably flat surface to gather good data)
Time, manner, and cause of death (state of decay of corpse)
Identification of state of remains to aid GPR interpretation
Recovery procedures of remains

1.7 AVALANCHE AND EARTHQUAKE VICTIMS

Radar has been used for research into the detection of victims of earthquakes and avalanches. Tests in the French Pyrenees by Daniels [2] and in the Italian Alps by ERA Technology showed that GPR [299] has a potentially useful role in this application. Detection of humans buried in earthquake rubble has also been the subject of much research, and Doppler radar systems are an obvious choice to detect humans by means of their respiration and heartbeat, while simultaneously eliminating the clutter that ranging radar would also detect.

1.8 CONCEALED HUMANS

A prime concern of immigration authorities in most developed countries is the detection of stowaways in road vehicles. Such stowaways may be opportunistic or facilitated and hide in the cargo zones of heavy-goods vehicles. It is difficult to establish the number of illegal entrants to any country, but it is clearly a problem for the developed countries of the world and a variety of estimates have been made in the United Kingdom, Europe, the United States, and Canada as to the scale of the problem. The legal and ethical issues are not the subjects of this volume, but the RF and microwave techniques for detection will be considered. Stowaways are found concealed in the cargo zones of heavy-goods vehicles. These cargo zones may be encased by fabric or metal and may carry a variety of goods from electrical white goods through to cargos of fruits and vegetables. Stowaways are found in a variety of concealed positions.

1.9 CONCEALED TARGETS ON HUMANS

There is clearly a major need to detect concealed targets on humans, and this can range from the less dramatic knife and gun carriers to the hardened terrorist. The targets can range from drugs inside body cavities, through metallic and nonmetallic weapons, knives, guns, as well as explosives hidden under a suspect's clothes or in luggage. Note that radar has not been used to investigate searches of targets inside human beings. Following the tragic events of 911 and the attempt by a shoe bomber to compromise safety, air travel is now extensively protected both in terms of baggage screening and person screening and by techniques ranging from X-ray scanning through to passive millimeter-wave imaging. However, the airport is a controlled environment, and, while the measures have undoubtedly made life for the traveler more difficult, it is now a harder target for the terrorist to attack. Other means of mass transportation as well as major public events are softer targets, so concealed weapons and explosives on humans are still a continued threat to national security. This is evidenced by recent events in Europe.

On March 11, 2004, in Madrid four packed commuter trains heading into Madrid from working-class neighborhoods were bombed using 10 backpacks filled with dynamite and nails. This killed 191 and wounded more than 1800 people. In the United Kingdom a series of bomb attacks on London's transport network on Thursday July 7, 2005, caused significant casualties. Three explosions on the Underground system left 35 dead, while 2 victims died in a blast on a double-decker bus, a further 700 plus people were injured. These soft targets need better protection, and sensors to detect concealed targets are an essential means to provide this.

1.10 RADIOLOGICAL CONSIDERATIONS

Public exposure to nonionizing electromagnetic fields such as those associated with radar, broadcast transmitters, mobile phones, power lines, and domestic equipment come under guidelines incorporated into various national and international standards.

In the United Kingdom the Health Protection Agency's Radiation Protection Division (RPD) [7] [formerly the National Radiological Protection Board (NRPB)] advises on risks from radiation including electromagnetic fields. The UK Health Protection Agency (HPA) is an independent body that protects the health and well-being of the UK population and has published a series of reports [8–16]. There is a continuing program of research in the United Kingdom relevant to health concerns about exposures to RF fields. The RPD has responsibility for providing advice on appropriate restrictions on the exposure of people to electromagnetic fields and radiation. These include static, power frequency (50 Hz), and other extremely low-frequency (ELF) electric and magnetic fields, and RF fields and radiation. Guidance limiting exposure to time-varying electric and magnetic fields is issued at regular intervals. NRB recommendations are based on an assessment of the possible effects on human health derived from biological information [17–19], from dosimetric data [20, 21], and from studies of exposed human populations [22, 23]. They apply equally to workers and to members of the public but not to people who are exposed to electromagnetic fields and radiation for medical diagnostic or therapeutic purposes.

The RPD advise that for electromagnetic radiation of frequencies greater than 100 kHz and less than 300 GHz, both induced current and thermal effects are relevant, but above 100 MHz only thermal effects need to be considered. The specific absorption rates (SAR) are measured in units of watts per kilogram (W kg^{-1}) and are given in Table 1.1.

Guidance for the protection of patients and volunteers during clinical magnetic resonance diagnostic procedures has been issued separately by the American National Standards Institute (ANSI) [24]. These recommendations are intended to provide a framework for a system of restrictions on human exposure to these fields and radiation as described by Elder and Cahill [25].

In Europe the European Recommendation (EC/519/1999) sets a framework that deals with limiting public exposure, providing public information, and undertaking research. The European Commission (EC) supports an ongoing program of further research on the topic.

The U.S. Department of Labor, Occupational Safety and Health Administration (OSHA) General Industry [Part 1910.97(a) (2) (i)] [26] defines the following standard for normal environmental conditions and for incident electromagnetic energy of frequencies from 10 MHz to 100 GHz: The radiation protection guide is 10 mW cm^{-2} (milliwatt per square centimeter) as averaged

TABLE 1.1 RPD Permitted Absorption Levels in United Kingdom

Frequency Range	Quantity	Basic Restriction
100 kHz to 10 MHz	Current density in the head and trunk	$F/100$ mA m^{-2}
100 kHz to 10 GHz	SAR averaged over the body and over any 15-min period	0.4 W kg^{-1}
100 kHz to 10 GHz	SAR averaged over any 10 g in the head and fetus and over any 6-min period	10 W kg^{-1}
100 kHz to 10 GHz	SAR averaged over any 100 g in the neck and trunk and over any 6-min period	10 W kg^{-1}
100 kHz to 10 GHz	SAR averaged over any 100 g in the limbs and over any 6-min period	20 W kg^{-1}

TABLE 1.2 Recommended Levels of Electromagnetic Fields: A Comparison of NRPB and ICNIRP Guidelines

1993 NRPB Guidelines (the same for occupational and public)			1998 ICNIRP Guidelines	
			Occupational	Public
Basic restriction (the quantity which must not be exceeded)	Induced current density in the central nervous system	10 mA m^{-2}	10 mA m^{-2}	2 mA m^{-2}
Reference level	Magnetic field	1600 μT	500 μT	100 μT
	Electric field	12 kV m^{-1}	10 kV m^{-1}	2 kV m^{-1}

over any possible 0.1 h period. This is expanded as follows: The acceptable absorption is equal to a power density of 10 mW cm^{-2} for periods of 0.1 h or more with an energy density of 1 mW-h cm^{-1} (milliwatt hour per square centimeter) during any 0.1 h period.

The available evidence indicates that human beings can tolerate moderate absorption rates of approximately 1 W kg^{-1}. Some standards also provide data on maximum allowable partial body exposures and criteria for avoiding RF shocks and burns. It should be noted that SAR criteria do not apply to exposures at low frequencies (less than 100 kHz) for which nerve stimulation (shock) occurs, or at frequencies higher than 6 GHz for which surface heating prevails.

The Institute of Electrical and Electronic Engineers (IEEE) has published a number of studies dealing with the issue of biological effects, and further details can be gathered from the references provided by the IEEE/ANSI [27]. The International Commission on Non-Ionizing Radiation Protection (ICNIRP) [28] is a body of independent scientific experts consisting of a main commission of 14 members, 4 scientific standing committees covering epidemiology, biology, dosimetry, and optical radiation, and a number of consulting experts. This expertise is brought to bear on addressing the important issues of possible adverse effects on human health of exposure to nonionizing radiation and their recommendations are given in Table 1.2.

1.11 LICENSING CONSIDERATIONS

Although compared with the number of mobile phones, Global Positioning System (GPS) users, and other occupants of similar bands, the number of radar/detection systems is very small with a very low potential for interference, and equipment must comply with regulatory requirements. A brief summary of the situation is provided in this section.

The International Telecommunication Union (ITU), based in Geneva, Switzerland, is the international organization within which governments coordinate global telecom networks and services. The United States is a member of the ITU. Therefore, the Federal Communications Commission (FCC) in the United States, the European Telecommunications Standards Institute (ETSI), and the national authorities in Region 3 define the key licensing requirements.

In the United States the FCC and the National Telecommunications and Information Administration (NTIA) jointly regulate radio spectrum use in the United States. Within the FCC, the Office of Engineering and Technology (OET) provides advice on technical and policy issues pertaining to spectrum allocation and use. OET also maintains the FCC's Table of Frequency Allocations. Further information can be found in EURO GPR [29] and Olhoeft [30].

In Europe the following requirements are relevant:

ETSI EN 302 066-1 Electromagnetic Compatibility and Radio Spectrum Matters (ERM); Short Range Devices (SRD); Ground- and Wall-Probing Radar Applications; Part 1: Technical Characteristics and Test Methods

ETSI EN 302 066-2 Electromagnetic Compatibility and Radio Spectrum Matters (ERM); Short Range Devices (SRD); Ground- and Wall-Probing Radar Applications; Part 2: Harmonized EN under Article 3.2 of the R&TTE Directive

ETSI EN 301 489-32 Electromagnetic Compatibility and Radio Spectrum Matters (ERM); Electromagnetic Compatibility (EMC) Standard for Radio Equipment and Services; Part 32: Specific conditions for Ground- and Wall-Probing Radar Applications

In addition to radio transmission regulations, all equipment in Europe must be CE marked to demonstrate that it satisfies the relevant directives of the European Union. (EU). The CE mark may only be applied when the requirements of all other relevant EU Directives, such as safety, have also been demonstrated. Europe has defined electromagnetic environments and agreed to European Norm (EN) specifications with test levels defined. These specifications have product, product family, and generic requirements with the application in that order of precedence. For commercial electromagnetic compatibility (EMC), there are emissions and immunity requirements. The emission requirements have been defined through the CISPR international committees and the immunity through the IEC committees.

In the United States the FCC document (*Federal Register* Vol. 67, No. 95/ Thursday, May 16, 2002/Rules and Regulations) defines the licensing requirements for various categories of equipment as listed below. Imaging systems under Part 15 of the FCC's rules are subject to certain frequency and power limitations.

GPR, which must be operated below 960 MHz or in the frequency band 3.1 to 10.6 GHz

Wall imaging systems, which must be operated below 960 MHz or in the frequency band 3.1 to 10.6 GHz

Through-wall imaging systems, which must be operated below 960 MHz or in the frequency band 1.99 to 10.6 GHz

Surveillance systems, which must be operated in the band 1.99 to 10.6 GHz

Medical systems, which must be operated in the band 3.1 to 10.6 GHz

Vehicular radar systems, which must be operated in the band 22 to 29 GHz

In response to various petitions, the FCC amended the rules to facilitate the operation of through-wall imaging systems by law enforcement, emergency, rescue, and firefighter personnel in emergency situations; eliminated the requirement that GPRs and wall-imaging systems operate with their −10dB bandwidths below 960 MHz or above 3.1 GHz; specified the limitations on who may operate GPR systems and wall-imaging systems and for what purposes; eliminated the requirement for nonhandheld GPRs to employ a dead man switch which requires the user to physically operate a switch for the equipment to transmit; clarified the coordination requirements for imaging devices; and clarified the rules regarding emissions produced by digital circuitry used by UWB transmitters.

The FCC also proposed additional new rules to address issues raised regarding the operation of low pulse repetition frequency (PRF) UWB systems, including vehicular radars, in the 3.1- to 10.6–GHz band; the operation of frequency hopping vehicular radars in the 22- to 29-GHz band as UWB devices; the establishment of new peak power limits for wideband Part 15 devices that do not operate as UWB devices; and the definition of a UWB device.

The designer, developer, or user of equipment must ensure compliance with existing national regulations. The changing nature of much of the ITU, European (ETSI), and U.S. standards is such that up-to-date information should be obtained. Other countries will have specific requirements, and proper advice should be sought from the appropriate national authorities.

1.12 STATISTICS OF THE DETECTION PERFORMANCE OF A SENSOR

A key issue in assessing the performance of any detection system is its probability of detection (PD), the probability of false alarm (PFA), as well as the confidence that can be placed in the claimed PD and PFA. This section describes the basic measures that are commonly used to describe the performance of a particular sensor and explains the most effective and commonly used description of sensor performance, which is the receiver operating characteristic (ROC).

A basic approach to describing a methodology for classifying the performance of a sensor is described by the confusion matrix reported by Provost and Kohavi [31], who use the descriptions as given in Table 1.3.

TABLE 1.3 Confusion Matrix

		Predicted	
		Negative	Positive
Actual	Negative	a	b
	Positive	c	d

a = number of correct predictions that an instance is negative
b = number of incorrect predictions that an instance is positive
c = number of incorrect of predictions that an instance negative
d = number of correct predictions that an instance is positive

Equation (1.1) gives the accuracy (AC), which is the proportion of the total number of predictions that were correct:

$$\mathrm{AC} = \frac{a+d}{a+b+c+d} \tag{1.1}$$

Equation (1.2) gives the recall or true positive (TP) rate, which is the proportion of positive cases that were correctly identified:

$$\mathrm{TP} = \frac{d}{c+d} \tag{1.2}$$

Equation (1.3) gives the false positive (FP) rate, which is the proportion of negative cases that were incorrectly classified as positive:

$$\mathrm{FP} = \frac{b}{a+b} \tag{1.3}$$

Equation (1.4) gives the true negative (TN) rate, which is defined as the proportion of negative cases that were classified correctly:

$$\mathrm{TN} = \frac{a}{a+b} \tag{1.4}$$

Equation (1.5) gives the false negative (FN) rate, which is the proportion of positive cases that were incorrectly classified as negative:

$$\mathrm{FN} = \frac{c}{c+d} \tag{1.5}$$

Equation (1.6) gives the precision (P), which is the proportion of the predicted positive cases that were correct, as calculated using the equation:

$$P = \frac{d}{b+d} \tag{1.6}$$

The ROC curve is another way of understanding the performance of a sensor and plots the true positive rate as a function of the false positive rate for different levels of sensitivity of the sensor. Consider two populations, one due to true reports or detections and one due to false alarms, which are shown in Figure 1.5 and labeled true and false, respectively.

Their Gaussian population distributions have identical standard deviations but different mean values. If a detection threshold value were set to 5, then the majority of the true positive reports would be detected, and a small proportion of true negative (false positive) reports would be included. This threshold can also be plotted as a pair of true positive/true negative reports or sometimes termed sensitivity/specificity parameters, and this generates an ROC curve. The ROC curve for Figure 1.5 is shown in Figure 1.6, and it can be seen that at a 0.95 true positive detection the false positive proportion is 0.35.

A test with perfect discrimination (no overlap in the two distributions) has an ROC plot that passes through the upper left corner (100% sensitivity, 100% specificity). Therefore, the closer the ROC plot is to the upper left corner,

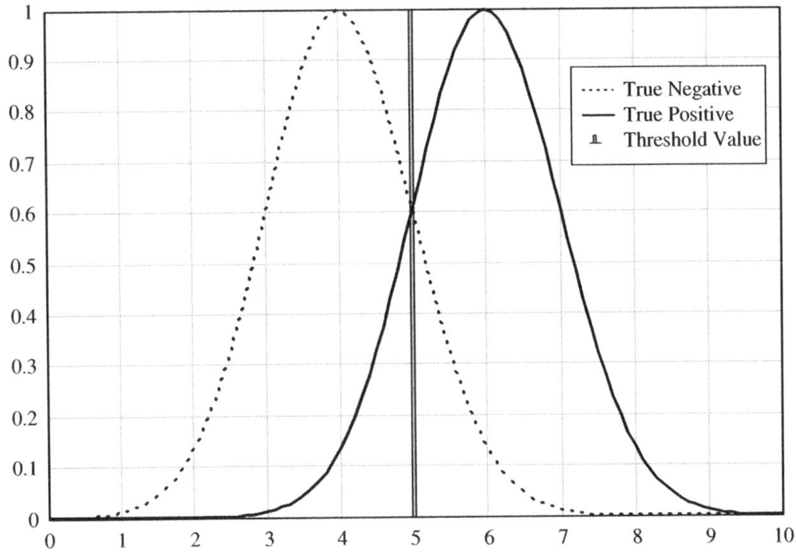

FIGURE 1.5 Distribution of true and false reports from a sample population.

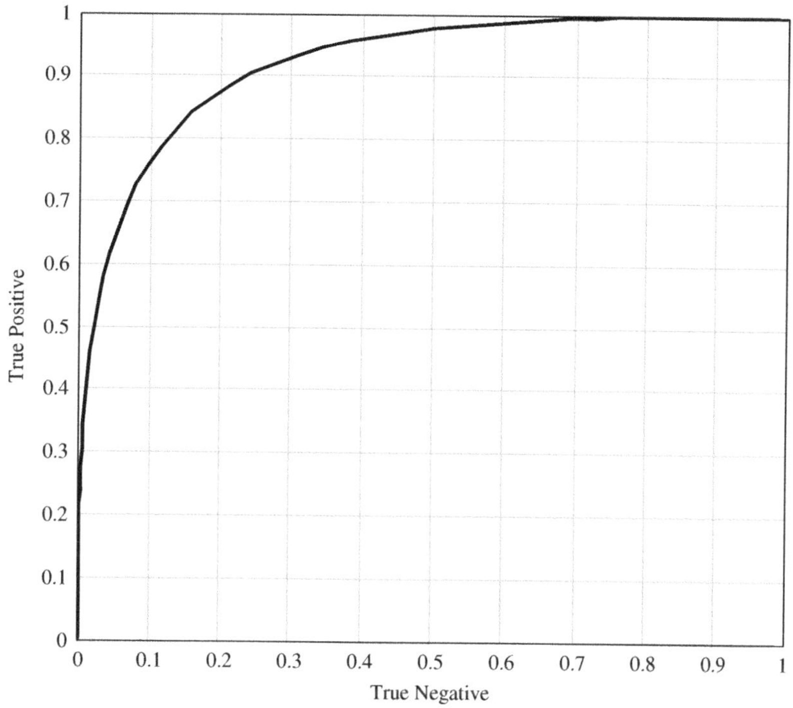

FIGURE 1.6 Receiver operating characteristic for distributions in Figure 1.5.

the higher the overall accuracy of the test. Examples of these are shown in Figure 1.7 for increasing separation of the mean values and hence less overlap, and Figure 1.8 shows decreasing separation of the mean values and hence more overlap.

The size of the sample must also be known in order to determine a confidence level in the result. Elementary statistical sampling theory can be used to show that the confidence that can be placed in a test of a limited sample set is fundamentally related to the size of the sample set. If 10 sensors are tested on a task and even if all provide a positive report (a probability of detection of 100%), the statistical confidence in the claim is limited by the number in the set. At the 95% limit, the upper and lower confidence bounds can be derived from the binomial distribution to show that with a sample set of 10, the bounds as shown in Figure 1.9 exist. The x axis shows the proportion of the sample set detected and the y axis shows the probability of detection.

In contrast, the limits for a sample set of 100 are much closer and are shown in Figure 1.10. These values are based on the small sample interval for calculating confidence intervals.

A fully detailed consideration of the issues is given by Simonson [32], who considers the difference between the small sample approach based on the

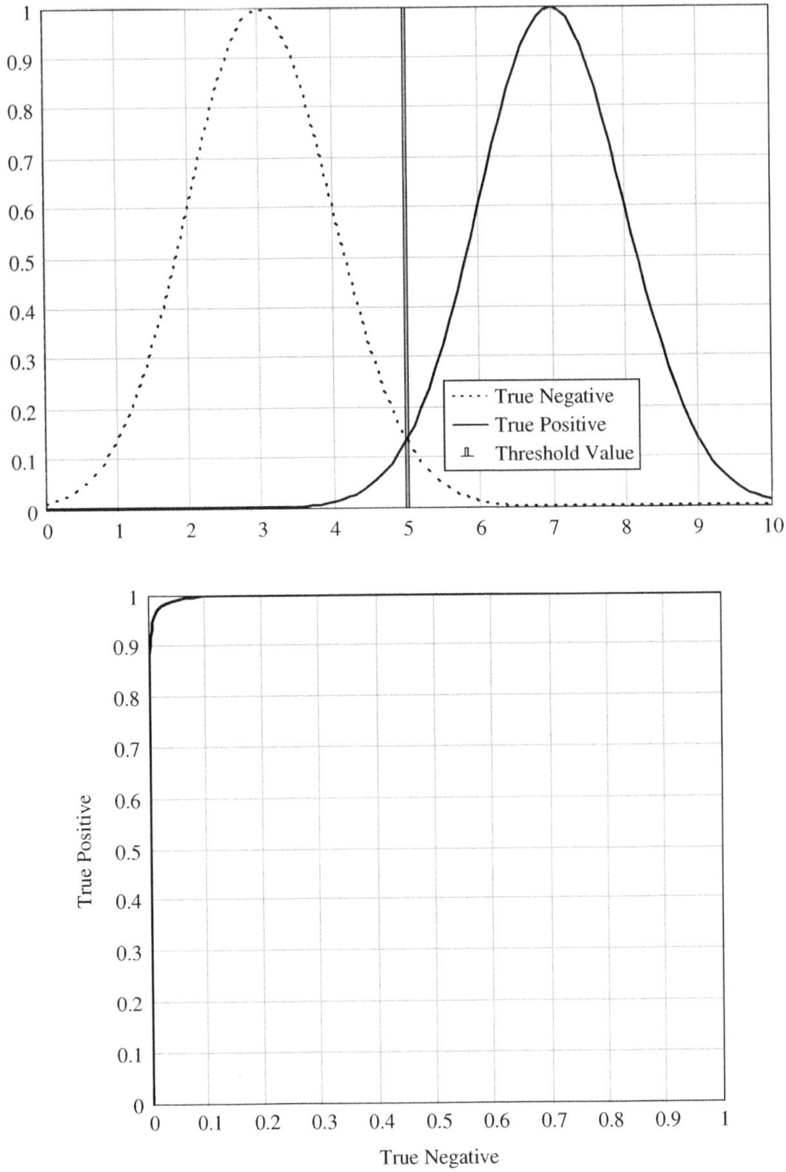

FIGURE 1.7 Effect of increased separation of information statistics on the ROC curve.

binomial distribution and a large sample approach using the normal distribution with respect to the detection of land mines. Note that the large sample distribution suggests a closer interval between the bounds. The detection of land mines is a useful example as a measure of the performance of a sensor as the consequences of a missed detection are significant for the operator.

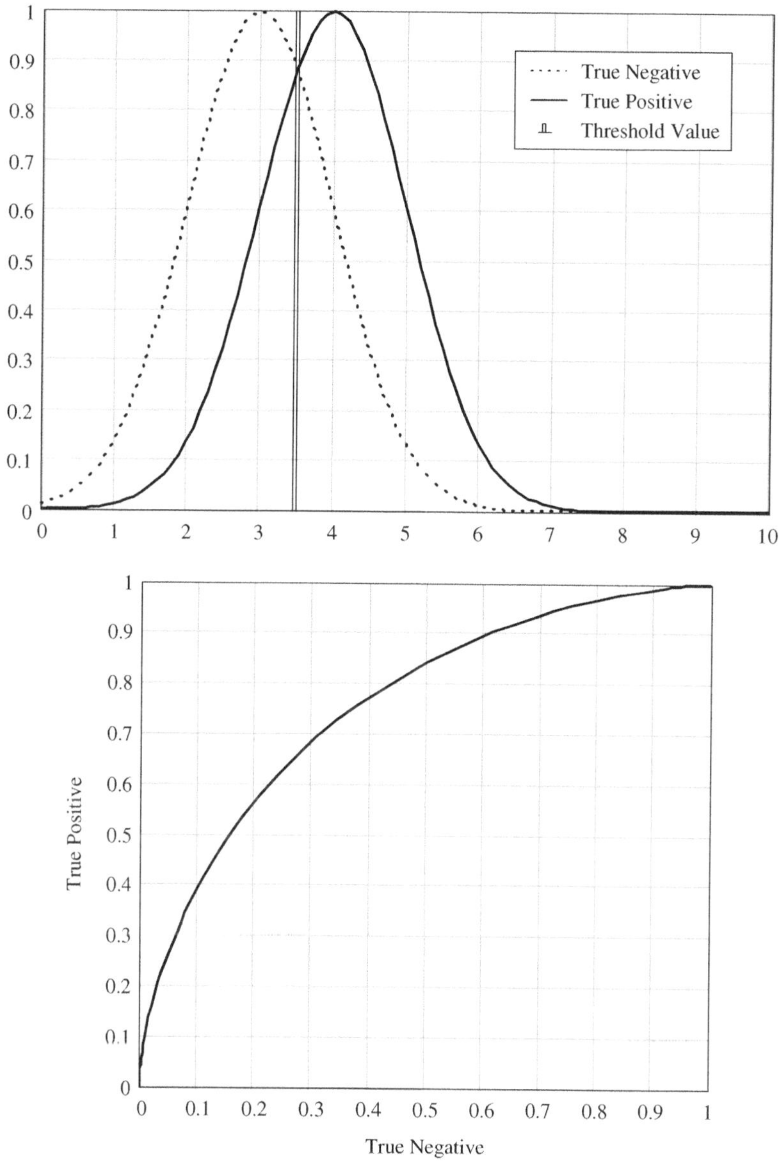

FIGURE 1.8 Effect of decreased separation of information statistics on the ROC curve.

Voles [33] considered this issue and showed that based on a Poisson distribution, even if no mines were missed by a sensor in a test of 100, then at the 95% confidence limit the highest value of probability of detection that can be claimed is 97%. Voles also showed that to achieve a 99.6% probability of detection at a confidence level of 95% would require a test of 750 mines and none should be missed.

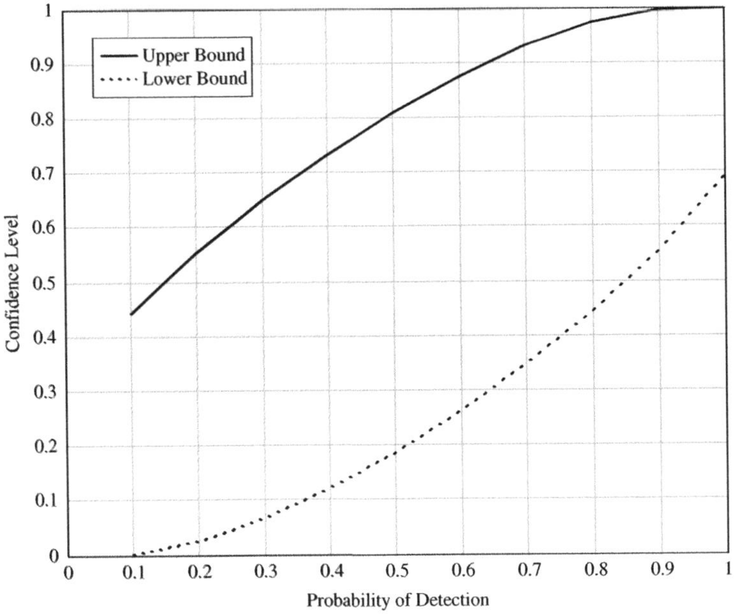

FIGURE 1.9 Upper and lower confidence bounds for a sample set of 10 (binomial distribution).

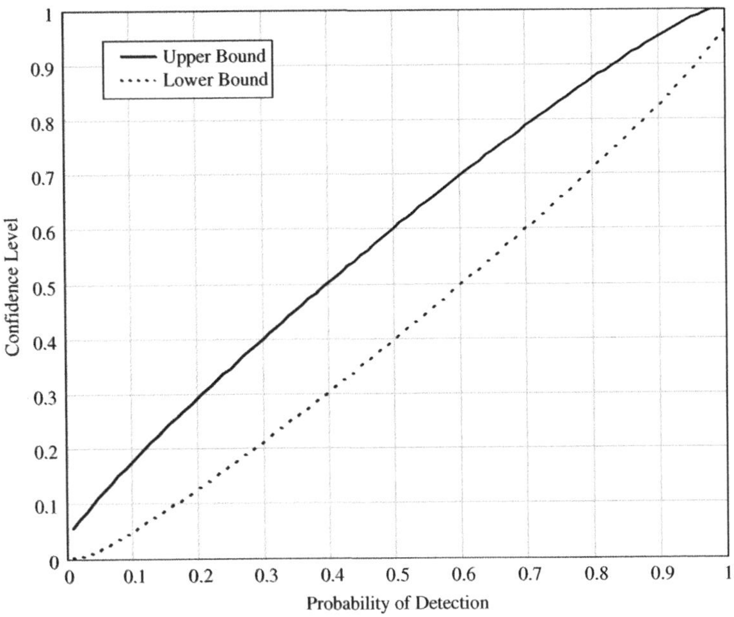

FIGURE 1.10 Upper and lower confidence bounds for a sample set of 100 (binomial distribution).

1.13 SUMMARY

The market for electromagnetic sensors for the detection of concealed targets has attracted considerable interest by potential end users from the military, security, and industrial bases. Considerable investment has been made in developing products for niche markets, and there is a growing interest in the various application areas. The threat from terrorist activities has galvanized research and development in this area, but the challenge of the basic physics in what is often an uncontrolled environment where high levels of clutter still remain. The most successful developments operate in a more controlled situation, and this may be one of the precursors to success. The opportunities are considerable, but issues of licensing, public acceptance of electromagnetic radiation, albeit at very low levels, as well as the reliability of detection and reduction of false alarms may be just as much a challenge to overcome as those related to the basic physics.

CHAPTER 2

Physics of Propagation

2.1 INTRODUCTION

This chapter provides an introduction to the key parameters and physics of propagation of electromagnetic waves through free space or a dielectric to the target and back to the receive antenna. The treatment is aimed to develop a basic insight and understanding and should enable the reader to estimate an order of magnitude assessment of propagation losses. As there is a wealth of literature available using wave equation methods, it has not been considered necessary to repeat such approaches. In any case, the variability of the situations encountered is such that accurate description and hence modeling is extremely difficult and a more general approach is often more cost effective. Total system loop path losses of 100 dB or more with variations of many tens of decibels are often encountered, and great accuracy in estimation of an individual component losses may not help to achieve the best perspective. The author is grateful for the kind permission of the IET to use copyright material from Daniels [2] relating to the propagation in dielectrics for this chapter.

The chapter introduces the topic with a basic description of propagation in free space using a small electric dipole and considers near and far fields. Polarization and basic scattering from canonical targets is described and is followed by a section on radar cross section. The reflection and refraction of electromagnetic waves is considered, as is a generic description of clutter. As many targets may be dielectric or concealed in dielectric, the propagation of energy in a dielectric is also considered. The chapter then reviews propagation losses in free space as well as the characteristics of various materials such as clothing, building materials, and the human body. A short section on the characteristics of explosives is included.

2.2 PROPAGATION OF ELECTROMAGNETIC FIELDS IN FREE SPACE

Many electromagnetic techniques for the detection of concealed targets use systems that operate in what is termed the near field. It is therefore useful to review the effect of this situation. When the target is some distance from the radar and is in the far field of the radar antenna, the spatial distribution of radiated energy if fully described by radiated field models, whereas radars operated in near field or what can be termed proximal mode are more properly described by the quasi-stationary and induction fields.

A small electric dipole radiates electric and magnetic fields as described in Eqs. (2.1) to (2.3):

$$E_r = \frac{2Z_0 Idl\pi\cos\theta}{\lambda^2}\left[\left(\frac{\lambda}{2\pi}\right)^3\cos\psi + \left(\frac{\lambda}{2\pi}\right)^2\sin\psi\right] \tag{2.1}$$

$$E_\theta = \frac{Z_0 Idl\pi\sin\theta}{\lambda^2}\left[-\left(\frac{\lambda}{2\pi}\right)^3\cos\psi - \left(\frac{\lambda}{2\pi}\right)^2\sin\psi + \left(\frac{\lambda}{2\pi}\right)\cos\psi\right] \tag{2.2}$$

$$H_\phi = \frac{Idl\pi\sin\theta}{\lambda^2}\left[\left(\frac{\lambda}{2\pi}\right)^2\sin\psi + \left(\frac{\lambda}{2\pi}\right)\cos\psi\right] \tag{2.3}$$

where dl = length of the current element

$\psi = 2\pi\phi/\lambda - \omega t$

ω = radian frequency of the signal

Z_0 = free space impedance (= 377 Ω)

I = current in the element

θ = zenith angle to radial distance r

λ = wavelength of the signal

r = distance from the element to the point of observation

Equations (2.1) to (2.3) can be considered to fall into three basic components:

1. The far field is described by a term proportional to r^{-1}, which is called the radiation term. This term represents the flow of energy away from the conducting element of the antenna.
2. The near field contains a term proportional to r^{-2}, which is called the induction term and represents energy stored in the field during one quarter of a cycle and then returned to the antenna in the next.

3. The near field contains a term proportional to r^{-3}, which is called the quasi-stationary term, or the electrostatic field term, and results from the accumulation of charge at the ends of the element.

At the distance $r = \lambda/2\pi$ all of these terms are equal, and this distance represents the boundary between the near fields and far fields where the contributions from the radiation, induction, and the quasi-stationary terms are all of the same magnitude. The regions may also be described as the reactive near field, the radiating near field, and the far field.

A graph of the various contributions is shown in Figure 2.1, the vertical axis relative field strength in units of $Z_0 Idl\pi \sin(\theta)/\lambda^2$ and the horizontal axis distance in multiples of $\lambda/2\pi$. The near field lies between 0.1 and 1 and the far field at distances greater than 1. Note that the total field = radiated field term + induction field term + quasi-stationary field term.

2.2.1 Reactive Fields

In the reactive near field, energy is stored in the electric and magnetic fields, very close to the source but not radiated from them. Instead, energy is exchanged between the signal source and the fields. If a target is capable of

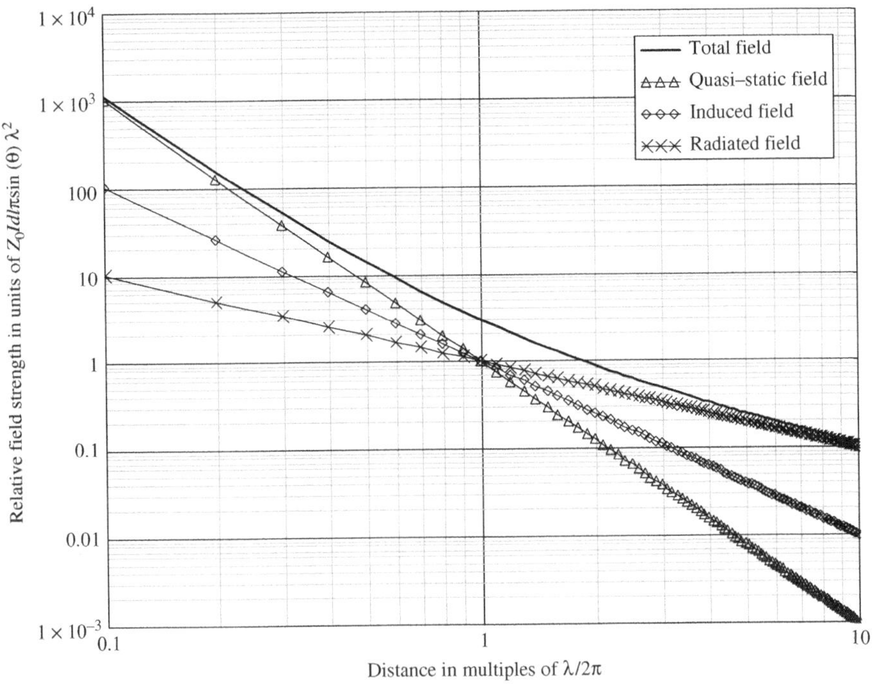

FIGURE 2.1 Signal level versus distance for the components of a radiated dipole field.

TABLE 2.1 Approximate Field Boundaries

Electrically Small-Antenna Reactive Near Field	Electrically Large-Antenna Reactive Near Field	Electrically Large-Antenna Radiating Near Field
$r < \lambda/2\pi$	$r < 0.62\,(D^3/l)^{1/2}$	$r < 2D^2/\lambda$

coupling energy from the source fields, a signal will be coupled or received in the antenna. The approximate field boundaries are given in Table 2.1.

Any consideration of the signal detected in a radar receiver should therefore fully account for the physical proximity of the antenna and target. In the case of an antenna on the surface of a dielectric and radiating frequencies centered at 0.5 GHz ($\lambda_m = 30$ cm for a dielectric whose $\varepsilon_r = 4$), it can be seen that targets closer than 5 cm will have increased field contributions, whereas radar radiating shorter wavelengths will lose out on the increased field contributions. Yarovoy et al. [34] and Lenler-Eriksen [35] have studied techniques of accurate near-field measurement antennas.

2.2.2 Near Fields

The near field, also called the reactive near-field region, is the region closest to the transmitting antenna and for which the reactive field dominates over the radiative fields.

2.2.3 Far Fields

The Fresnel zone, also called the radiating near field, is that region between the reactive near-field and the far-field regions and is the region in which the radiation fields dominate and where the angular field distribution depends on distance from the transmitting antenna. The far field or Fraunhofer zone is the region where the radiation pattern is independent of distance from the transmitting antenna. In the far field, the power received per unit area from an isotropic antenna is calculated from the following equation:

$$P_r = \frac{P_t}{4\pi R^2} \tag{2.4}$$

Equation (2.4) is also referred to as the inverse square law, since doubling the range gives a fourfold reduction in signal power. The region where the near field becomes the far field is a gradual transition as shown in Figure 2.2; however, for practical applications, radio and optical engineers have defined the maximum radius to be r_{min}. This is also sometimes known as the Rayleigh distance.

A simplified representation of the near-far field boundary can be shown using a ray diagram of an antenna pattern, and it can be seen that the antenna aperture always defines the extent of the field.

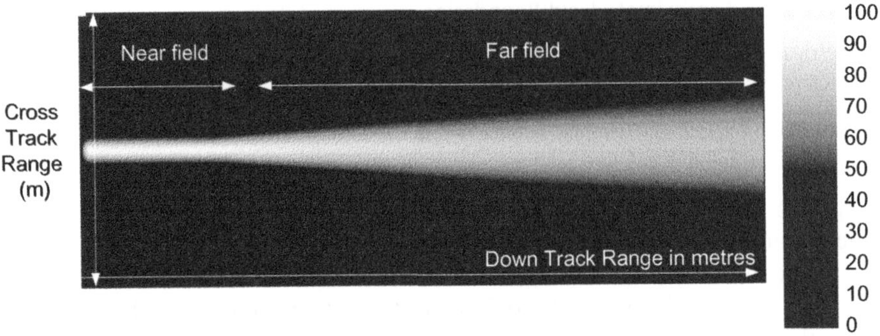

FIGURE 2.2 Representation of near–far field boundary showing divergence in the far field.

2.2.4 Polarization

The polarization of a radiated wave can be defined as the orientation of the transmitted (or received) electric field (*E* field). A complete description of the radar scattering cross section of a target includes a description of its polarization scattering characteristics (not the same as molecular polarization). The polarizing properties of targets are described by the Stokes parameters, and the polarization coordinates can be represented on the Poincaré sphere. All of these are well described in standard texts on optics and electromagnetic theory. The optimum polarization for a radar system depends on the polarization scattering properties of the target. Fundamentally all polarizations can be derived from circularly polarized radiation, with suitable selection of the vector amplitudes and phase. Circular polarization is defined as right-hand circular polarization (RHCP) or left-hand circular polarization (LHCP) in which the direction of rotation of the polarization vector will be handed as shown in Figure 2.3. Elliptical polarization can be used deliberately, but is often the result of imperfections in the transmit-and-receive antennas being used to radiate nominally circular polarization. The quality of the polarized wave is often defined by the cross-polar ratio in the case of linear polarization or the axial ratio of the antenna radiation pattern at a defined angle from boresight in the case of circular polarization. Practically, most antennas provide a cross-polar ratio of around −25 dB on bore sight and with careful design and manufacture a value of better than −35 dB can be achieved. Off bore sight the cross-polar ratio degrades and an antenna radiating nominally circular polarization on bore sight will radiate elliptical polarization in other off-bore-sight directions. In summary, these descriptions allow the state of an electromagnetic wave to be described in terms of linear either vertical, horizontal, or slant linear polarization (45°) and elliptical and circular polarization (left handed or right handed).

The IEEE [36] convention of the description of polarization used here is that polarization is right handed when an observer looking in the direction of

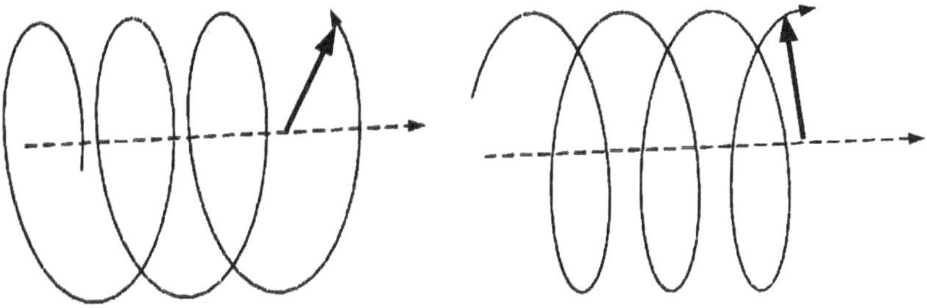

FIGURE 2.3 Circularly polarized vectors left-hand rotation (lhs) and right-hand rotation (rhs).

propagation observes the electric field vector rotating clockwise and left handed for the electric field vector rotating anticlockwise. Note that alternative definitions particularly related to optics exist.

If the locus of the electric field vector is described as

$$E_x = a_1 \cos(\tau + \delta_1) \tag{2.5}$$

$$E_y = a_2 \cos(\tau + \delta_2) \tag{2.6}$$

where $2a_1$ and $2a_2$ are the sides of a rectangle inscribing the ellipse and δ_1 and δ_2 are the respective phase angles of each component.

Born and Wolf [37] showed that

$$\frac{E_y}{E_x} = \frac{a_2}{a_1} e^{-i\delta} \tag{2.7}$$

This is illustrated in Figure 2.4, which shows the locus swept out by an elliptically polarized vector for a 2 : 1 ratio of major to minor axis and at angles of 0°, 30°, 45°, and 60° to 90°.

When the ratio of major to minor axis equals unity, the polarization becomes circular and δ becomes $\pi/2$ for right-handed and $-\pi/2$ for left-handed polarization. When $\delta = n\pi$ ($n = 0, \pm 1, \pm 2, \pm 3, \ldots$) and a_2/a_1 is finite, a linearly polarized wave exists.

Linear targets, such as wires, pipes, or thin cylinders, act as depolarizing features, although the received signal will always encounter a polarization loss. An RHCP wave will be reflected back as an LHCP wave from a planar surface, but some proportion of RHCP will be reflected from a thin pipe or wire. This may enable the surface reflection to be reduced while enhancing that from a thin pipe, wire, or cylindrical target concealed beneath a material or buried in a soil. The relationship between the transmit-and-receive antenna polarizations is important to ensure that the polarization is received (or rejected) and is given

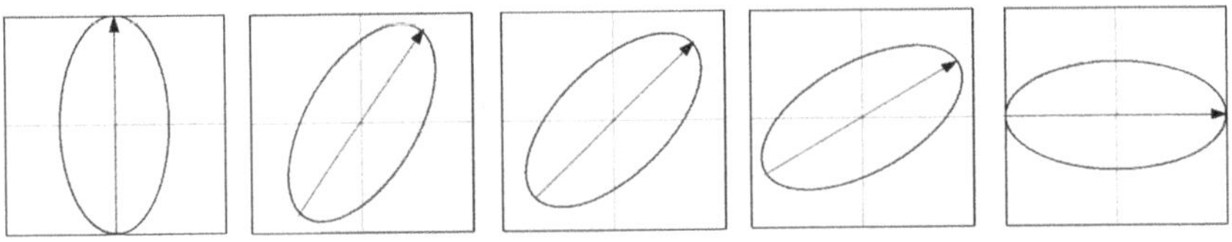

FIGURE 2.4 Elliptical polarizations from vertical to horizontal.

TABLE 2.2 Relationship between Transmit and Receive Antenna Polarization Loss

Transmit Antenna Polarization	Receive Antenna Polarization	Theoretical Ratio
Vertical	Vertical	0 dB
Vertical	Slant $\pi/4$ or $3\pi/4$	−3 dB
Vertical	Horizontal	$-\infty$ dB
Vertical	RHCP or LHCP	−3 dB
Horizontal	Horizontal	0 dB
Horizontal	Slant $\pi/4$ or $3\pi/4$	−3 dB
Horizontal	RHCP or LHCP	−3 dB
RHCP	RHCP	0 dB
RHCP	LHCP	$-\infty$ dB
LHCP	LHCP	0 dB
LHCP	RHCP	$-\infty$ dB
RHCP or LHCP	Slant $\pi/4$ or $3\pi/4$	−3 dB

in Table 2.2, and Table 2.3 lists the parameters that influence target discrimination by means of polarization.

When the target is a linear feature such as a cylinder, then the backscattered field exhibits a polarization characteristic that is independent of the state of polarization of the incident field. For linear targets it is possible to use orthogonally disposed transmit-and-receive antennas as a means of preferential detection. Essentially, the received signal varies sinusoidally with angle between antenna pair and the target. As it is inconvenient to physically rotate the antenna, it is also possible to electronically switch (commutate) the transmit/receive signals to a set of multiple co-located antenna pairs. Alternatively, circular polarization, which is essentially a means of automatically rotating the polarization vector in space, may be used. However, circular polarization inherently requires an extended time response of the radiated field, and consequently either hardware or software deconvolution of the received signal is needed. This section has not considered bistatic radar where the scattered signal from the target may be forward scatter rather than backscatter, and the scattering mode should also be considered in choosing a suitable polarization. The size of the target in wavelengths should also be taken into account as, for example, it has been found that very large diameter pipes exhibit depolarizing effects, not from the crown of the pipe but from the edges. The choice of polarization-dependent schemes should thus be considered very

TABLE 2.3 Relationship between Target Polarizations for Canonical Targets

Transmit Antenna Polarization	Target	Reflected Polarization
Vertical	Planar reflector	Vertical
Horizontal	Planar reflector	Horizontal
Slant $\pi/4$ or $3\pi/4$	Planar reflector	Slant $\pi/4$ or $3\pi/4$
RHCP	Planar reflector	LHCP
LHCP	Planar reflector	RHCP
Vertical	Dihedral	Vertical
Horizontal	Dihedral	Horizontal
Slant $\pi/4$ or $3\pi/4$	Dihedral	Slant $\pi/4$ or $3\pi/4$
RHCP	Dihedral	RHCP
LHCP	Dihedral	LHCP
Vertical	Trihedral	Vertical
Horizontal	Trihedral	Horizontal
Slant $\pi/4$ or $3\pi/4$	Trihedral	Slant $\pi/4$ or $3\pi/4$
RHCP	Trihedral	LHCP
LHCP	Trihedral	RHCP
Vertical	Vertical thin cylinder	Vertical
Horizontal	Vertical thin cylinder	Zero
Slant $\pi/4$ or $3\pi/4$	Vertical thin cylinder	Vertical (−3 dB) and horizontal (−3 dB)
RHCP	Vertical thin cylinder	Vertical ($\phi = -\pi$, max amplitude = −3 dB)
LHCP	Vertical thin cylinder	Vertical ($\phi = -\pi$, max amplitude = −3 dB)
Vertical	Horizontal thin cylinder	Zero
Horizontal	Horizontal thin cylinder	Horizontal
Slant $\pi/4$ or $3\pi/4$	Horizontal thin cylinder	Vertical (−3 dB) and horizontal (−3 dB)
RHCP	Horizontal thin cylinder	Horizontal ($\phi = -\pi$, max amplitude = −3 dB)
LHCP	Horizontal thin cylinder	Horizontal ($\phi = -\pi$, max amplitude = −3 dB)
Vertical	Sphere	Vertical
Horizontal	Sphere	Horizontal
Slant $\pi/4$ or $3\pi/4$	Sphere	Slant $\pi/4$ or $3\pi/44$
RHCP	Sphere	LHCP
LHCP	Sphere	RHCP

carefully as it may not be possible to cover all possible sizes of targets with one antenna/polarization scheme.

2.2.5 Radar Cross Section

The radar cross section σ (RCS) of a target is fundamentally defined by the product of its geometric cross section × reflectivity × directivity (all as functions of polarization) and is the effective surface area that intercepts the incident wave and that scatters the energy isotropically in space. It is often expressed in decibels relative to the radar cross section of 1 m^2; hence a target with a 1 m^2 RCS would be 0 dBsm = dB square metre. The RCS seen by radar will also depend on whether the radar is monostatic or bistatic in which case the bistatic RCS must be considered. The main methods used to predict RCS are exact methods based on Maxwell's equations, approximate methods usually applied to the optical and Rayleigh regions (which are considered later), geometric optics, physical optics, geometric theory of diffraction (GTD), physical theory of diffraction (PTD), incremental length diffraction coefficient (ILDC), and method of equivalent currents, depending on the complexity of the target and the accuracy required from the prediction.

In this section the basic geometric cross section is considered, and Skolnik [38] gives the most frequently used definition for the radar cross section in Eq. (2.9) as σ = incident power density/4π power reflected toward source per unit solid angle

$$\sigma = \frac{\lim}{R \to \infty} 4\pi R^2 \left| \frac{E_r}{E_i} \right|^2 \tag{2.8}$$

Alternatively, the radar cross section may be defined as

$$\sigma = 4\pi \frac{P_{\text{backscatter}}}{P_{\text{intercepted}}} \tag{2.9}$$

The components of the energy reradiated from a target can be considered to be the specular reflection, diffracted reflections, creeping waves, evanescent waves, and in the case of nonmetallic targets, glory waves. Ruck et al. [39] describe these in detail. The complexity of the analysis of real targets leads to an analytic approach based on canonical targets such as spheres, cones, cylinders, and the like. The sphere, being the simplest case, has been considered in detail, and the RCS of a sphere can be considered to be composed of backscattering by two main components, specular reflection and a creeping wave. The RCS of a sphere as a function of the sphere circumference in wavelengths is shown in Figure 2.5; note that the value of RCS is normalized to the echo area divided by πa^2.

The main regions of the RCS are the *Rayleigh region* where the RCS rises as the fourth power of the sphere radius for $k_0 a < 0.7$, the r*esonance region* this is also known as the Mie region between $0.7 < k_0 a < 10$, and the o*ptical region* where $k_0 a > 10$. The oscillations in the resonance region are due to the creeping wave adding and subtracting in phase with the specular reflection as a function of sphere circumference from the front face since the phase depends on the wavelength. Real targets show a more complex RCS, which is due to a range of scattering mechanisms. Although complex RCS signatures are more commonly associated with larger targets such as ships, vehicles, or aircraft, it is useful to note that in addition to specular reflections, creeping waves, and diffracted waves other mechanisms such as cavity reflections (the intakes of jet engines

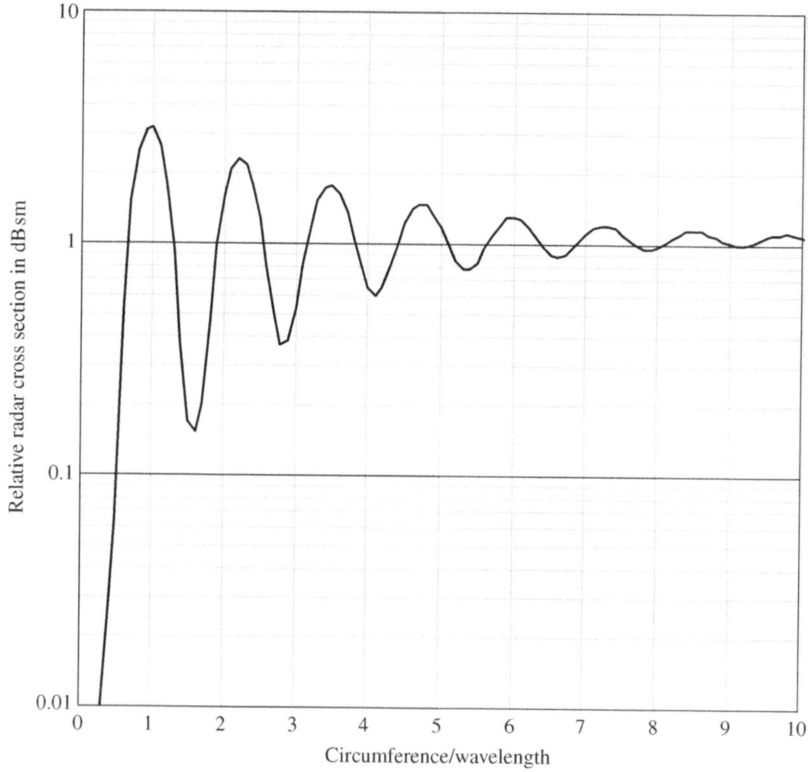

FIGURE 2.5 RCS of a sphere as a function of circumference in wavelengths.

being a typical example), traveling-wave reflections, and surface discontinuities will all contribute to target RCS. The RCS of a range of canonical targets is given in Table 2.4. Blore [40] provides graphs of the RCS in various planes of such targets.

Where the target is a dielectric the RCS is changed from a metallic target and is the target is concealed under clothing or embedded in soil, by the properties of that material. When a target is buried in the ground, the effect of the RCS on the radar range equation, which assumes a point source scatterer, needs to be considered. The radar range equation may need adjusting for the different types of targets as shown in Table 2.5 as the nature of the target influences the magnitude of the received signal. The following approximate relationships apply for targets, which extend across the zone illuminated by an antenna (i.e., its footprint).

The RCS will vary with angle and one phenomenon, which is often seen, is a scintillation in the target RCS as a function of angular position. Associated with deep fading amplitude events, the scintillations cause large and sudden variations in the apparent position of the target and are also known as glint. Borden [41] provides further details.

TABLE 2.4 Table of RCS Approximations for Canonical Targets

Scatterer	Aspect	RCS	Symbols
Sphere		$\sigma = \pi a^2$	a = radius
Flat plate	Normal	$\sigma = 4\pi \frac{A}{\lambda^2}$	A = area of plate
Cylinder	Broadside in line	$\sigma = \frac{a\lambda \cos\theta \sin^2(kl \sin\theta)}{2\pi \sin^2\theta}$	a = radius, l = length
Dihedral	Axis of symmetry	$\sigma = \frac{4\pi a_d^2}{\lambda^2}$	a_d = area contributing to reflection
Trihedral	Axis of symmetry	$\sigma = \frac{4\pi a_t^4}{3\lambda^2}$	a_t = area contributing to reflection
Prolate spheroid	Axis of symmetry	$\sigma = \frac{\pi b_0^2}{a_o^2}$	a_0 = major axis
			b_0 = minor axis
Curved edge	Edge ⊥ to bore sight	$\sigma = \frac{a\lambda}{2}$	a = radius of edge contour
Cone tip	Axial on bore sight	$\sigma = \lambda^2 \sin^4(\alpha/2)$	α = half angle of cone

TABLE 2.5 Adjustment of Range Law for Different Types of Target

Nature of Target	Magnitude of Received Signal
Point scatterer (small void)	$(\text{Target depth})^{-4}$
Line reflector (pipeline)	$(\text{Target depth})^{-3}$
Planar reflector (smooth interface)	$(\text{Target depth})^{-2}$

2.2.6 Reflection

In any estimation of received signal level, it is necessary to consider the coefficients of reflection and transmission, as the wave passes through the dielectric to the target, and Snell's laws describe the associated angles of incidence, reflection, transmission, and refraction. Note that the polarization of the incident wave is important as discussed earlier in Section 2.2.4. When lossy materials are involved, complex angles of refraction may occur unlike the simple classical case, and polarization and the Stoke's matrix may also be required for oriented high aspect ratio features such as pipes, wires, and fractures.

The intrinsic impedance Z of a medium is the relationship between the electric field E and the magnetic field H and is a complex quantity given by

$$Z = \sqrt{\frac{-j\omega\mu}{\sigma - j\omega\varepsilon}} \tag{2.10}$$

At the boundary between two media, some energy will be reflected and the remainder transmitted. For normal incidence the reflected field strength is described by the reflection coefficient, r:

$$r = \frac{Z_2 - Z_1}{Z_2 + Z_1} \tag{2.11}$$

where Z_1 and Z_2 are the impedances of medium 1 and 2, respectively.

The reflection coefficient has a positive value when $Z_2 > Z_1$, such as where an air-filled void exists in a dielectric material. The effect on a pulse waveform is to change the phase of the reflected wavelet so that targets with different relative dielectric constants to the host material show different phase patterns of the reflected signal. However, the propagation parameters (relative dielectric constant and loss tangent) of the host material, the geometric characteristics of the target, and its dielectric parameters affect the amplitude of the reflected signal.

2.2.7 Refraction

A plane wave incident on a dielectric will propagate into the dielectric but will be refracted while passing through the dielectric according to Snell's law, which states that

$$\sin\alpha = \frac{\eta_1}{\eta_2}\sin\theta \tag{2.12}$$

where α is the angle of incidence, θ is the angle of refraction, η_1 is the refractive index of the first dielectric, η_2 is the refractive index of the second dielectric. The relative permittivity is equal to the square root of the refractive index so that for radar or microwave applications it is more usual to write

$$\sin\alpha = \frac{\sqrt{\varepsilon_{r1}}}{\sqrt{\varepsilon_{r2}}}\sin\theta \tag{2.13}$$

One of the effects of refraction is to modify the radiation patterns of antennas when they are placed in proximity to dielectrics. The effect of a dielectric will be to reduce the beam spreading of the antenna radiation pattern, as is shown schematically in Figure 2.6 where a $\pi/2$ radiation pattern is effectively reduced by the refraction in the dielectric.

2.2.8 Brewster Angle

At a dielectric boundary and at a particular angle of incidence when the electric field of the incident wave is in the same plane as the plane of incidence, the impedances between the two media can become equal, and this angle is known as the Brewster or polarizing angle. The reflection coefficient is plotted in Figure 2.7 for the cases when the field is in the plane of incidence $r_{//}$. (i) where the suffix $_{//}$ denotes incidence parallel to the plane and with the field normal to the plane of incidence r_{n} (i) where the suffix $_{n}$ denotes incidence normal to the plane. This shows that the field in the plane has

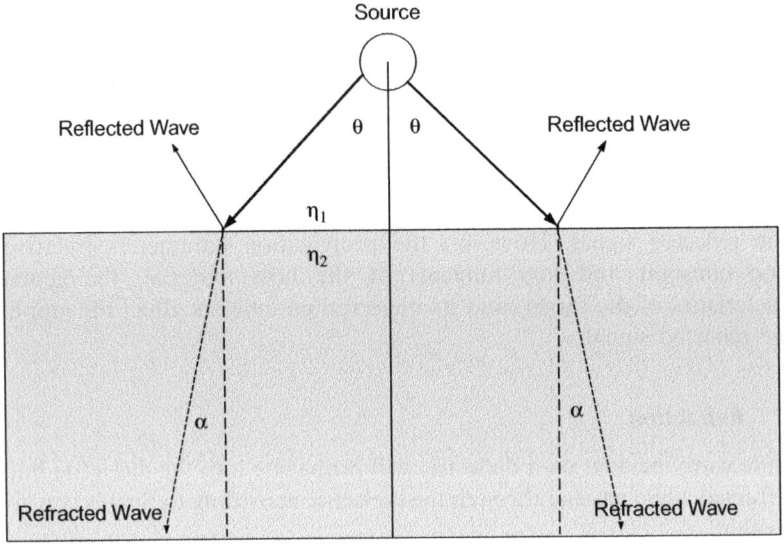

FIGURE 2.6 Effect of a dielectric on the ray paths from an isotropic radiator.

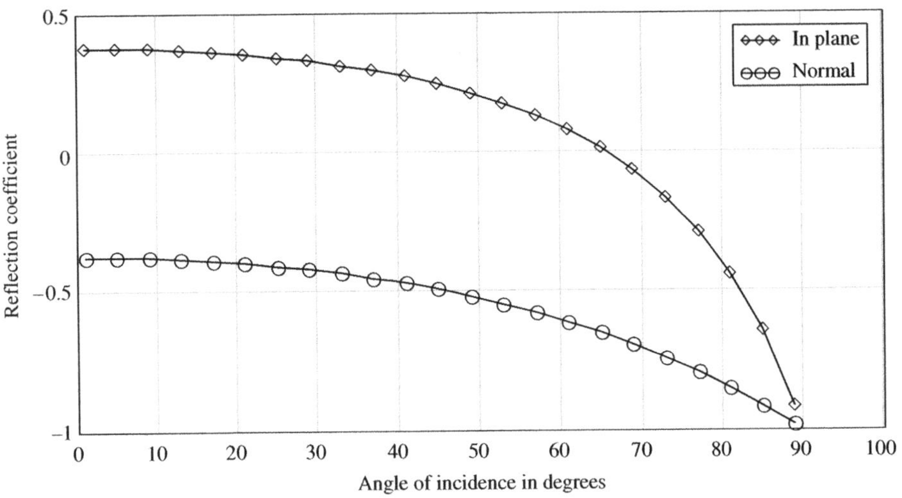

FIGURE 2.7 Reflection coefficients as a function of angle of incidence (i) for fields in and normal to the plane of incidence from a dielectric of 1 to a dielectric of 4.

zero reflection coefficient at 76° (the Brewster angle), but the field normal to the plane has no angle at which the wave impedances are equal. This reflection coefficient for the field normal to the plane causes a 180° phase change. For an unpolarized wave some of the field components will be reflected at the Brewster angle, that is, only those normal to the plane of incidence.

2.2.9 Dispersion

The frequency-dependent nature of the dielectric properties of the material causes the phase velocity of the component frequencies of a wideband signal to suffer differential propagation values. Hence, there will be variation in the velocity of propagation with frequency. Dielectrics exhibiting this phenomenon are termed dispersive. In this situation, the different frequency components within a broadband radar pulse would travel at slightly different speeds, causing the pulse shape to change with time. However, the propagation characteristics of octave band radar signals in most earth materials remain largely unaffected by dispersion. In many instances, the potential variation in the velocity of wave propagation over the frequency range of interest is small and can be ignored.

2.2.10 Anisotropy

An isotropic material is one in which its properties are uniform in every direction, whereas an anistropic material has properties that are direction dependent. For example, a seam of coal has bedding planes, which exhibits different attenuations depending on whether the direction of propagation is normal to or in line with the plane.

2.2.11 Clutter

The clutter that affects a radar or detector system can be defined as those signals that are unrelated to the target scattering characteristics but occur in the same time range and have similar spectral characteristics to the target reflection. This is a somewhat different definition from conventional radar clutter and should be borne in mind when considering conventional methods of clutter filtering such as MTI, which would be inappropriate to apply. Short-range systems are often limited by self-generated clutter, which is caused by internally generated reflections within the antenna, transmitter, and receiver as well as leakage between the transmit-and-receive antennas and multiple reflections between the antenna and the surface of the material under investigation. Clutter will vary according to the type of antenna configuration, co-polar arrangements having generally higher levels of leakage than cross-polar arrangements. Screening bistatic antennas reduces their leakage, but when antennas are operated very close to the surface of the material, small variations in spacing will affect the leakage quite significantly. For GPR systems the parallel planar dipole arrangement is one where the stability of the level of breakthrough is most constant.

The planar, crossed dipole antenna can be designed and manufactured to provide very low levels of breakthrough (> -70 dB). However, it then becomes very susceptible to "bridging" by dielectric anomalies on the near surface, which can degrade the breakthrough in a random manner as the antenna is moved over the ground surface. The variability of the breakthrough is unfortunate as it is not usually amenable to signal processing. The same problem is encountered with planar spiral antennas. Local variations in the characteristic impedance of the ground can also cause clutter, as can inclusions of groups of small reflection sources within the material. In addition, reflections from targets in the side lobes of the antenna, often above the ground surface, can be particularly troublesome. This problem can be overcome by careful antenna design and incorporating radar absorbing material to attenuate the side and back lobe radiation from the antenna.

In general, self-generated clutter is more significant at short time range and reduces at longer times. It is possible to quantify the rate of change of this as a

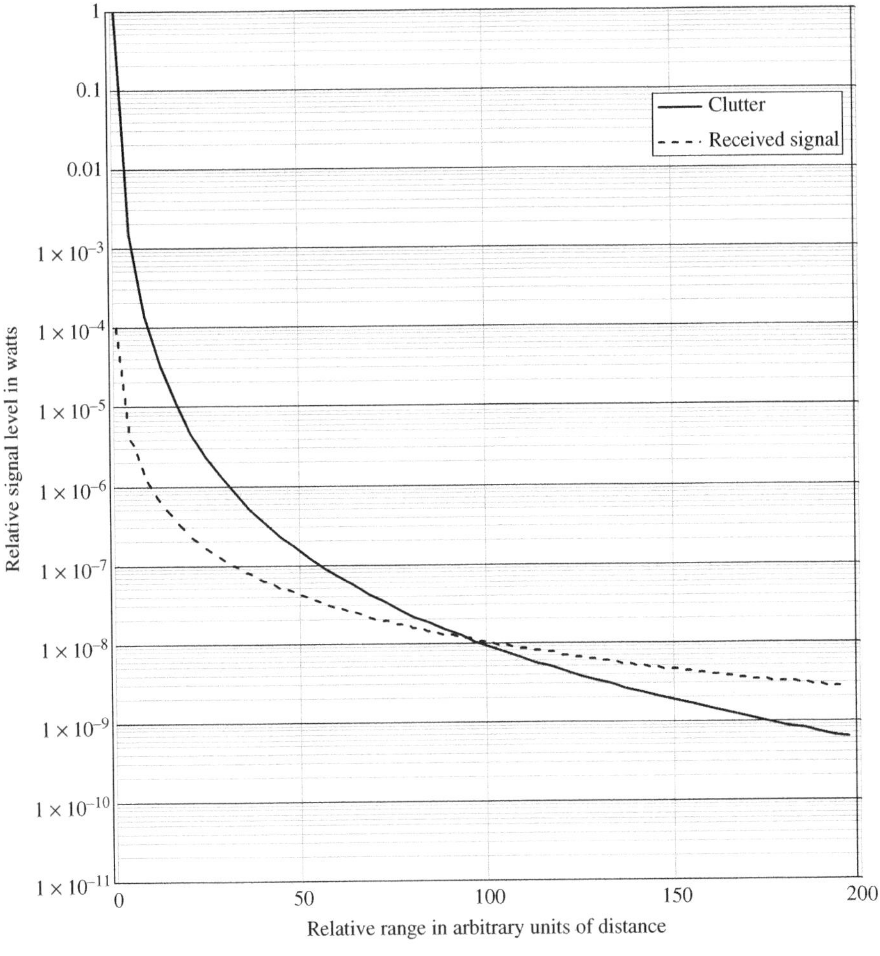

FIGURE 2.8 Generic signal and clutter characteristics.

function of time as in many cases this parameter sets a limit to the detection capability of the radar system. The effect of clutter on system performance is illustrated in Figure 2.8, which illustrates the consequent limitation on near-range radar performance when the self-generated clutter exceeds the level of the received signal at a range of 100 units. It may be possible to recover the signal provided suitable subclutter signal processing is used, but the performance of this is critically dependent on the stability of the clutter with range.

Various techniques have been investigated in the search for a method of reducing clutter. In the case of impulse radars using transverse electromagnetic (TEM) horns or frequency-modulated continuous-wave (FMCW) radars using ridged horns and reflectors, it has been found possible to angle the bore sight of the horn antennas to take advantage of the critical angle, thereby suppressing to some extent the ground surface reflection.

When the target is buried, the path losses to and from the target may be of the same order of magnitude as reflections from free-space targets at longer ranges. The system design must consider what sources of clutter could affect the system performance, and appropriate measures must be implemented to reduce their effect.

2.3 PROPAGATION OF ENERGY IN A DIELECTRIC

2.3.1 Introduction

This section provides a brief introduction to the topic with acknowledgment to the IET for permission to publish material from Daniels [2]. Targets concealed in or beneath lossy dielectric material will be verified by detecting the electromagnetic energy that is reflected or emitted by the target. In the case of radar, whether monostatic or bistatic, the energy will be radiated by the antenna of the radar, whereas for radiometers the energy is received by the receiver antenna from energy emitted from both the target by virtue of its emissivity and reflected by the target from a source of energy. In the case of radiometers operated in the open, the sky, which has a low temperature, acts as a cold source of temperature several tens of kelvins, whereas inside buildings the source may be heat and may have a temperature of several hundreds of kelvins. This section considers earth material as well as man-made materials such as clothing, although the latter may be constructed from natural substances such as cotton, wool, or animal skin.

While Maxwell's equations are the foundation for understanding the propagation of energy, simpler start points for considering the coupling mechanisms of EM energy are the coefficients of reflection and transmission, as the wave passes through whatever dielectric obscures the target. Snell's laws describe the associated angles of incidence, reflection, transmission, and refraction. When a system is operated very close to the intervening material, the efficiency of the coupling process is generally high, but this is not the case for standoff systems since, where lossy materials are involved, complex angles of refraction may occur. Concealed targets pose a difficult detection problem for standoff sensors such as radar, and their performance is strongly influenced by the properties of the material. With vertical polarization at incidence angles less than the Brewster angle, transmission losses at the air–material interface are relatively small, but at larger incidence angles than the Brewster angle the losses increase more rapidly.

This section considers the mechanisms that affect the energy that is coupled into a material to reflect from a concealed target. The main mechanisms are path absorption losses, material absorption losses, and transmission and reflection losses. When a system operates at standoff range, then the path absorption losses will be due to atmospheric absorption, whereas at close range these losses are relatively small and are usually ignored.

Plane waves are good approximations to real waves in many practical situations, particularly in low loss and resistive media such as clothing. However, when targets are buried in earth materials much higher losses are encountered. More complicated electromagnetic wave fronts can be considered as a superimposition of plane waves, and this method may be used to gain an insight into more complex situations.

In free space the magnetic susceptibility and electric permittivity are constants that are independent of frequency and the medium is not dispersive. In a perfect dielectric no propagation losses are encountered, and hence there is no consideration of the attenuation, which occurs in real dielectric media.

As a starting point, electromagnetic wave propagation can be represented by a one-dimensional wave equation of the following form. Propagation is taken along the z axis, with perpendicular electric (E) and magnetic (H) fields and is given in Eqs. (2.14) to (2.16).

$$\frac{\partial^2 E}{\partial z^2} = \mu\varepsilon \frac{\partial^2 E}{\partial t^2} \tag{2.14}$$

where the velocity of propagation is

$$v = \frac{1}{\sqrt{\mu\varepsilon}} \tag{2.15}$$

and the velocity of light in free space is

$$c = \frac{1}{\sqrt{\mu_0\varepsilon_0}} \tag{2.16}$$

where $\mu_0 = 1.26 \times 10^{-6}$ H m^{-1} absolute magnetic susceptibility of free space
$\varepsilon_0 = 8.86 \times 10^{-6}$ F m^{-1} absolute electric permittivity of free space
$\mu = \mu_0, \mu_r$, absolute magnetic susceptibility of medium
$\varepsilon = \varepsilon_0, \varepsilon_r$, absolute electric permittivity of medium

ε_r = relative permittivity, having a value in the range 1 to 80 for most materials

μ_r = relative magnetic susceptibility, being 1 for nonmagnetic materials

2.3.2 Velocity in a Dielectric

The velocity of propagation of electromagnetic waves in free space is approximately 3×10^8 m s^{-1} but slows in a material depending on its relative permittivity and relative magnetic permeability. The velocity of propagation of electromagnetic waves in a material with a value for ε_r of 9 would be slowed by a factor of $\sqrt{9}$ to 1.10^8 m s^{-1}.

In general, it is not possible to make a reliable estimate of propagation velocity or relative permittivity in a medium from a single measurement without performing a calibrated measurement. When the material is man-made and known to be homogeneous, then the relative dielectric constant can be derived from measurement of the phase of the reflection coefficients at several different frequencies. However, for earth materials and rocks, even in the case where a measurement is carried out at one location, it is often found that significant variations in velocity will occur within comparatively short distances from the original position. This can lead to significant errors in the estimation of depths of targets. One procedure that overcomes this limitation is known as common depth point surveying, which utilizes two antennas in bistatic operation at a number of transmit-and-receive positions.

The velocity of propagation is given by $v = (\mu_0\mu_\rho\varepsilon_0\varepsilon_\rho)^{-1/2}$. Hence in a material with $\mu_r = 1$ the velocity becomes $v = c/\sqrt{\varepsilon_r}$. The phase velocity is given by $v = \omega/\beta$ as given in Eq. (2.17):

$$\beta = \omega\sqrt{\left[\frac{\mu\varepsilon'}{2}\left(\sqrt{1+\left(\frac{\varepsilon''}{\varepsilon'}\right)^2}+1\right)\right]} \tag{2.17}$$

the phase velocity is also dependent on the factor $\varepsilon''/\varepsilon'$, which is also known as $\tan\delta$ also known as the loss tangent.

It is also possible to derive velocity from multiple measurements scanning over a target, but this works well only in relatively uncluttered situations where the media has no anisotropic characteristics. The velocity of propagation is also slowed by increase of loss tangent as well as relative dielectric constant as in Eq. (4.18),

$$v = c\left[\frac{\varepsilon_e'}{2\varepsilon_0}\left(\sqrt{(1+\tan^2\delta)}+1\right)\right]^{1/2} \tag{2.18}$$

However, tan δ must be significantly greater than 1 for any slowing to occur, and it is reasonable to assume that for $\tan\delta < 1$

$$v = \frac{c}{\sqrt{\varepsilon_r}} \tag{2.19}$$

2.3.3 Impedance of a Dielectric

The intrinsic impedance of a medium is the relationship between the electric field E and the magnetic field H:

$$\eta = \frac{E}{H} \tag{2.20}$$

A wave propagating in the positive z direction in a perfect dielectric can be described by the following:

$$E(z) = E_0 e^{-jkr} \tag{2.21}$$

where the phase constant

$$k = \frac{\omega}{v} = \omega\sqrt{\mu\varepsilon} \tag{2.22}$$

This describes the change in phase per unit length for each wave component; it may be considered as a constant of the medium for a particular frequency and is then known as the wave number. It may also be referred to as the propagation factor for the medium.

The wavelength, λ, is defined as the distance the wave propagates in one period of oscillation. It is then the value of r, which causes the phase factor to change by 2π:

$$k\lambda = 2\pi \tag{2.23}$$

Rearranging

$$\lambda = \frac{2\pi}{\omega\sqrt{\mu\varepsilon}} = \frac{v}{f} \tag{2.24}$$

This is the common relationship between wavelength, phase velocity, and frequency.

The intrinsic impedance (the ratio of the electric to the magnetic field) of the medium is

$$Z_i = \sqrt{\frac{\mu}{\varepsilon}} \tag{2.25}$$

The intrinsic impedance of the medium becomes

$$Z_i = \sqrt{\frac{\mu}{\varepsilon}} = \sqrt{\mu}\left[\varepsilon'\left(1 - j\frac{\varepsilon'}{\varepsilon''}\right)\right]^{-1/2} \tag{2.26}$$

2.3.4 Propagation Loss in a Dielectric

Electromagnetic waves propagating through natural media experience losses to either the electric (*E*) or magnetic (*H*) fields or both. This causes attenuation of the original electromagnetic wave. For most materials of interest the magnetic response is weak and need not be considered as a complex quantity, unlike the permittivity and conductivity. In the case of lossy dielectric materials, both conduction and dielectric effects cause absorption of electromagnetic radiation. It is not possible to distinguish, by measurement at a single frequency, the separate components of loss for such materials.

If an alternating electric field is applied to a material, the individual molecules will be induced to rotate in an oscillatory manner about an axis through their centers, the inertia of the molecules preventing them from responding instantaneously. Similar translational effects can occur. The polarization produced by an applied field (such as a propagating radar wave) is closely related to the thermal mobility of the molecules and is, therefore, strongly temperature dependent. Note that polarization in this context is different from the polarization of EM waves. In general, the relaxation time (which may be expressed as a relaxation frequency) depends on activation energy, the natural frequency of oscillation of the polarized particles, and on temperature. Relaxation frequencies vary widely between different materials.

The physical models, which are used to predict the propagation of electromagnetic waves in dielectric materials, have two main sources: electromagnetic wave theory and geometrical optics. The latter method is only relevant when the wavelength of the electromagnetic radiation employed is considerably shorter than the dimensions of the object or medium being illuminated and when the materials involved can be considered to be electrical insulators. Optical theory is therefore most relevant to dry materials. Materials containing appreciable amounts of moisture will behave as conducting dielectrics, especially if the water contains ions. Most naturally occurring waters have some degree of ionic conduction and so act as aqueous electrolytes.

For example, maximum absorption occurs at very low frequencies in ice (10^3 Hz), whereas it takes place in the microwave region in water (10^6 to 10^{10} Hz), thus the effects of this phenomenon can have a direct bearing upon the dielectric properties of materials at the frequencies employed by radar, especially if moisture is present within a material. There are a number of other mechanisms that cause a separation of positively and negatively charged ions resulting in electric polarization. These mechanisms can be associated with ionic atmospheres surrounding colloidal particles (particularly clay minerals), absorbed water, and pore effects, as well as interfacial phenomenon between particles.

The general form of the model that describes the frequency dependence of such systems is the well-known Debye [42] relaxation Eq. (2.27):

$$\varepsilon' - i\varepsilon'' = \varepsilon_\infty + \frac{\varepsilon_s - \varepsilon_\infty}{1 + i\omega\tau} \tag{2.27}$$

where ε' = real part of the dielectric permittivity
ε'' = imaginary part of the dielectric permittivity
ε_∞ = high-frequency limiting value of permittivity
ε_s = low-frequency limiting value of permittivity
ω = radian frequency ($= 2\pi f$)
τ = relaxation time constant

The frequency of maximum movement and loss occurs at $\omega = 1/\tau$.

In general, single relaxations are rarely observed in natural systems. Instead, there are distributions of relaxations corresponding to distributions of size scales that influence movement of charge. There are several equations describing such distributed systems, with the most common experimental observations in agreement with the model from Cole and Cole [43] as in Eq. (2.28):

$$\varepsilon' - i\varepsilon'' = \varepsilon_\infty + \frac{\varepsilon_s - \varepsilon_\infty}{1 + (i\omega\tau)^\alpha} \tag{2.28}$$

where α describes the breadth of the time constant distribution from a single relaxation, $\alpha = 1$, to an infinitely broad distribution, $\alpha = 0$, with a common process. Different polarization processes may be described by a series of Cole–Cole equations with different values of α and other parameters.

Free-space radar systems need only consider propagation phenomena through the atmosphere as described by Currie [44]. For most soils, the magnetic response is weak and need not be considered as a complex quantity, unlike the permittivity and conductivity. However, in certain soil types, such as those derived from volcanic rocks or otherwise high in iron content, full consideration of the magnetic properties is necessary. In the case of lossy dielectric materials, both conduction and dielectric effects cause absorption of electromagnetic radiation.

In general, the complex permittivity and the complex conductivity may be expressed as

$$\varepsilon = \varepsilon' - j\varepsilon'' \tag{2.29}$$

and

$$\sigma = \sigma' - j\sigma'' \tag{2.30}$$

where σ' and σ'' are real parts and imaginary parts of the conductivity, ε' and ε'' are real parts and imaginary parts of the permittivity, and j or $i = \sqrt{-1}$.

The nature of the parameter ε' relates to the electric permittivity, which may also be expressed in terms of relative permittivity. The parameter ε'' relates to losses associated with both conductivity and frequency. For practical purposes at frequencies up to 1 GHz and conductivities below 0.1 S m^{-1}, the effect of the ε' term will be small and is commonly disregarded (i.e., ε taken as the real components ε'') in such circumstances.

When measurements are made on a conducting dielectric, the parameters measured are the apparent permittivity and apparent conductivity:

$$\bar{\varepsilon} = \varepsilon_e' - j\varepsilon_e'' \tag{2.31}$$

$$\bar{\sigma} = \sigma_e' - j\sigma_e'' \tag{2.32}$$

The behavior of a material may be specified either by its apparent permittivity or, equivalently, by its apparent conductivity since

$$\bar{\sigma} = j\omega\bar{\varepsilon} \tag{2.33}$$

In terms of wave propagation equations ε and σ always occur in the combination

$$\sigma + j\omega\varepsilon = \sigma_e' - j\omega\varepsilon_e' \tag{2.34}$$

where σ_e' = real effective conductivity
ε_e = real effective permittivity

$$\sigma_e' = \sigma' - \omega\varepsilon'' \tag{2.35}$$

$$\varepsilon_e' = \varepsilon_e' - \frac{\sigma''}{\omega} \tag{2.36}$$

From Eq. (2.14) the propagation of an electromagnetic field E_0 originating at $z = 0, t = 0$ in a conducting dielectric can be described by $E(z, t)$ at a distance z and time t by

$$E(z, t) = E_0 e^{-\alpha z} e^{j(\omega t - \beta z)} \tag{2.37}$$

where the first exponential function is the attenuation term and the second the propagation term.

From the first exponential function it is seen that at a distance $z = 1/\alpha$ the attenuation is $1/e$. This distance is known as the skin depth, d, and provides an indication of the penetration depth of the energy. However, there are a number of other factors that influence the effective penetration depth, notably the strength of reflection from the target sought and the degree of clutter suppression of which the system is capable. These may reduce the calculated performance and must also be considered.

In general, the parameters of interest for radar applications are the attenuation and velocity of wave propagation. In a conducting dielectric the phase constant is complex and is given by

$$k = \omega\sqrt{\mu(\varepsilon' - j\varepsilon'')} \tag{2.38}$$

The wave number may be separated into real and imaginary parts:

$$jk = \alpha + j\beta = j\omega\sqrt{\mu\varepsilon'\left(1 - j\frac{\varepsilon''}{\varepsilon'}\right)} \tag{2.39}$$

where α = attenuation factor

β = phase constant

The parameters α and β can be related to σ and $j\omega\varepsilon$ and give expressions for α and β as shown in Eqs. (2.40) and (2.41):

$$\alpha = \omega\sqrt{\frac{\mu\varepsilon'}{2}\sqrt{1 + \left(\frac{\varepsilon''}{\varepsilon'}\right)^2} - 1} \tag{2.40}$$

$$\beta = \omega\sqrt{\frac{\mu\varepsilon'}{2}\sqrt{1 + \left(\frac{\varepsilon''}{\varepsilon'}\right)^2} + 1} \tag{2.41}$$

The dimensionless factor $\varepsilon''/\varepsilon'$ is more commonly termed the material loss tangent. The above expressions may be rearranged to provide the attenuation constant α in decibels/meter^{-1} and the wave velocity v in meters/second.

It can be seen from the above expressions that the attenuation constant of a material is, to a first order, linearly related (in decibels/meter^{-1}) to frequency. It is not sufficient to consider only the low-frequency conductivity σ_0 when attempting to determine the loss tangent over the frequency range of 1×10^7 to 1×10^{10} Hz. In the case of a material that is dry and relatively lossless, it may be reasonable to consider that $\tan\delta$ is constant over that frequency range. However, for materials that are wet and lossy such an approximation is invalid as

$$\tan\delta = \frac{\sigma' + \omega\varepsilon''}{\omega\varepsilon' - \sigma''} \tag{2.42}$$

It is important to consider the magnitude of σ'' and ε'' in attempting to determine the value of tan δ. In general for lossy materials tan δ is large at low frequencies, exhibits a minimum at around 1×10^8 Hz, and increases to a maximum at several gigahertz, remaining constant thereafter. Accurate determination of attenuation must therefore consider all the coefficients that comprise the expression for tan δ. It should be noted that the complex dielectric constant, and hence the loss factor, of a material is affected by both temperature and water content. The general effect of increasing the temperature is to reduce the frequency of the dielectric relaxation, while increasing the water content also increases the value of the loss factor while shifting its peak frequency down. It is observed that the frequency of the maximum dielectric loss of the water relaxation in materials is reduced and occurs over a more limited frequency range when compared with conductive water tan δ can increase with frequency over the range of 1×10^8 to 1×10^{10} Hz as the dipolar losses associated with the water content of the material become more significant and the conductivity losses reduce.

$$\tan \delta = \text{conductivity losses} + \text{dipolar losses}$$

$$\tan \delta = \frac{\sigma_{\text{dc}}}{\omega \varepsilon_o \varepsilon_r} + \frac{\varepsilon''}{\varepsilon'} \tag{2.43}$$

An approximation, which enables an order of magnitude indication, is that when σ is small

$$\tan \delta \approx \frac{\sigma'}{\omega \varepsilon'} \tag{2.44}$$

The attenuation loss can be expressed in decibels/meter as

$$L_a = (8.686 \times 2R \times 2\pi f)\sqrt{\left(\frac{\mu_0 \mu_r \varepsilon_0 \varepsilon_r}{2}\left(\sqrt{(1 + \tan^2 \delta)}\right) - 1\right)} \tag{2.45}$$

where f = frequency in Hz
tan δ = loss tangent of material
ε_r = relative permittivity of material
ε_0 = absolute permittivity of free space
μ_r = relative magnetic susceptibility of material
μ_0 = absolute magnetic susceptibility of free space

2.3.5 Coupling Losses into Materials

At the boundary between two media, some energy will be reflected and the remainder transmitted. The reflected field strength is described by the reflection coefficient, Γ:

$$\Gamma = \frac{Z_2 - Z_1}{Z_2 + Z_1} \tag{2.46}$$

where Z_1 and Z_2 are the impedances of medium 1 and 2, respectively.

In a low-loss nonconducting medium and when considering only a single frequency of radiation, the above expression may be simplified and rewritten as

$$\Gamma' = \frac{\sqrt{\varepsilon_{r2}} - \sqrt{\varepsilon_{r1}}}{\sqrt{\varepsilon_{r2}} + \sqrt{\varepsilon_{r1}}} \tag{2.47}$$

where ε_r is the relative permittivity of the medium.

The reflection coefficient has a positive value when $\varepsilon_{r2} > \varepsilon_{r1}$, such as when an air-filled void exists in a dielectric material. The effect on a pulse waveform is to change the phase of the reflected wavelet so that targets with different relative dielectric constants to the host material show different phase patterns of the reflected signal. However, the amplitude of the reflected signal is affected by the propagation dielectric of the host material, the geometric characteristics of the target, and its dielectric parameters.

In general, the transmission loss from air into a material is given by

$$L_{t1} = -20\log_{10}\left(\frac{4Z_m Z_a}{|Z_m + Z_a|^2}\right) \quad \text{dB} \tag{2.48}$$

where Z_a = characteristic impedance of air (377 Ω)
Z_m = characteristic impedance of the material given by

$$Z_m = \left(\sqrt{\frac{\mu_0\mu_r}{\varepsilon_0\varepsilon_r}}\right)\frac{1}{(1+\tan^2\delta)^{1/4}}\left(\cos\frac{\delta}{2} + j\sin\frac{\delta}{2}\right) \tag{2.49}$$

The retransmission loss from the material to the air on the return journey is given by

$$L_{t2} = -20\log_{10}\left(\frac{4Z_m Z_a}{|Z_m + Z_a|^2}\right) \tag{2.50}$$

These losses must be accounted for when estimating the total path losses because as the impedance of the material decreases compared with free space the coupling losses increase. These expressions do not consider the frequency-dependent characteristics of materials so that wideband signals may suffer additional and frequency-dependent losses while propagating. Proper consideration of both ε_r and tan δ as a function of frequency is necessary to calculate wideband effects

2.4 DIELECTRIC PROPERTIES OF SOILS AND ROCKS

When targets are buried in soil, rocks, or immersed in water, it is necessary to understand the propagation characteristics of such materials. It is an complex subject mainly because natural earth materials cover an enormous range from clays, which readily absorb moisture and whose fine-grain size allows good electrical contact from particle to particle thus increasing the conductivity, to granites and marbles, which when not fractured are largely impermeable to moisture.

The dielectric properties of soils and rocks are discussed in detail by De Loor [45], Hoekstra and Delaney [46], Hipp [47], Wang and Schmugge [48], Hallikainen et al. [49], Wobschall [50], and extensive data is provided by Parkhomenko [51], Keller [52], Fuller and Ward [53], Campbell and Ulricks [54], and Endres and Knight [55] and the reader can gain a sound understanding from these authors.

All of these authors consider in detail the theoretical effect of microscopic fluid distribution on the dielectric properties of partially saturated soils and rocks. The electromagnetic characteristics of soils and rocks can include anisotropy, an enormous range (20 orders of magnitude) of dc conductivity, and an order of magnitude range in permittivity between about 2 or 3 for dry sand or sandstone to 40 or so for wet porous soils and rocks. The use of radar techniques to penetrate soils and rock is usually intended to provide information about deep geological features where resolution is not usually critical. This permits the use of lower frequencies to aid penetration; however, for more shallowly buried targets it is still useful to understand the main parameters that influence material impedance and propagation.

As values of permittivity and velocity of propagation are strongly influenced by the presence of moisture, there is clearly a relationship between these parameters and porosity of the medium. Work upon rocks has shown the following relationships:

For layered material Parkhomenko [56] showed that with the electric field applied parallel to the bedding, for a particular frequency f:

$$\varepsilon_r = (1 - p)\varepsilon_m + p\varepsilon_\omega \tag{2.51}$$

where ε_r = permittivity of layered material

ε_m = permittivity of matrix

ε_ω = permittivity of water

p = porosity ratio

For layered material with electric field applied perpendicular to the bedding, for a particular frequency λ:

$$\varepsilon_r = \frac{\varepsilon_m \varepsilon_\omega}{(1-p)\varepsilon_m + p\varepsilon_\omega} \tag{2.52}$$

from which velocity of propagation can be calculated. Alternatively, porosity can be calculated from

$$p = \frac{c}{v^2} = \frac{\varepsilon_m}{\varepsilon_m - \varepsilon_\omega} \tag{2.53}$$

Endres and Knight [55] have put forward a theoretical treatment of the effect of microscopic fluid distribution on the dielectric properties of partially saturated rocks. Hanai [300] has also made attempts, among others, to model the dielectric behavior of soils by extending the theory of dielectric mixtures produced by Wagner and others for emulsions for the oil-in-water type, establishing relationships of the following type:

$$\frac{\varepsilon^* - \varepsilon_p^*}{\varepsilon_m^* - \varepsilon_D^*}\left(\frac{\varepsilon_m^*}{\varepsilon^*}\right)^{0.33} = 1 - \phi \tag{2.54}$$

where ϕ = volume fraction of the dispersed phase
ε_D^* = permittivity of dispersed phase
ε_m^* = permittivity of dispersing medium
ε^* = permittivity of mixture

It is necessary to determine the dipolar losses from a consideration of the dielectric relaxation spectrum of the soil–water mixture. Typically, for a mixture of sandy–clay soils, (60% sand and 15% clay) with a water content of 25%, the losses shown in Figures 2.9 and 2.10 could be expected. A dielectric mixture model as described by Peplinski et al. [57] is strictly applicable over the range of 0.3 to 1.3 GHz, although the original model developed by Hallikainen et al. [58] covered the range of 1.4 to 18 GHz. Therefore, its accuracy is limited to that frequency range. However, it gives a good insight into the wideband characteristics of soils. A graph of dielectric properties of ε' and ε'' for a medium loss soil is shown in Figure 2.9. The attenuation loss in decibels/meter as is shown in Figure 2.10, and after 1 GHz a steady rise is observed until reaching a maximum value at 100 GHz.

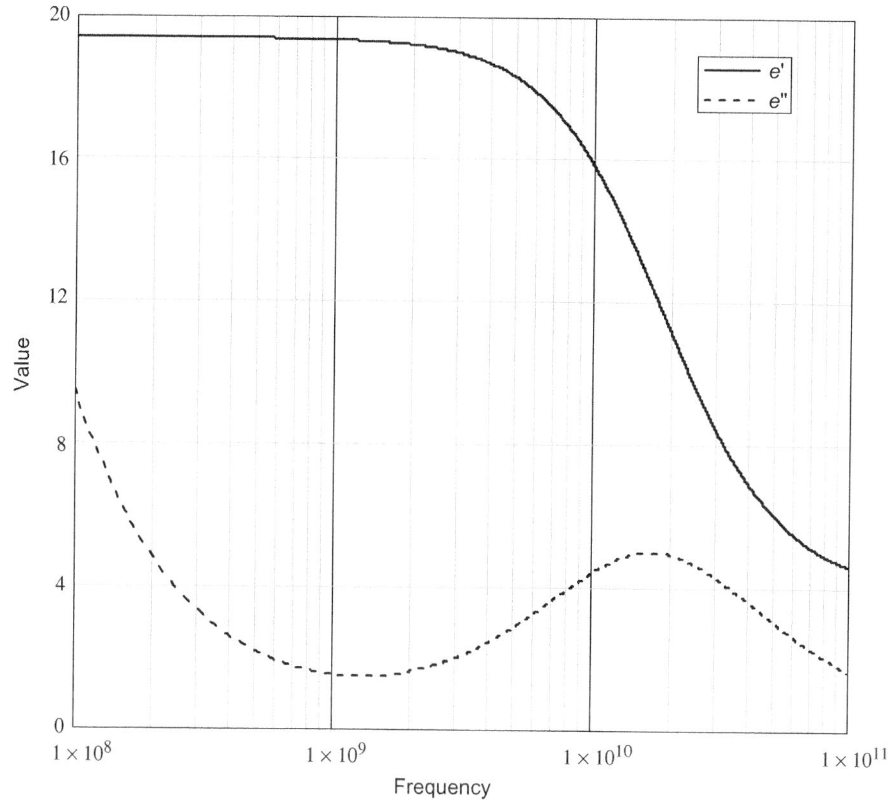

FIGURE 2.9 Dielectric properties ε' and ε'' of lossy soil as a function of frequency.

The dielectric properties of soils have been studied for many years, and there is now a large body of experimental data available as well as a range of theoretical models. Considerable difficulties are posed by the variability of the material, and none of the models developed are universally applicable. Simple models for dielectric loss tend to be deficient, with the major discrepancy between theory and experiment being the frequency dependence of the observable effects. The principal errors are understood to relate to the representation of the energy absorption by moisture, although there are numerous other factors that have a bearing upon the matter.

Typical changes in relative permittivity with moisture content are shown in Figure 2.11 for several types of rock. These data were obtained at frequencies in the range of 0.1 to 1 MHz and represent a combined estimation of the effect of water on type of rock. The key points are that the relative dielectric constant of a particular rock type is given by an approximate equation as shown in Eq. (2.58):

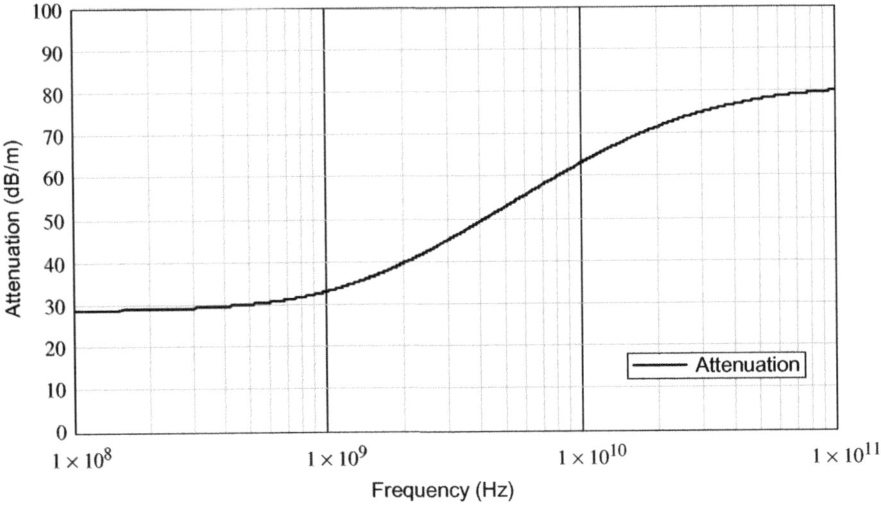

FIGURE 2.10 Material attenuation as a function of frequency for a medium loss soil.

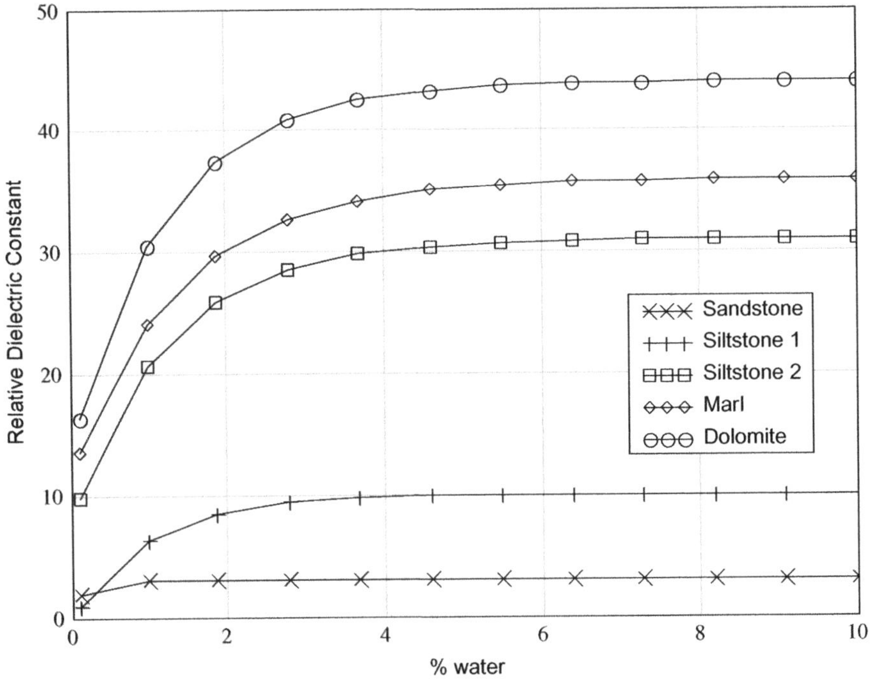

FIGURE 2.11 Effect of moisture content of rock on relative permittivity.

$$\varepsilon_r = \varepsilon\{\text{rocktype}\}_r[\text{sat}](1 - e^{-kw}) \tag{2.55}$$

where $\varepsilon\{\text{rock type}\}_r[\text{sat}]$ = saturated value of ε_r
k = coefficient related to rock type
w = percentage water content

Experimental studies by Hipp [59] soils have shown a rise in permittivity with water content and, at a given water content, a fall in permittivity with increasing frequency. Effective conductivity, and hence attenuation, rose both with frequency and water content. Figure 2.12 illustrates the dielectric behavior of a silty clay soil at a moisture content of 15 wt % over a wide range of frequencies after Hoekstra and Delaney [60]. Note that below a frequency of 1×10^9 Hz the values are extrapolated.

Field measurements have produced a wide scatter of results primarily due to the inherent variability of the "natural" environment caused by the presence of stones, boulders, and localized regions of higher conductivity within the "ground" mass. Such variations cause the dielectric parameters to change in a statistically unpredictable way as the radar antenna is scanned over the ground surface, introducing clutter into the received signal. There is no simple parameter, such as water content or low-frequency conductivity, which can be used as a convenient measure of dielectric loss in the frequency range of 100 MHz to 1 GHz.

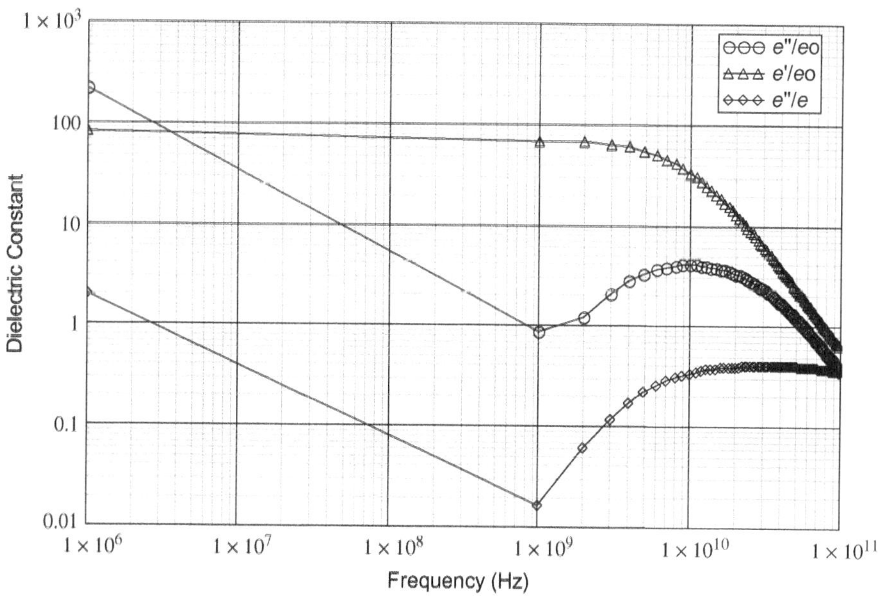

FIGURE 2.12 Complex relative permittivity and loss tangent of a silty clay soil at a water content of 15% wt.

Various workers have sought to employ a modified Debye model for describing the dielectric properties of moist soils. Bhagat and Kadaba [61] indicate that a relaxation mechanism alone is adequate at frequencies above about 1 GHz, but at lower frequencies the soil structure affects the results. De Loor [62] summarizes the chief loss mechanisms occurring in soil and their frequency range of applicability. In comparing the relaxation of water in bulk and soil water, it is observed that the frequency at maximum dielectric loss is displaced to a much lower frequency in soils and that relaxation occurs over a narrower frequency band in soils than in bulk water.

Soil is not only a mixture of dielectrics, but even when the composition of a given sample is known in terms of its components and their individual properties, that alone is not sufficient to define its nature dielectrically. The particle sizes, the electrochemical nature of their boundaries, and the way in which the water is distributed, both physically and chemically, among the matrix also affect the behavior. Wobschall [50] distinguishes, for example, between free water and pore water and developed a theory of the complex dielectric permittivity of soils based on the semidisperse model.

Systematic attempts have been made to produce models of the dielectric behavior of soils, which use parameters obtained independent of the dielectric measurements to be predicted. A number of models are believed to show reasonable agreement with experimental values over certain frequency ranges, but further work needs to be done on the subject of soil dielectric properties.

Wobschall's model is based on the Hanai / Bruggelman / Wagner theory of mixtures and considers the ionic conducting water as partly dispersed and partly the dispersing medium, an important distinction with this theory. Wobschall [50] considered soil to consist of irregular particles containing micro/macroscopic air-filled voids (pores) and crevices, which become increasingly filled by water as the percentage of water content increases.

Figure 2.13 shows the schematic composition of the soil suggested by Wobschall [50]. The key element of the semidisperse model is a phase of composition in which the solid particles contain disperse water filled with micropores and around each particle is a coating of water. These particles and the entrapped air are dispersed throughout the remaining water. The procedure for calculation takes into account 10 parameters including the frequency, the volume fraction of water, the volume fraction of voids, conductivity of pore water and microcrevice water, the dispersed water fraction, as well as the crevice water fraction. There is, however, a sufficient body of information to enable an assessment to be made of the likely range of dielectric properties, which might be encountered, assuming there is a detailed knowledge of the site in question.

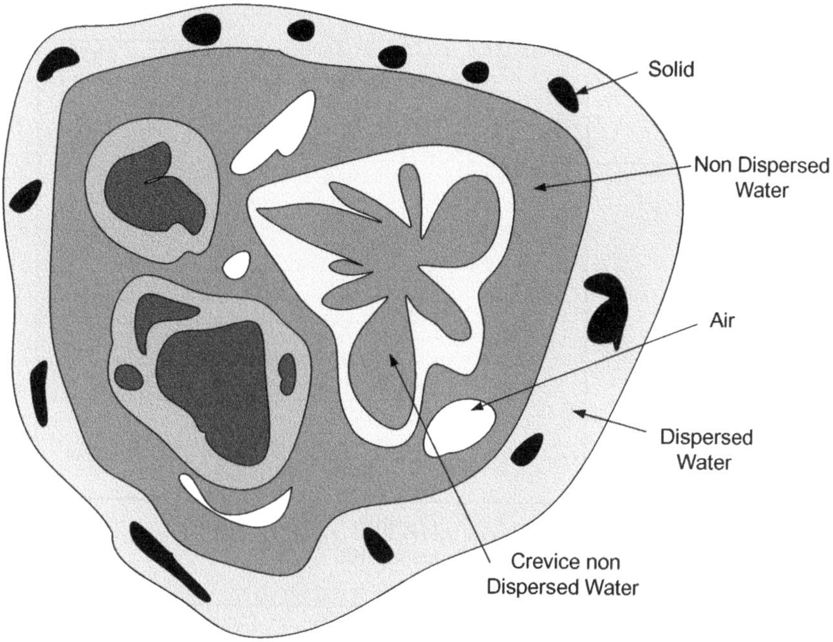

FIGURE 2.13 Particles and water dispersed in soil.

2.5 PROPAGATION IN WATER

The principal loss mechanism in many natural materials, certainly at frequencies over 500 MHz, is the absorption of energy by water present in the pores. The dielectric relaxation properties of water are described well by a Debye relaxation with a single relaxation time [59]. Pure liquid water at 0°C exhibits a maximum absorption at 9.0 GHz; this increases as temperature rises, with the maximum absorption occurring at 14.6 GHz at 10°C. Figure 2.14 shows the dielectric relaxation spectrum for water at two temperatures proposed by Hoekstra and Delaney [60].

King and Smith [63] provide a useful introductory summary of the properties of water, both liquid and solid, while Pottel [64] reviews the topic of aqueous electrolytes. All water encountered in earth materials has some degree of ionic conduction, and Pottel's treatment gives useful quantitative information to enable calculations to be made of the frequency dependence of dielectric properties. Cole and Cole [65] in a classic study proposed a modified Debye relaxation equation for liquid water containing free ions. An indication of the variation in permittivity for a solution of NaC1 and pure water is for values of ε_s between 63 and 73 and values of ε_∞ between 5 and 6. It has been found that the quoted range of values for ε_∞ is common to water and all aqueous solutions.

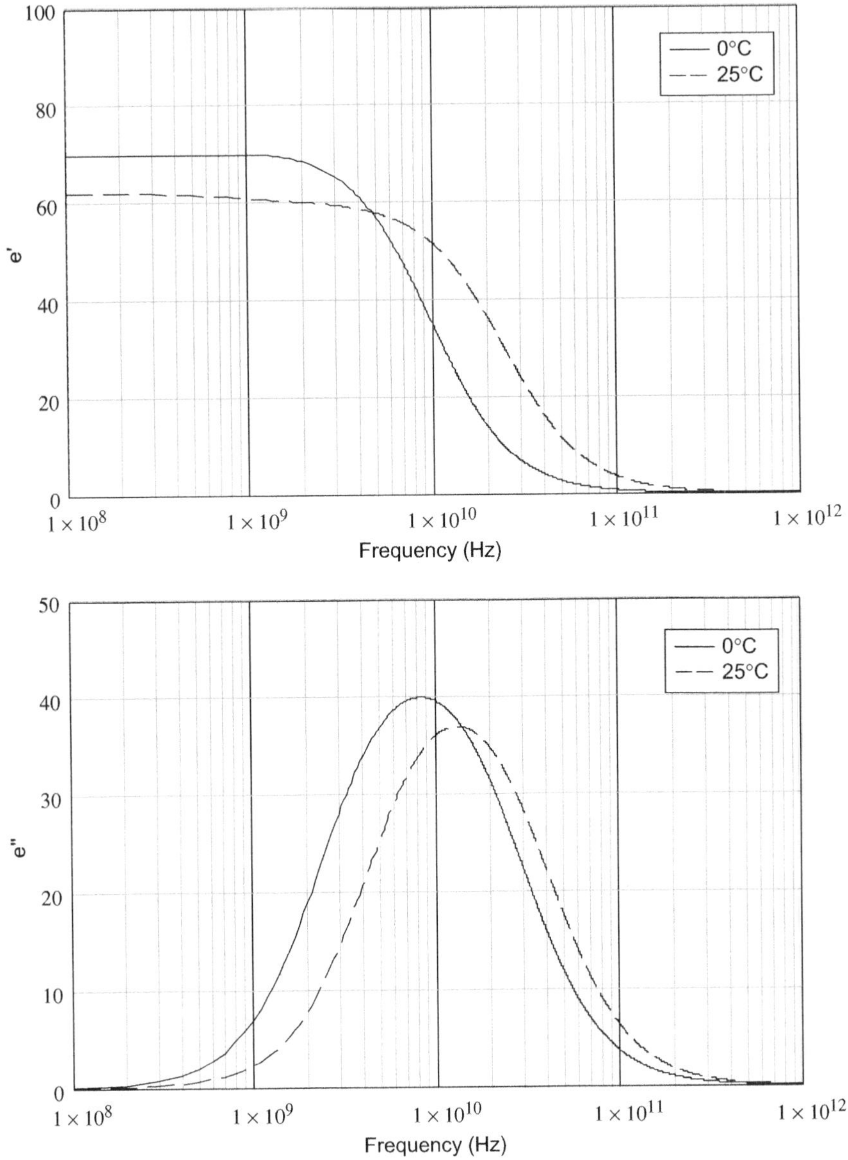

FIGURE 2.14 Dielectric relaxation spectrum of water at two temperatures.

The dielectric properties of naturally occurring water can be adequately described by the Debye theory or its modification when there is knowledge of its temperature and low-frequency conductivity; with the other constants needed in the calculation being available in the literature. The velocity of wave propagation is slowed substantially with increasing ionic concentrations.

This has the most pronounced influence at frequencies below 200 MHz, and above this frequency the wave velocity closely approaches that of pure water. At high conductivities the attenuation rises with frequency, and there is no plateau in the loss curve before the rise due to resonance absorption takes place [62].

2.6 ATMOSPHERIC ABSORPTION OF ELECTROMAGNETIC WAVES

The atmosphere will attenuate electromagnetic waves, due primarily to absorption by water vapor, oxygen, and nitrogen gases. Brussaard and Watson [66] provide a detailed discussion of all of the factors influencing atmospheric path losses and give details of path losses for dry air and increasing levels of water vapor. Note that altitude, density of the water vapor temperature, as well as the contribution due to nitrogen are all factors that influence path loss.

Liebe [67] developed a model that formed the basis for an approximate characteristic, which is given in Figure 2.15. Note that the absorption peaks at around 22 and 60 GHz are O_2 lines, whereas all others are water vapor absorption resonances. For either radar or radiometer systems operation at the minima points on the graph optimizes the range unless covert operation is required. The graph relates to water vapor at a concentration of 7.5 g m^{-3}, a temperature of 293 K, and an altitude of 250 m. For most sensor systems for the detection of concealed targets the ranges are relatively short and in the order of 1 to 20 m. In this situation atmospheric path losses are unlikely to be

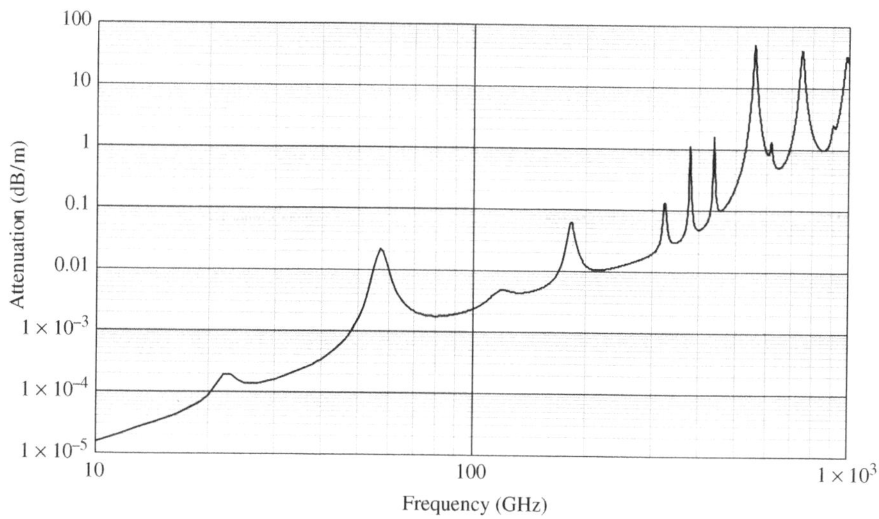

FIGURE 2.15 Atmospheric attenuation in dBm^{-1} from 10 GHz to 1 THz (with kind permission from Springer Science + Business Media and Dr. Liebe).

significant at millimeter wavelength but may become more significant at submillimeter wavelengths or if the range increases to several hundred meters.

The path propagation loss models developed by Crane [68] have been accepted by the ITU as its standard method for calculating attenuation. However, there are reservations about the models, particularly when the points of minima are concerned, as the predictions do not fully match the measured values. Appleby and Wallace [69] note the more recent work to refine the various models and refer to the model developed by Wallace using several models to provide a prediction, as shown in Figure 2.16. From 10 GHz to 1 THz the atmospheric absorption is predicted using the ITU standards, which are primarily based on the MPM89 model developed by Liebe [67].

From 1 to 10 THz the atmospheric absorption is predicted using the Fast Atmospheric Signature Code (FASCODE) and Line-by-Line Radiative Transfer Model (LBLRTM) in conjunction with the AFRL and JPL line databases. From 10 THz through to the visible region the atmospheric absorption is predicted using a propagation model, MODTRAN 4, maintained by the Air Force Research Laboratory.

Figure 2.16 relates to rain at 4 mm h^{-1}, fog visibility of 100 m, water vapor at a concentration of 7.5 g m^{-3}, a temperature of 293 K and an altitude

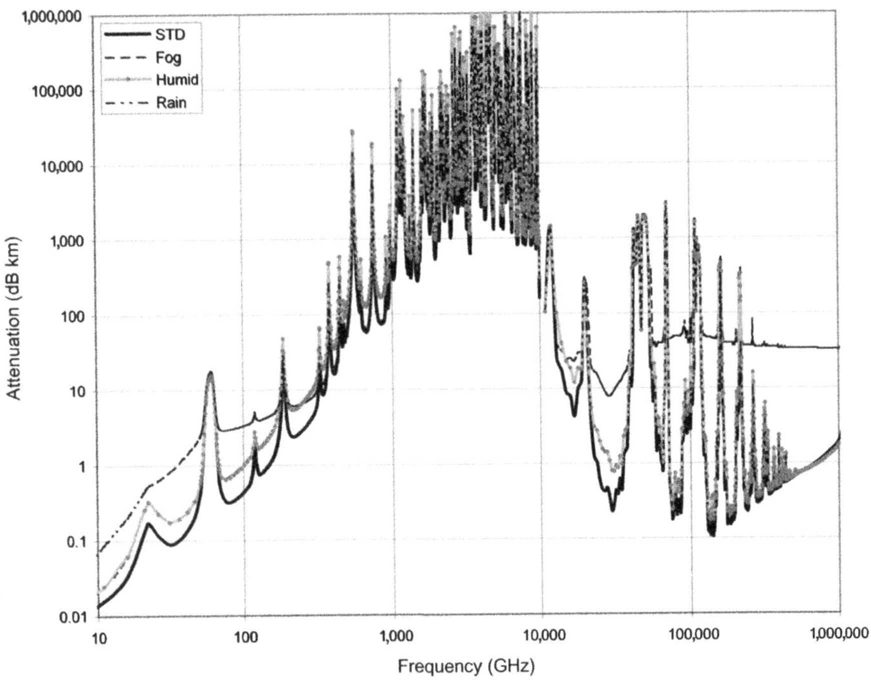

FIGURE 2.16 Atmospheric path losses in different water conditions (with kind permission © 2007 IEEE).

of 250 m. The data for Figure 2.16 were based on the AT model and used the same databases for the line structures but calibrated the attenuations and line shapes by contrasting them with MPM89 in the 500-GHz to 1-THz region. From Figure 2.16 it is apparent that, at sea level, above 500 GHz the attenuation is primarily a function of the atmospheric water vapor, and the effects of rain or fog are not dominant contributors to the total attenuation. In order to maintain a reasonable level of detail, the attenuation was averaged over a 1% frequency window. The rain and fog attenuations were calculated using the method of ITU standards.

The main factor affecting the range suitability of microwave, millimeter-wave and submillimeter-wave technology for standoff detection is its attenuation by water vapor, which significantly limits the operating range of the system. The trade-off is between the greater transmission offered by the microwave and millimeter-wave region versus the resolution achievable in the submillimeter-wave region at the target distances of interest (up to 100 m) and any system size constraints relative to the specification for the equipment. Higher frequencies offer greater resolution and reduced glint, thus enabling better discrimination at greater standoff distances. There is also the potential benefit of target specificity at frequencies above 400 GHz, due to the presence of some absorption bands of certain explosives in this region, where explosives show high absorption. As already indicated, the effect of fog and rain is to increase the path loss in proportion to the density of the water vapor and the rainfall rate and is discussed in the following section.

2.6.1 Rain and Fog

The scattering cross section of water spheres is discussed by Brussaard and Watson [66], who note the four main analytical models for scattering by spherical particles derived from Mie theory, Rayleigh scattering, the Born approximation, and the WKB wave number approximation. The normalized cross section of a 1-mm dielectric sphere is shown in Figure 2.17, with the contributions due to scattering and attenuation both contributing to the total cross section. As the diameter increases, the cross section increases. Hence a 4-mm dielectric sphere has a total cross section of 16 times that of the 1-mm sphere with a peak value at 20 GHz. Thus, the water droplet size will have a significant effect on path propagation loss.

The attenuation for six different atmospheric conditions described by Appleby and Wallace [69] is shown in Figure 2.18. This represents the extremes of loss from very low atmospheric attenuation in the winter to very high levels in tropical climates at sea level. Interestingly the highest losses are encountered in humid conditions, which are typical of tropical and semitropical latitudes.

A more detailed prediction of the attenuation of moist air and fog was also developed by Liebe [67] and is shown in Figure 2.19. Note that the general trend is an increase of several orders of magnitude from the dry air case shown as the baseline attenuation. Interestingly, the prediction shows little increase in

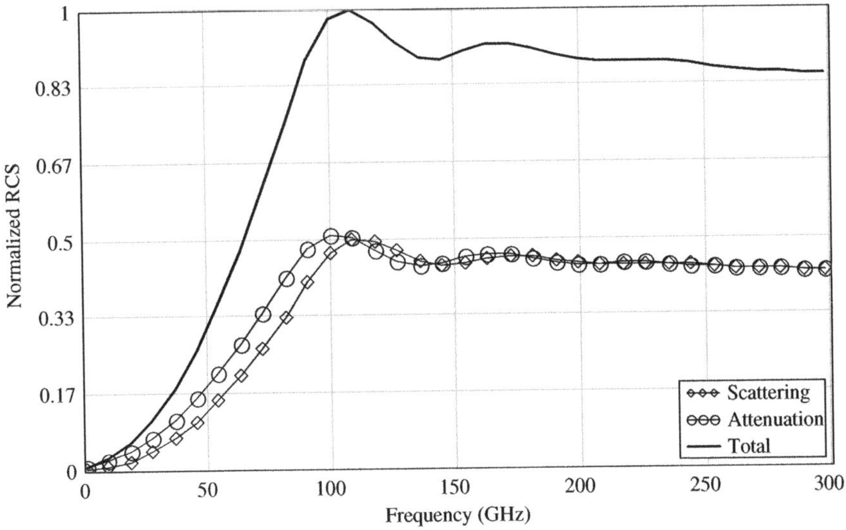

FIGURE 2.17 Normalized cross section of a 1-mm-diameter dielectric sphere.

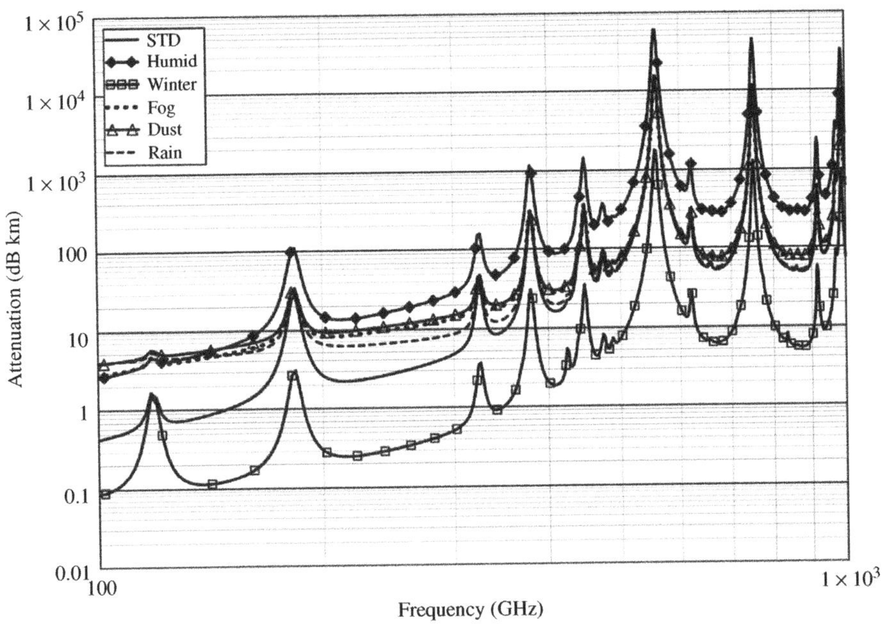

FIGURE 2.18 Path attenuation from 100 GHz to 1 THz for different atmospheric conditions (with kind permission © 2007 IEEE).

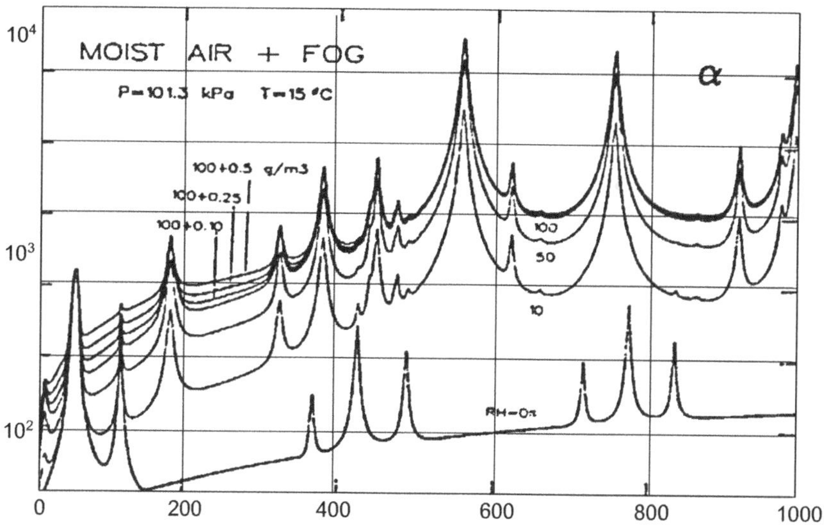

FIGURE 2.19 Predicted path attenuation in dB km^{-1} for three fog cases 0.1, 0.25, and 0.5 g m^{-3} and air at 0, 10, and 50% relative humidity (with kind permission from Springer Science + Business Media and Dr. Liebe).

the oxygen lines but a convergence to a maximum value of around 5 dB km^{-1} at 100 GHz, increasing to around 100 dB km^{-1} at 500 GHz. Goldstein [70] modeled the attenuation due to fog as

$$K_\alpha = \frac{0.438 M_w}{\lambda^2} \tag{2.56}$$

where M_w is one-way attenuation.

The effect of rain is, unsurprisingly, even more marked as shown in Figure 2.20, where a rainfall rate of 100 mm hr^{-1} is predicted to cause an attenuation of 40 dB km^{-1} at 100 GHz. Mechanisms such as scattering may also cause path losses.

2.6.2 Dust, Smoke, and Sand Storms

Particles such as sand have a relatively small effect on the propagation path loss, but dust can have more effect mainly as a result of scattering. Studies by the U.S. Army reported by Knox [71] indicated that seven levels of obscurant were tested from white and red phosphorus as well as hexachloroethane provided different burning rates and concentrations of smoke, fog-oil, and dust dispersed by HE detonation. Only the dust affected the transmission path, and the highest attenuation measured was 5×10^{-4} dB m^{-1} at 140 GHz. The other constituents showed lower values and at frequencies of 94 and 35 GHz the attenuation was never greater than 3.75×10^{-4} dB m^{-1}. Ghobrial et al. [72] using 2- and 7.5-GHz microwave links operating in Khartoum reported that

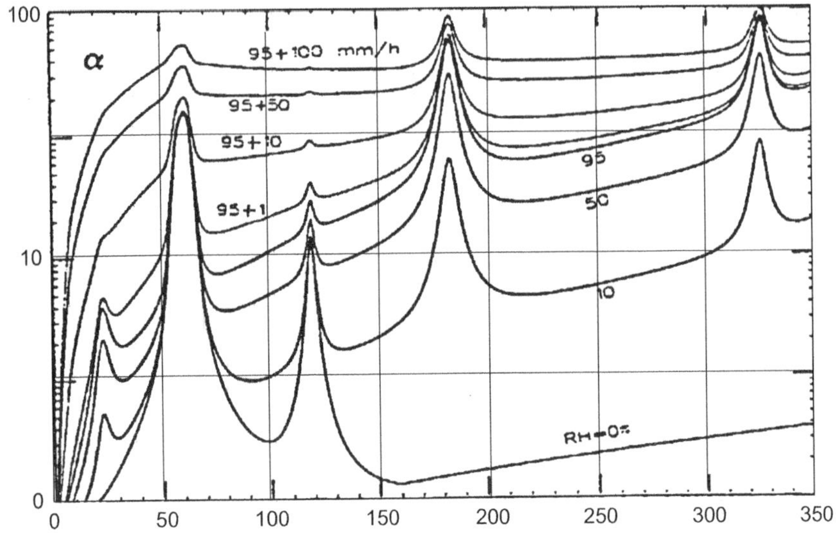

FIGURE 2.20 Predicted path attenuation in dB km^{-1} for four rain cases 1, 10, 50, and 100 mm h^{-1} and air at 0, 10, and 50% relative humidity all at sea level (with kind permission from Springer Science + Business Media and Dr. Liebe).

the attenuation effect of dust storms is negligible. Attenuations of less than 0.5 dB were recorded during storms with visibilities of less than 200 m. This confirms theoretical findings made by a number of workers.

Path loss can also be encountered due to multipath fading. Ali and Alhaider [73] carried out a 4-year field study in the city of Riyadh, Saudi Arabia, of clear air fading encountered on millimetric wave and infrared radio links. Ali and Alhaider showed a statistical characterization of fading for both the millimetric wave links operating near 40 GHz and the infrared radio link at a wavelength of 0.88 μm. Fading was dominated by multipath, having a Rayleigh amplitude distribution and an occurrence factor similar to the microwave band. Fades were highly correlated for the two links sharing the same path and separated in frequency by 1%. Multipath fading was also measured on the infrared link. Even at the short hop length at 0.75 km, the occurrence factor was about 1%. An assessment of millimeter-wave and Terahertz technology [74] by the National Academies noted that attenuation due to dust modeled by Goldhirsh [75] had been validated at 10 GHz and represents the attenuation from a storm that has a visibility of 10 m. This can also been seen in fig 2.18 which shows the atmospheric attenuation under various conditions up to 100 GHz. The curve labeled "humid" represents what would be expected when the atmosphere may contain up to five times the water vapor contained by the standard atmosphere.

2.7 ATTENUATION OF ELECTROMAGNETIC FIELDS BY MATERIALS

2.7.1 Human and Animal

Much work has been carried out on the electromagnetic properties of humans and animals. Gabriel and Gabriel [76] documented the properties of many organs such as the heart and brain and blood, skin, fat, and other tissues over a wide range of frequencies. The Italian National Research Council Institute for Applied Physics also published data on body tissue electrical properties for skin, fat, cartilage, and muscle with derived impedance, reflection coefficient (with respect to air), and emissivity values. A set of summary graphs are shown in Figures 2.21 to 2.26, for conductivity, relative permittivity, loss tangent, reflection coefficient, skin penetration, and emissivity, respectively, over the frequency range of 3 to 30 GHz. There are some general observations on the properties of such material. With the exception of fat, which shows a lower conductivity than muscle, dry skin, wet skin, and cartilage, the general trend is that conductivity and hence loss tangent increase with frequency, while relative dielectric constant decreases. The reflection coefficient thus gradually decreases with frequency. The penetration depth rapidly decreases to only a few millimeters above 30 GHz. The emissivity gradually increases with frequency.

The results documented by the Italian National Research Council Institute for Applied Physics can be compared with measurements described by Daniels [77]. A total of 80 scattering parameter measurements on two male subjects, over all potential areas of the body that could be used for concealment, were carried out. The objective of the measurements was to determine whether above 15 GHz the skin has a substantially constant reflectivity. The measurements were taken using a calibrated, open-ended waveguide (WG20) pressed against the skin of the target. The complete measurement was performed using a Wiltron Vector Network Analyser Type 360. An example is shown in Figure 2.27, which shows the voltage reflection coefficient over the frequency range of 18 to 26 GHz.

Computational models of the human body were also reported, and the predicted voltage reflection coefficient for several layers, comprising skin, fat, and then bone, are shown in Figure 2.28, which predicts an overall voltage reflection coefficient of around 0.68 at 24 GHz and compares reasonably well with results shown in Figure 2.27. The parameters of the model are as follows: The thickness of the skin is 3 mm with $\varepsilon_r = 31.3$ and $\tan\delta = 0.459$, the thickness of the fat layer was 20 mm with $\varepsilon_r = 4.6$ and $\tan\delta = 0.2288$, and the thickness of the bone was15 mm with $\varepsilon_r = 8$ and $\tan\delta = 0.161$.

If a layer of dielectric, such as explosive, is placed in front of the skin, the effect is to create a periodicity in the characteristic of the reflection coefficient in frequency domain, as shown in Figure 2.29. This is a method of detecting whether a layer of dielectric material is in front of human skin, but requires care in measurement.

The values given in Table 2.6 produce the reflection coefficient as shown in Figure 2.30. The absorption coefficient and refractive index of skin can be

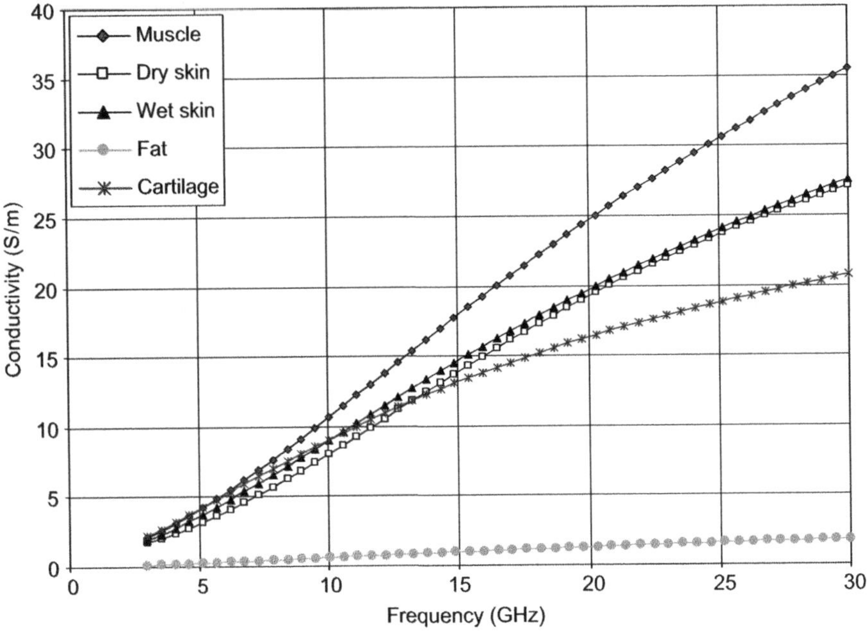

FIGURE 2.21 Conductivity of human body versus frequency in gigahertz.

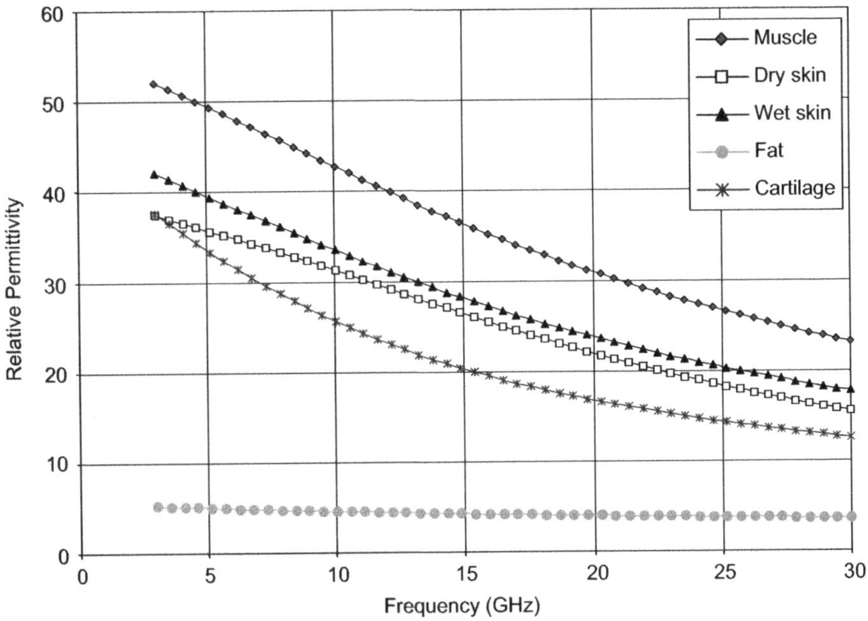

FIGURE 2.22 Relative permittivity of human body versus frequency in gigahertz.

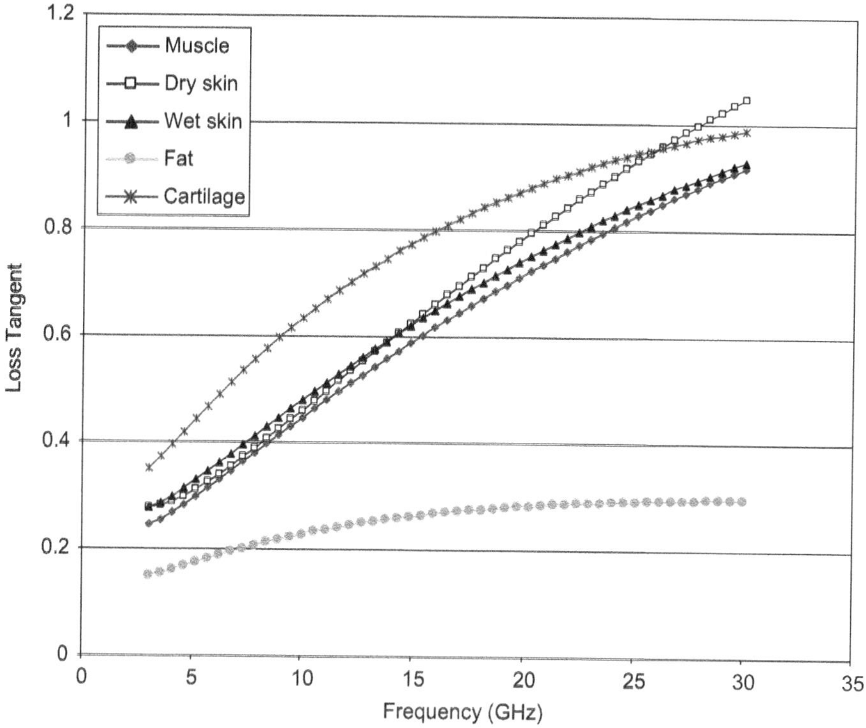

FIGURE 2.23 Loss tangent of human body versus frequency in gigahertz.

approximated by a dielectric mixture equation of dry skin and water, as the frequency response of skin is similar to water over the millimeter and submillimeter wave region, since skin contains approximately 70% water. Below 300 GHz a single Debye model suffices. Above 300 GHz the double Debye equation must be used for modeling skin. A simple one-dimensional model of skin in reflection was developed by Woodward [78] in 2003 and has shown good agreement with experimental results for the study of skin cancer.

Measurements by Daniels [77] at 24 GHz are summarized in Table 2.7, which shows the variation between the two male volunteers.

2.7.2 Heartbeat

In normal adults, the mass of the heart is 250 to 350 g, or about 75% the size of a clenched fist, but extremely diseased hearts can be up to 1000 g due to hypertrophy. The heart consists of four chambers, the two upper atria and the two lower ventricles. The typical heartbeat rate for a resting subject is 72 beats per minute, rising to 140 beats per minute (b min^{-1}) after exercise. Note that the maximum heart rate is 220 minus the age of the subject. Assume a rest rate

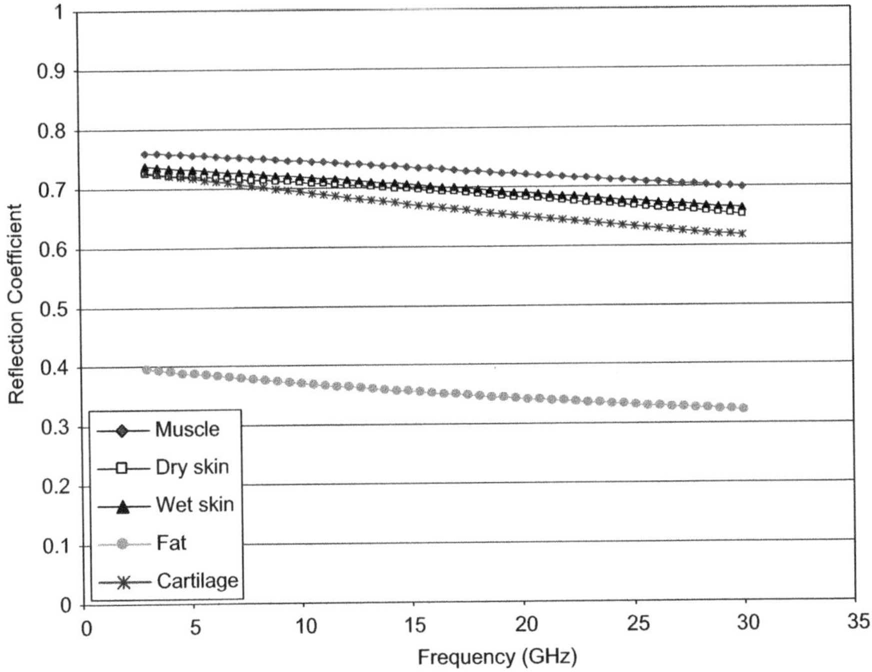

FIGURE 2.24 Reflection coefficient of human body versus frequency in gigahertz.

of 72 b min^{-1} gives a frequency of 1.2 Hz. When the heart muscle contracts or beats (systole), it pumps blood out of the heart. The heart contracts in two stages; in the first stage the right and left atria contract at the same time, pumping blood to the right and left ventricles; then the ventricles contract together to propel blood out of the heart. The heart muscle then relaxes (diastole) before the next heartbeat, and this allows blood to fill up the heart again. The first heartbeat sound is caused by the acceleration and deceleration of blood and vibration of the heart at the time of the closure of the tricuspid and mitral valves. The second heartbeat sound is caused by the same acceleration and deceleration of blood and vibrations at the time of closure of the pulmonic and aortic valves. Blood leaves the heart at a rate of approximately 0.45 m s^{-1}. The movement of the heart muscle can be detected using Doppler radar techniques, and these are discussed in later chapters.

2.7.3 Respiration

A normal human at rest breathes at a rate of around 12 breaths per minute or 0.2 Hz with a velocity of 10^{-2} m s^{-1} and up to 25 breaths per minute or 0.4 Hz when exercising. The mechanism for breathing is as follows. During inhalation,

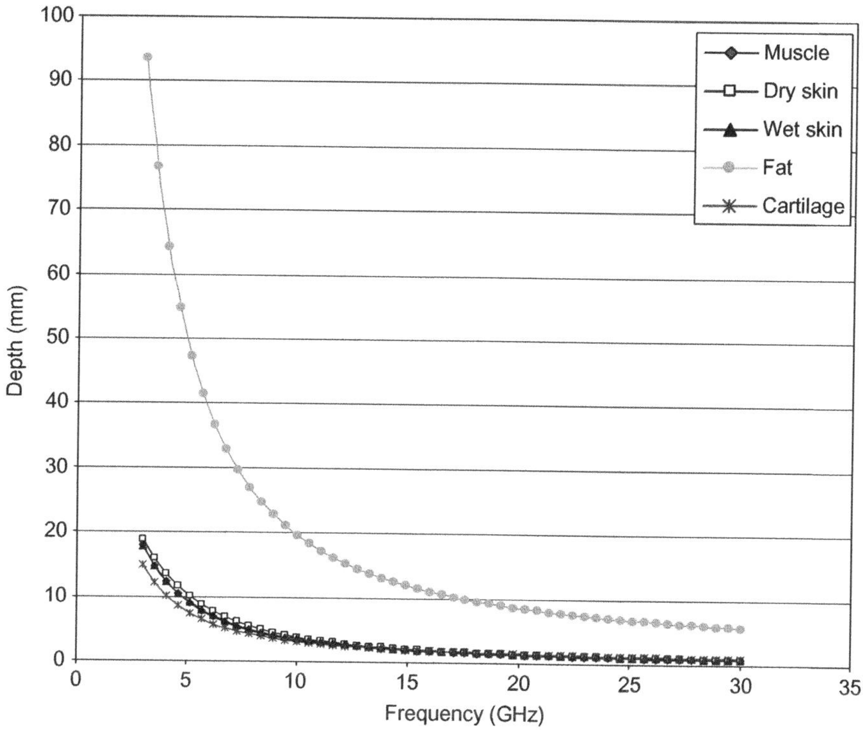

FIGURE 2.25 Skin penetration depth of human body versus frequency in gigahertz.

the intercostal muscles contract, the sternum moves upward and outward, the ribs move upward and outward, the diaphragm flattens, and there is an increase in volume of the thoracic cavity. The pressure in the thoracic cavity decreases and air is drawn in. During exhalation, the intercostal muscles relax, the sternum moves downward and inward, and the ribs move downward and inward. The diaphragm relaxes, forming a bell shape, and the volume in the thoracic cavity decreases. Pressure in the thoracic cavity increases and air is expelled. The volume of air exchanged during breathing is categorized into various components as follows. The minute volume is the amount of air drawn into the lungs during one minute while the person is at rest and at 12 to 16 breaths/minute is approximately 6 to 8 liters/minute. The residual volume is the air that remains in the lungs after expiration and is approximately 0.35 liters. The tidal volume is the amount of air taken into the lungs during one breath when the person is at rest and is approximately 0.5 liters. The vital capacity is the maximum volume of inspired breath, following maximum expiration and is approximately 3.5 to 4.5 liters. The chest cavity movement is approximately 3 to 5 cm during breathing. The movement of the chest and the diaphragm can be detected by Doppler radar and can be modeled as the superposition of a number of sine waves.

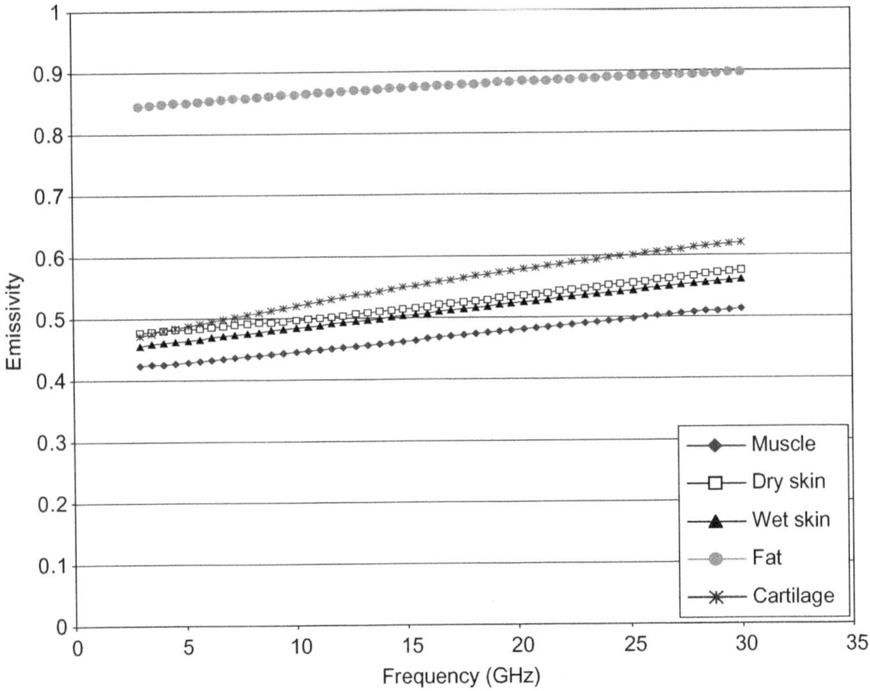

FIGURE 2.26 Emissivity of human body versus frequency in gigahertz.

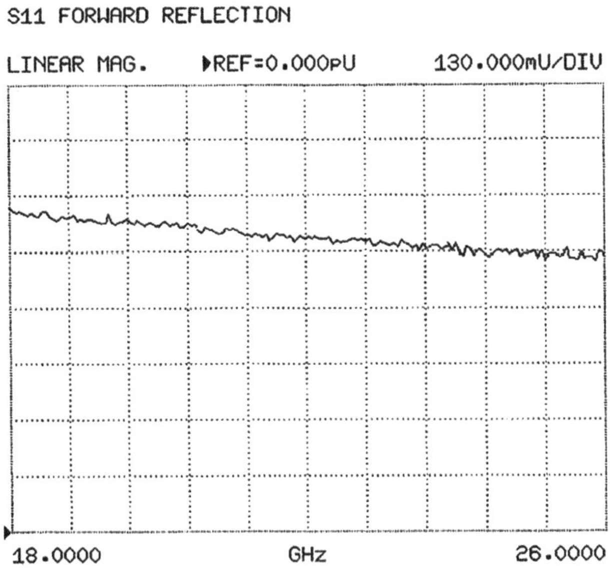

FIGURE 2.27 Voltage reflection coefficient 18 to 26 GHz, right upper arm inside, no target (with kind permission ©ERA Technology).

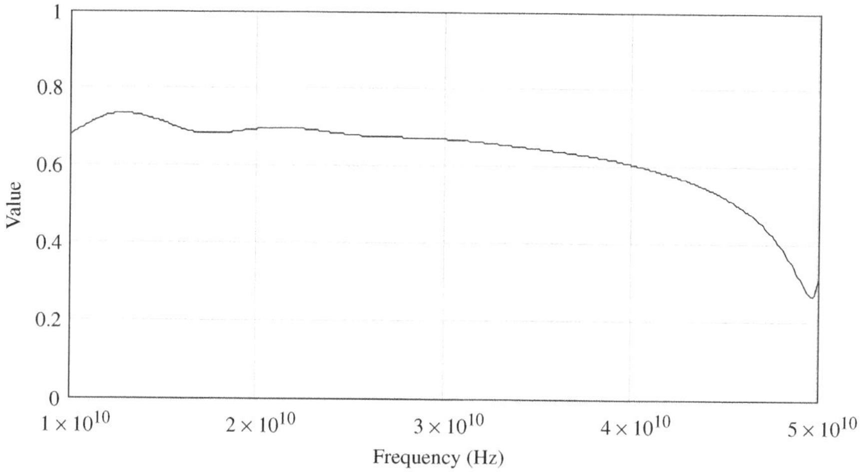

FIGURE 2.28 Voltage reflection coefficient for skin–fat–bone over the frequency range 10 to 50 GHz for the model values indicated.

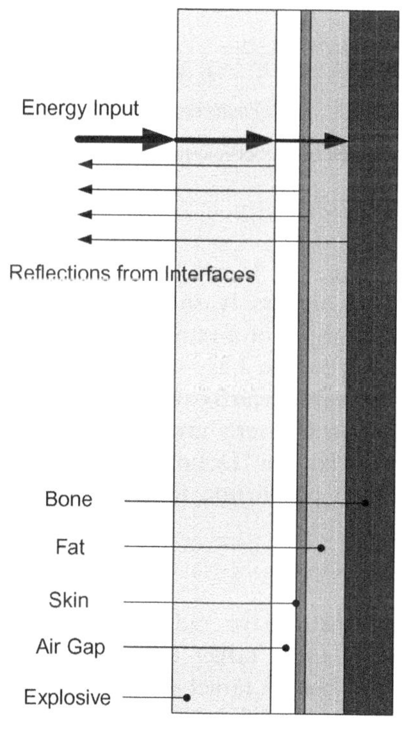

FIGURE 2.29 Model of explosive and skin.

TABLE 2.6 Average Values Used in Model Prediction

Material	Thickness (mm)	ε_r	tan δ
Explosive	10	2.5	0.01
Air gap	5	1	0
Skin	3	31	0.459
Fat/cartilage, etc.	20	4.6	0.224
Bone	20	8	0.613

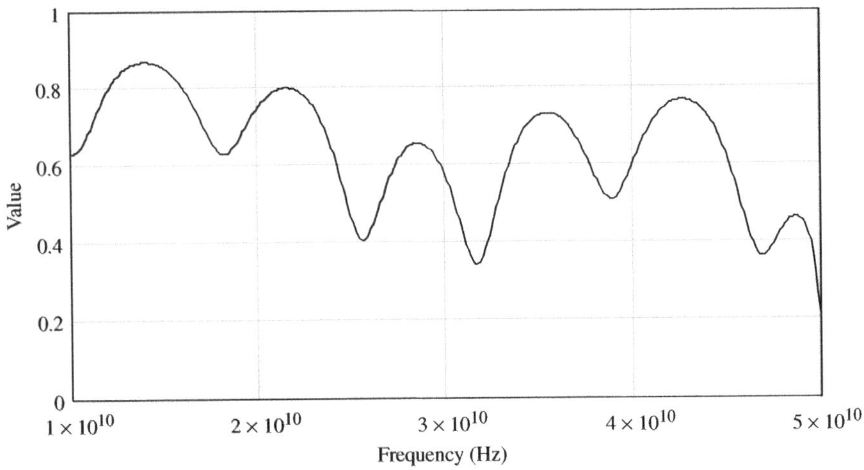

FIGURE 2.30 Wideband reflection coefficient of sheet explosive in front of human skin.

The model shown in Figure 2.31 can be compared with practical measurements taken by various researchers. Ivashov et al. [79] measured the respiration and pulse rate of a male subject at a range of 1.1 m through a brick wall and recorded data as shown in Figure 2.32.

Staderini [80] also measured heartbeat and respiration and showed similar responses. Doppler frequencies are given in Table 2.8 for radar with a center frequency of 1 GHz. The Doppler frequency can be calculated for other frequencies in direct proportion to the value at 1 GHz.

2.7.4 Clothing

Millimeter- and submillimeter-wave radiation is reasonably transparent to clothing. The wavelength used, target material composition and structural configuration, background clutter, target size, distance to target, angle to target, polarization state of the source, and detector (and hence system design) all play an important role in the transparency of the target material and hence system

TABLE 2.7 Summary of Voltage Reflection Coefficients at 24 GHZ

Zone of the body	Volunteer A	Volunteer B
Under left arm	0.63	0.72
Under right arm	0.64	0.74
Left collarbone	0.63	0.60
Right collarbone	0.63	0.56
Center chest	0.57	0.59
Left side chest	0.66	0.60
Right side chest	0.65	0.60
Center back	0.66	0.60
Solar plexus	0.62	0.56
Left side waist	0.64	0.59
Right side waist	0.63	0.57
Lower back	0.65	0.59
Right front lower leg	0.36	0.53
Left front lower leg	0.53	0.52
Right rear calf	0.41	0.50
Left rear calf	0.43	0.49
Right outer thigh	0.43	0.45
Left outer thigh	0.42	0.45
Right inner thigh	0.40	0.50
Left inner thigh	0.44	0.50
Average	0.55	0.56
Standard deviation	0.11	0.08

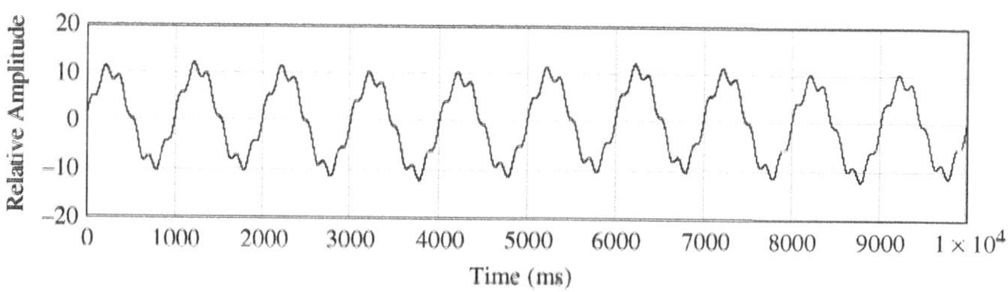

FIGURE 2.31 Modeled respiration and pulse record of human.

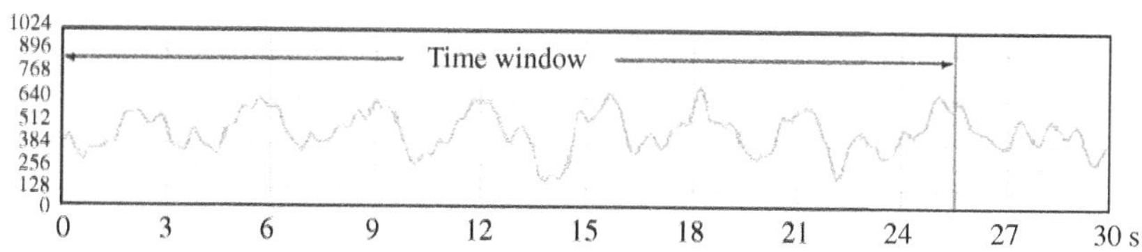

FIGURE 2.32 Respiration and pulse record of human behind wall (with kind permission ©PIERS).

TABLE 2.8 Doppler Frequencies at Center Frequency of 1 GHZ for Human Movement

Parameter	Velocity (m s^{-1})	Doppler Frequency (Hz)
Hand movement	5	33
Rapid body movement	3	20
Walking	0.5	3.3
Heart movement	0.45	1–3
Breathing	0.01	0.1–0.5

performance. Work by numerous researchers in the millimeter-wave and submillimeter-wave/Terahertz regions driven by a need to detect suicide bombers has resulted in a good understanding of the properties of clothing. The U.S. National Research Council [81] published a major review of Terahertz technology and referenced its report with data from various sources. A major source of information came from the Terahertz Imaging Focal Plane Array Technology (TIFT) program of the Defense Advanced Research Projects Agency (DARPA), work carried out by Ohio State University (OSU) and the University of California at Santa Barbara (UCSB) characterized the submillimeter-wave transmission properties of various articles of clothing. The U.S. Army provided four items of clothing to OSU for measurement, and university students and staff from UCSB provided three items of clothing. The transmission measurements in decibels are shown in Figure 2.33. From this it can be seen that above 400 GHz attenuation increases with frequency to significant levels. The moisture content and composition of the materials is not stated.

Measurements on the published transmission properties of clothing by Dickinson et al. [82] are shown in Figure 2.34. The measurements cover the range of 300 GHz to 1.5 THz and show the same general pattern of increasing attenuation with frequency, particularly for materials such as wool.

Gatesman et al. [83] measured the transmittance and reflectance of an variety of building, fabric, and clothing materials at frequencies extending from millimeter waves (94 GHz) to approximately 2 THz. Measurements were made at 94 GHz using a Gunn diode source giving a dynamic range of 70 dB while measurements at 326 GHz, 584 GHz, and 1.04 THz were made using an optically pumped laser. The attenuation for the E field both parallel and perpendicular to a (approximate) symmetry direction in the sample was measured. For example, the symmetry direction was chosen to be parallel to the grain lines in the wood samples. The radiation output was linearly polarized in the vertical direction, so the measurements were made with the symmetry direction in the vertical direction (parallel attenuation) and with the sample rotated 90° (perpendicular attenuation). A summary of material attenuation values found using fixed frequency techniques is given in Table 2.9.

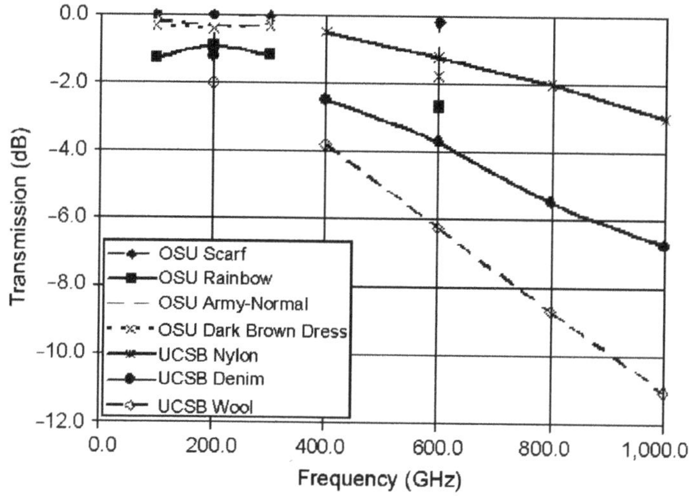

FIGURE 2.33 Submillimeter-wave attenuation of clothing materials (©OSU and UCSB).

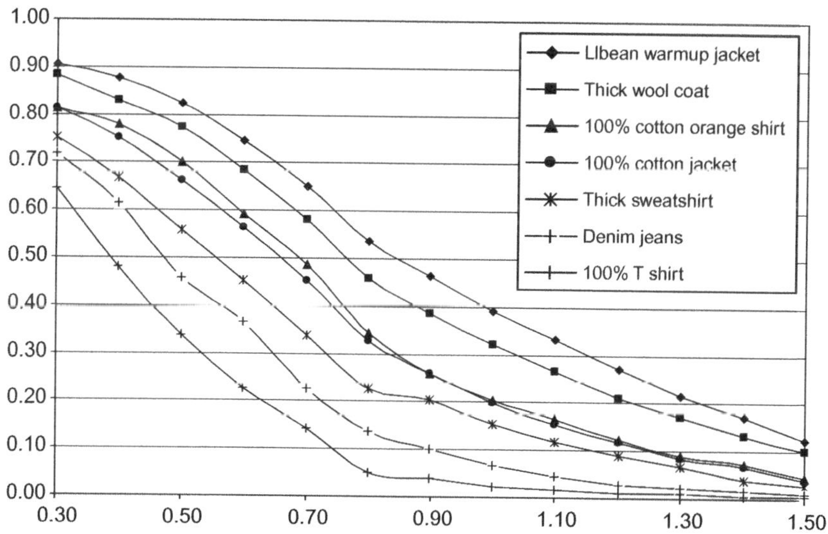

FIGURE 2.34 Submillimeter-wave attenuation of clothing materials (with kind permission ©IEEE).

From the results shown in Table 2.9 there is evidence of variation in the cross-polar attenuation characteristics of clothing in addition to the general trend of increasing attenuation with frequency. However, only certain materials show a consistent trend as a function of frequency, leather being the best

TABLE 2.9 Attenuation of Common Clothing Materials (all values in decibels)

Material	Thickness (mm)	94 GHz		Ratio	326 GHz		Ratio	584 GHz		Ratio	1042 GHz		Ratio
		∥	⊥		∥	⊥		∥	⊥		∥	⊥	
Cotton Shirt	0.3	0.2	0.1	2.0	0.3	0.5	0.6	1.0	1.1	0.9	3.1	3.2	1.0
Denim	0.6	0.7	0.7	1.0	1.3	1.4	0.9	3.4	2.9	1.2	10.0	7.9	1.3
Drapery	0.9	0.3	0.5	0.6	3.0	1.7	1.8	7.5	7.6	1.0	12.3	11.4	1.1
Leather	1.3	0.7	0.6	1.2	2.3	2.1	1.1	6.0	5.2	1.2	17.9	15.3	1.2
Sweater	2.1	0.4	0.4	1.0	3.8	4.0	1.0	14.5	13.7	1.1	19.1	21.4	0.9
Sweatshirt	2.1	0.3	0.2	1.5	0.8	1.1	0.7	4.3	3.8	1.1	14.3	13.9	1.0

n/t, no transmission; n/m, no measurement; ∥, parallel; ⊥, perpendicular.

With kind permission of Dr. Gatesman [83].

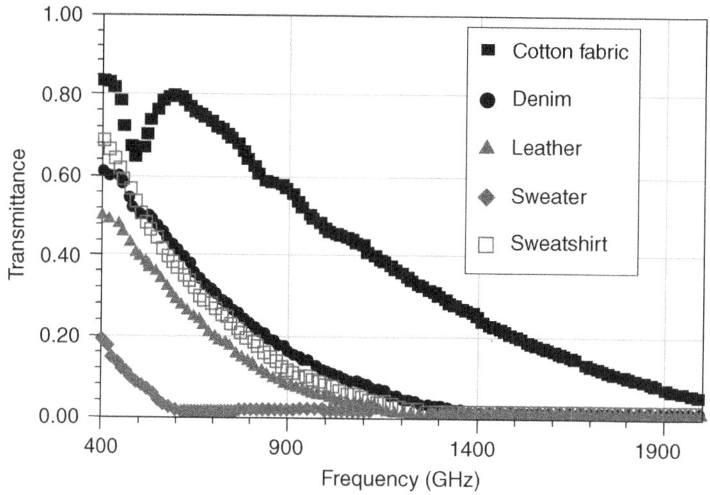

FIGURE 2.35 Terahertz transmittance of common clothing (reprinted with kind permission of Dr. Bjarnason and ©2004 American Institute of Physics, *Applied Physics Letters* [84]).

example. This anisotropic behavior is noteworthy insofar as it affects the reflectivity versus frequency characteristics of a particular material. It is also expected that diffraction will occur since the periodicity of the weave in many clothing materials is on the order of the wavelength of the radiated signal. As the wavelength becomes smaller, the period of the weave of the material in terms of reflection and transmission will show this effect. Figures 2.35 and 2.36 (detailed) show evidence of this for cotton at $\lambda = 0.56$ mm (535 GHz).

Bjarnason et al. [84] carried out measurements of the millimeter wave, Terahertz, and midinfrared transmission through common clothing at 1 THz and produced results as shown in Table 2.10. From this it can be seen that leather and wool show the highest attenuations at 1 THz.

Since individuals attempting to conceal objects may well wear leather coats, systems operating below 0.4 THz will outperform those at higher frequencies at longer standoff distances from the detection system.

2.7.5 Construction Materials

Targets may often be concealed behind walls, which can be constructed from a variety of materials such as wood, brick, concrete, and the like. Attenuation can be calculated from Eq. (2.45), and it can be seen that attenuation (decibels/meter) depends linearly on frequency. Attenuation versus frequency values for typical dry materials is given in Figure 2.37. However, these values are for dry indoor materials, and, when wet, the attenuation can rise dramatically. It is worth noting that concrete just after mixing in its "green" state may exhibit

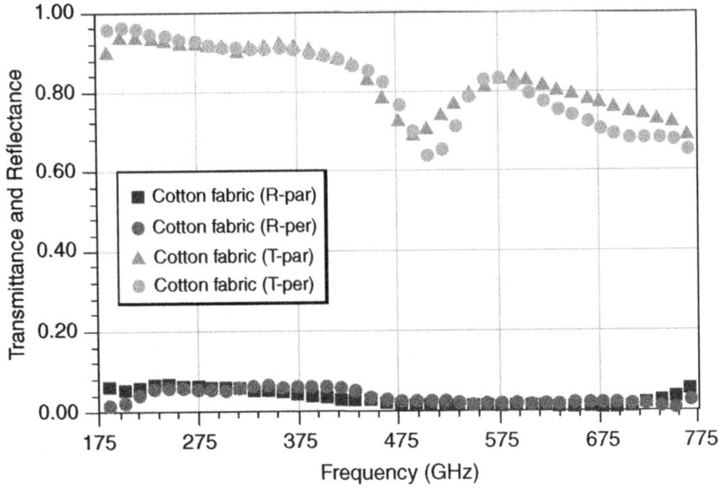

FIGURE 2.36 Terahertz transmittance and reflectance of cotton fabric displaying evidence of diffraction (reprinted with kind permission of Dr. Bjarnason and ©2004 American Institute of Physics, *Applied Physics Letters* [84]).

TABLE 2.10 Physical Properties and Transmission Results for Cloth Samples

Fabric	Thickness (mm)	Density ($kg\ m^{-3}$)	Frequency (THz) (−3 dB)	Attenuation in dB @ 1 THz	ε' @100 kHz
Wool	0.17	214	0.35	11.0	1.6
Linen	0.12	509	0.35	8.0	2.9
Leather	0.6	813	0.40	10.0	5.1
Denim	7.4	490	0.50	6.5	2.3
Naugahyde	8.3	800	0.70	5.5	2.6
Silk	15	256	1.0	3.0	1.4
Nylon	31	379	1.0	3.0	1.9
Rayon	34	733	>1.0	2.5	1.9

Source: Reprinted with kind permission of Bjarnasun and © 2004 American Institute of Physics, *Applied Physics Letters* [84].

high values of both relative permittivity (ε_r = 10 to 20), and attenuation losses in excess of 200 dB m^{-1} at 1 GHz can be expected. Even apparently dry concrete can have high values of attenuation (115 dB m^{-1} at 0.9 GHz) as reported by AvaLAN systems [85], and the attenuation loss and relative dielectric constant in concrete is dependent upon a number of factors including water content, mix constituents, and the curing regime employed. When hydration and curing are effectively complete, at periods of possibly up to

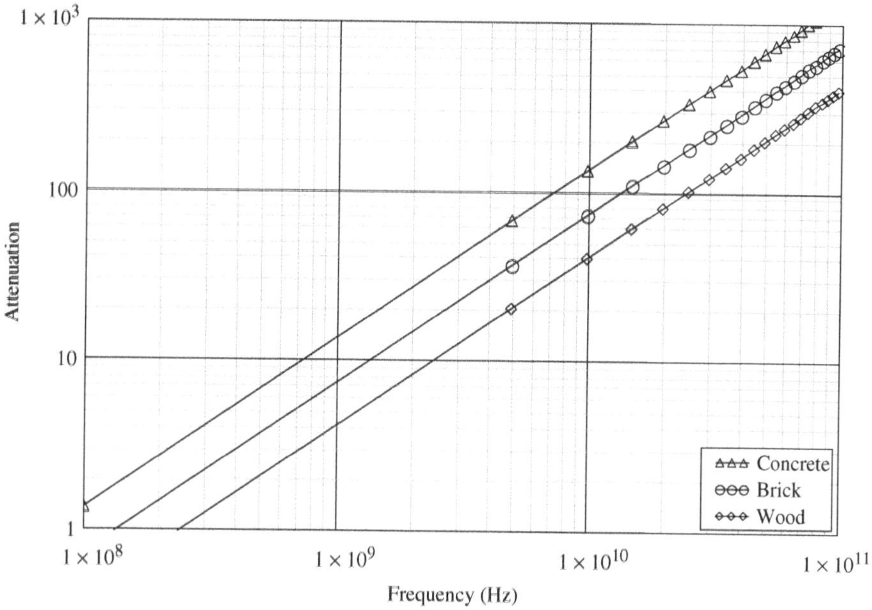

FIGURE 2.37 Estimated one-way attenuation of building materials as a function of frequency.

about 6 months, lower values may be measured such that $\varepsilon_r = 4$ to 10 and $\alpha = 5$ to 25 dB m^{-1} at 1 GHz. Note that reinforced concrete may have losses of 15 to 20 dB m^{-1} greater, and the size and orientation of the reinforcing may have significant effects on propagation.

It should be noted that the possibility of detecting or imaging objects through walls reduces as frequency increases, and, at submillimeter and Terahertz wavelengths, the only realistic situations in which this can be achieved are when the material is very thin (<10 mm) and very dry. Measurements made with millimeter-wavelength and Terahertz imaging systems of transmission through various building materials by Ohio State University are shown in Figure 2.38. As the thickness of the material is not specified, it is not possible to determine absolute attenuation, but clearly the losses at Terahertz wavelengths are such as to rule out Terahertz as a reliable detection regime.

Work by Gatesman et al. [86] on the attenuation of common materials, as shown in Tables 2.11 and 2.12, also shows the same general trend, and concrete shows an equivalent loss of nearly 900 dB m^{-1} at 94 GHz. Other materials show similarly high values of attenuation at 94 GHz: glass = 2000 dB m^{-1}, brick = 1000 dB m^{-1}, wood = 787 dB m^{-1}, cinder block = 1964 dB m^{-1}.

TABLE 2.11 Attenuation of Common Construction Materials (all Values in Decibels)

Material	Thickness (mm)	94 GHz		Ratio	326 GHz		Ratio	584 GHz		Ratio	1042 GHz		Ratio
		∥	⊥		∥	⊥		∥	⊥		∥	⊥	
Cardboard	3.9	1.2	1.3	0.9	2.8	3.2	0.9	4.4	5.0	0.9	9.0	9.4	1.0
Maple 1	3.2	2.6	1.8	1.4	8.4	5.3	1.6	16.2	11.0	1.5	n/m	n/m	n/m
Maple 2	6.4	5.9	4.0	1.5	20.1	16.0	1.3	31.4	22.7	1.4	65.7	52.4	1.3
Maple 3	12.7	10.8	7.1	1.5	32.4	22.5	1.4	62.6	45.9	1.4	n/t	n/t	n/t
Maple 4	19.4	16.9	10.4	1.6	48.0	33.1	1.5	n/t	68.4	n/t	n/t	n/t	n/t
Maple 5	19.4	16.6	9.5	1.7	46.5	31.0	1.5	n/t	62.7	n/t	n/t	n/t	n/t
OSB 1	6.4	6.4	7.1	0.9	33.2	33.4	1.0	47.8	48.2	1.0	n/t	n/t	n/t
OSB 2	12.7	20.6	18.9	1.1	59.4	55.8	1.1	n/t	n/t	n/t	n/t	n/t	n/t
Plywood 1	6.4	5.3	4.5	1.2	18.2	16.7	1.1	31.3	30.2	1.0	n/t	61.9	n/t
Plywood 2	12.7	8.7	10.8	0.8	30.3	30.4	1.0	31.3	30.2	1.0	n/t	n/t	n/t
SPF1	3.2	3.0	1.7	1.8	7.2	4.9	1.5	14.3	9.3	1.5	29.2	20.7	1.4
SPF2	6.4	4.8	2.8	1.7	14.0	8.5	1.6	24.5	15.9	1.5	56.5	38.4	1.5
SPF3	12.7	10.8	5.8	1.9	28.8	19.2	1.5	55.9	38.4	1.5	n/t	n/t	n/t
SPF4	18.8	15.8	9.6	1.6	42.6	27.3	1.6	53.3	72.0	0.7	n/t	n/t	n/t
SPF5	25.0	21.0	12.3	1.7	55.3	35.8	1.5	n/t	74.0	n/t	n/t	n/t	n/t
SPF6	36.3	30.2	18.7	1.6	70.4	54.4	1.3	n/t	n/t	n/t	n/t	n/t	n/t

n/t, no transmission; n/m, no measurement; ∥, parallel; ⊥, perpendicular.

With kind permission of Dr. Gatesman.

TABLE 2.12 Attenuation of Common Building Materials

Material	Thickness (mm)	94 GHz		Ratio	326 GHz		Ratio	584 GHz		Ratio	1042 GHz		Ratio
		‖	⊥		‖	⊥		‖	⊥		‖	⊥	
Concrete	11.1	9.8	10.5	0.9	47.7	49.2	1.0	n/t	n/t	n/t	n/t	n/t	n/t
Drywall	9.5	1.6	1.7	0.9	10.7	10.5	1.0	35.2	35.0	1.0	n/t	n/t	n/t
Drywall	12.7	2.2	2.8	0.8	12.8	13.1	1.0	49.1	50.4	1.0	n/t	n/t	n/t
Glass	2.2	4.5	4.3	1.0	10.8	11.0	1.0	25.3	25.4	1.0	n/t	n/t	n/t
Plastic blind	0.8	0.5	0.5	1.0	1.3	1.4	0.9	3.6	3.4	1.1	8.2	8.1	1.0
Vinyl siding	1.1	0.7	0.8	0.9	2.5	2.3	1.1	5.4	5.4	1.0	12.6	11.4	1.1
Vin. sid. & §″ Cdx	13.0	10.8	12.6	0.9	32.5	32.5	1.0	68.1	68.2	1.0	n/t	n/t	n/t
Wall section	121.1	17.9	22.1	0.8	69.0	70.6	1.0	n/t	n/t	n/t	n/t	n/t	n/t
Brick 1	11.5	8.7	8.9	1.0	62.7	64.3	1.0	n/t	n/t	n/t	n/t	n/t	n/t
Brick 2	15.1	15.1	15.3	0.9	n/t	n/t	n/t	n/t	n/t	n/t	n/t	n/t	n/t
Bricks	6.4	5.4	5.1	0.9	39.9	40.3	1.0	n/t	n/t	n/t	n/t	n/t	n/t
Brick 4	5.2	5.9	5.7	1.0	n/m	n/m	n/t	n/t	n/t	n/t	n/t	n/t	n/t
Bricks	8.6	7.1	6.7	1.1	46.9	50.0	0.9	n/t	n/t	n/t	n/t	n/t	n/t
Cinder block 1	27.7	45.9	48.3	1.0	n/m	n/m	n/t	n/t	n/t	n/t	n/t	n/t	n/t
Cinder block 5	8.4	16.5	16.5	1.0	53.1	53.1	1.0	52.8	56.8	0.9	n/t	n/t	n/t
Cinder block 7	9.8	17.7	17.4	1.0	54.1	52.7	1.0	62.3	60.6	1.0	n/t	n/t	n/t
Cinder block 9	17.0	26.6	26.9	1.0	74.7	75.6	1.0	n/t	n/t	n/t	n/t	n/t	n/t

n/t, no transmission; n/m, no measurement; ‖, parallel; ⊥, perpendicular.

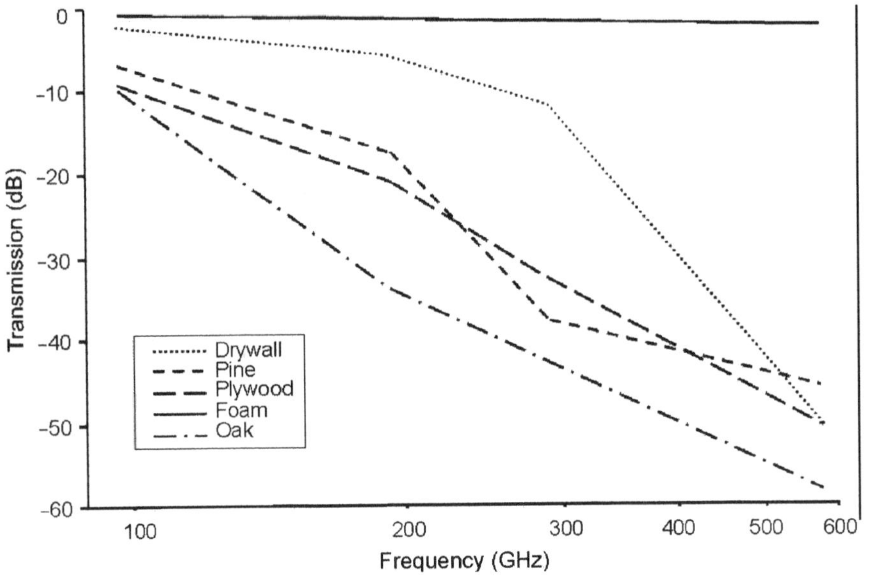

FIGURE 2.38 Submillimeter-wave attenuation of construction materials (with acknowledgment ©OSU).

2.7.6 Explosives

Most explosives, with the exception of nitroglycerin and ammonium nitrate, have a comparatively low relative dielectric constant, as shown in Table 2.13. The loss tangent of most explosives is in the region of 0.001, and this results in a reflection coefficient as a function of relative dielectric constant, as shown in Figure 2.39.

Considerable work has been carried out on the spectroscopic identification of explosives and other materials such as narcotics at Terahertz wavelengths. The literature provides many examples of predicted and measured spectra of both reflected and transmission spectra. Note that the latter is only relevant to systems that can be used two sided around the target and its concealing structure, and transmission properties of the concealing material may well be excessive at Terahertz wavelengths.

Kemp [87] first reported the spectra of common energetic compounds and commercial explosives based on these compounds (PE4, Semtex-H) in 2006. These results are shown in Figure 2.40 and have subsequently been validated and extended by Yamamoto et al. [88], Cook et al. [89], and Shen et al. [90] using time-domain Terahertz spectroscopy and Fourier transform infrared (FTIR). The strong absorption features, particularly of RDX-based explosives around 800 GHz, open up the possibility of material-specific detection of these

TABLE 2.13 Relative Dielectric Constant of Explosives at 1GHZ

Substance	Name	Molecular Weight	Density (g cm^{-3})	Relative Dielectric Constant
TNT	2,4,6-Trinitrotoluene	227.13	1.65	2.70
Deta sheet	PETN			2.72
PETN	Pentaerythritol tetranitrate	316.20	1.78	2.72
Comp B	RDX TNT			2.90
Octol	HMX TNT			2.90
Tetryl	2,4,6-Trinitrophenyl-*N*-methylnitramine	287.15	1.73	2.90
Semtex-H	RDX-PETN			3.00
HMX	Cyclotetramethylene-tetranitramine	296.16	1.96	3.08
Comp C-4	RDX			3.14
RDX	RDX Hexahydro-1, 3,5-trinitro-1, 3,5-triazine	222.26	1.83	3.14
AN	Ammonium nitrate	80.05	1.59	7.10
NG	Nitroglycerin	227.09	1.59	19.00

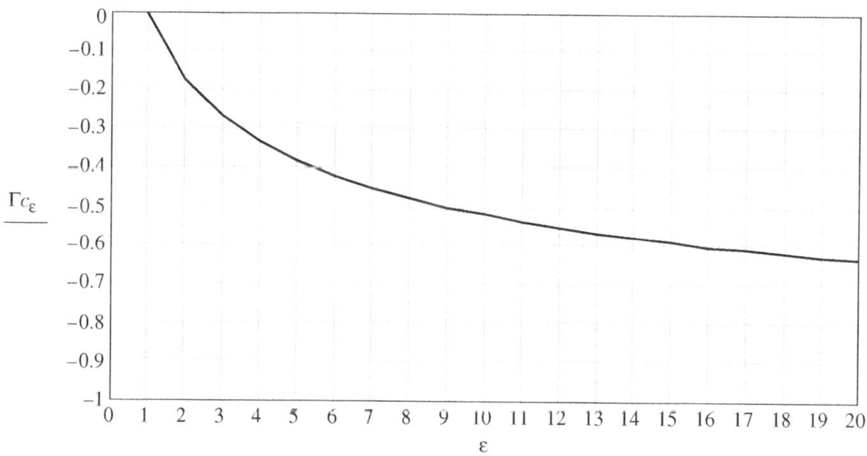

FIGURE 2.39 Reflection coefficient of explosive as a function of relative dielectric constant.

materials [cyclotrimethylenetrinitramine (RDX), Pentaerythritol tetranitrate (PETN), tetrahexamine tetra + nitramine (HMX), and trinitrotoluene (TNT)].

It should, however, be noted that practical detection systems will need to operate in reflection rather than transmission mode due to the high absorption coefficients of the explosives themselves and, for people screening applications, due to absorption by the body. Figure 2.41 shows the refractive index of RDX

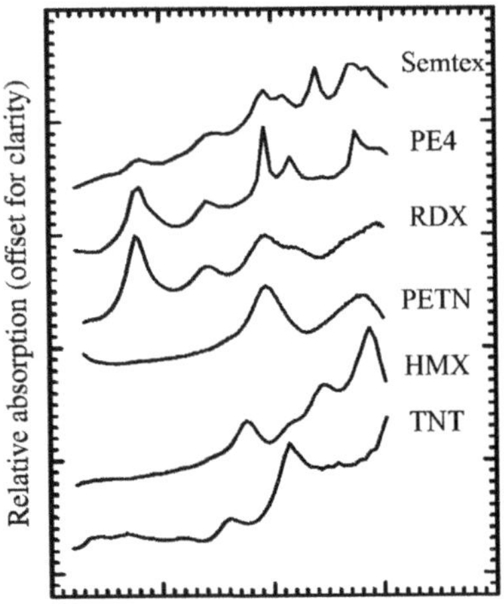

FIGURE 2.40 Terahertz absorption spectra for various explosives (©SPIE).

and PETN; the spectral features due to the resonances are still visible but are much less strong in reflection geometry measurements.

Kemp used pulsed Terahertz systems to demonstrate proof-of-principle standoff explosives detection starting with the materials characterization of explosives, barrier and confusion materials, work on reflection spectroscopy, and the development of laboratory prototype systems for standoff detection at a distance of 1 m as described by Baker et al. [91]. Real-time spectroscopic detection of RDX-based explosives (1/15-sec integration time) beneath several layers of clothing at a distance of 1 m, is reported by Kemp [87].

Liu et al. [92] reported measurements on RDX whose relative power reflection is shown in Figure 2.42. The diffusely reflected Terahertz pulses from the RDX and from the Teflon pellet surfaces were measured in a chamber purged with nitrogen to avoid the effect of water vapor absorptions in ambient air. Other materials that are commonly encountered are shown for comparison.

In the transmission measurements Liu [92] noted a strong absorption peak at 0.82 THz and three relatively weak absorption peaks at 1.05, 1.35, and 1.55 THz and concluded these can also be well identified by the diffuse reflection measurement. The RDX samples were covered with the following materials: polyethylene sheet (~0.1 mm, black), leather (~0.3 mm, yellow), and polyester cloth (~0.4 mm, green), and the main peak was still clearly observable. Liu notes that in the atmosphere, the water vapor absorptions affected the obtained Terahertz spectrum, especially in the range above 1.0 THz.

Cook at al. [93] as shown in fig 2.44 concentrated on examining the effect of the atmospheric absorption on the reflection resonance spectra as they considered that suitable windows must be used for reliability of detection.

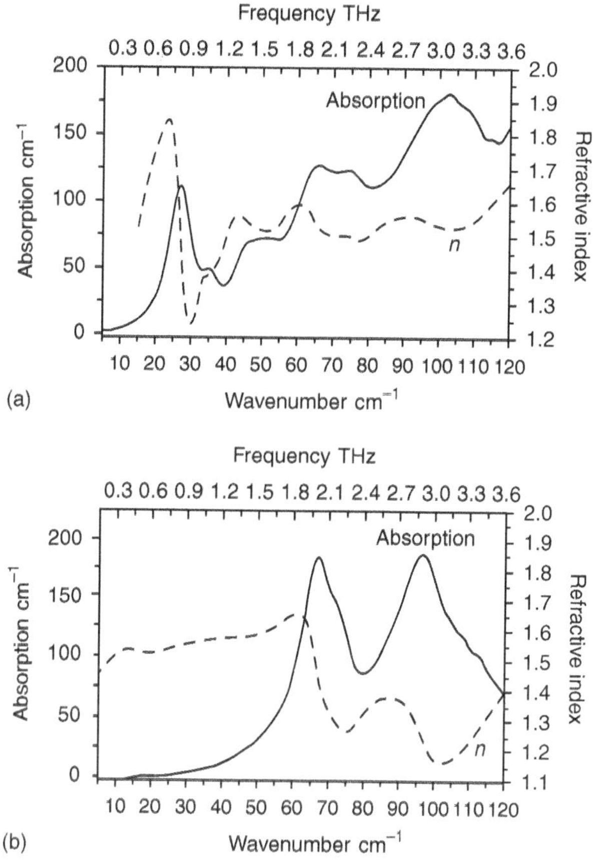

FIGURE 2.41 Terahertz absorption spectra (solid line) and refractive index (dashed line) of (*a*) RDX and (*b*) PETN (©SPIE).

They developed a standoff system and reported results for samples of PETN and HMX.

Work on the Terahertz detection of explosives by Federici et al. [94] also confirmed the absorption spectra of various explosives. Methods of detection considered by Federici [94] included operation over the 3 to 7-THz range, identifying the material based on the slope of the increase in Terahertz absorption in the 0.2 to 3-THz range or dual measurements one at 0.2 THz and another at roughly 2.5 THz. His paper reviews some of the fundamental physics of absorption process and also compares work predominantly by Kemp et al. [95] who produced spectra of various explosives as shown in Figure 2.40. Note that each has been offset, and the graphs show the relative relationship rather than absolute absorption.

Federici listed the various resonance frequencies for various explosives, and these are listed in Table 2.14.

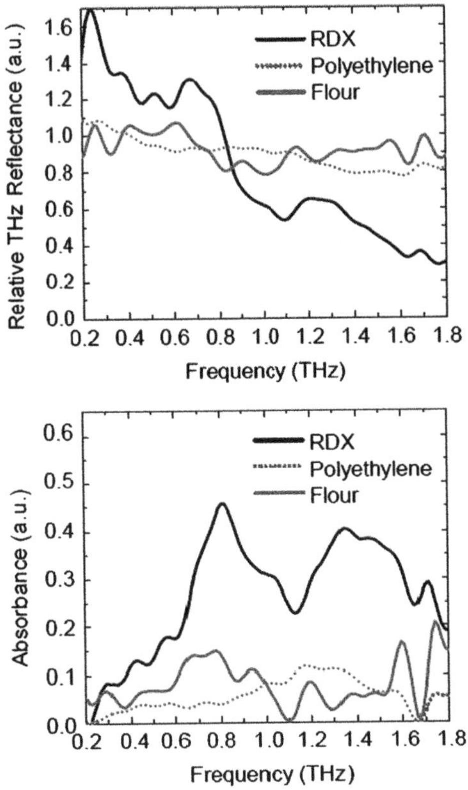

FIGURE 2.42 Relative power reflection spectra (lhs) and transmission spectra of RDX, polyethylene, and flour (rhs) (Liu ©Optics).

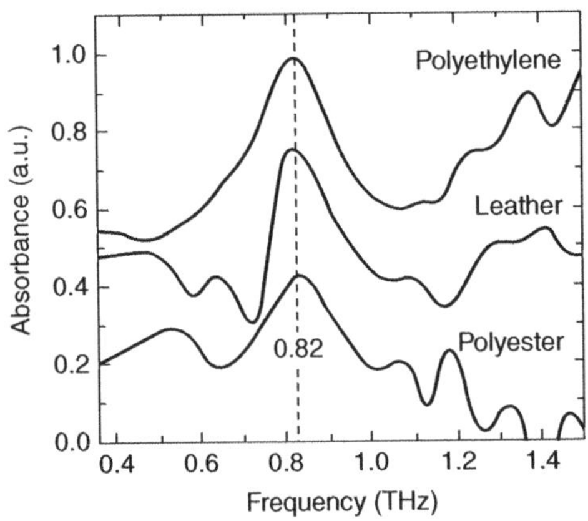

FIGURE 2.43 Absorption spectra of RDX using different coverings (Liu ©Optics).

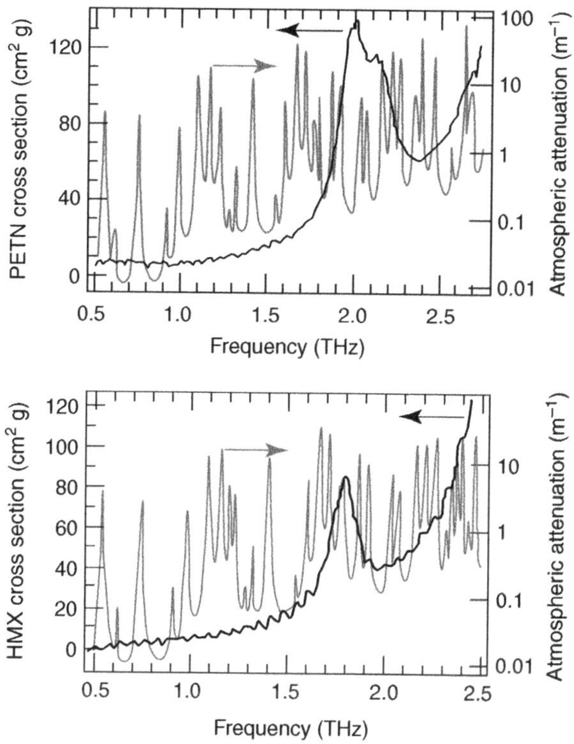

FIGURE 2.44 Spectra of PETN (lhs) and HMX (rhs) with overlaid atmospheric absorption spectra (Cook ©2005 IEEE).

TABLE 2.14 Resonant Absorption Frequencies for Various Explosives

Explosive	Frequency (THz)
Semtex-H	0.72, 1.29, 1.73, 1.88, 2.15, 2.45, 2.57
PE4	0.72, 1.29, 1.73, 1.94, 2.21, 2.48, 2.69
PETNa	1.73, 2.51
PETNb	2.01
HMXa	1.58, 1.91, 2.21, 2.57
HMXb	1.84
TNTa	1.44, 1.91
TNTb	1.7
TNT	5.6, 8.2, 9.1, 9.9
NH_4NO_3	4, 7

Source: With kind permission of © I.P.

While the spectral features of explosives do vary from type to type, many of the resonances are low Q and relatively small and are generally dominated by the gradually increasing attenuation as a general function of frequency. In addition the reflection spectra are less pronounced than the absorption spectra.

2.8 SUMMARY

The main physical phenomena that influence the detection of concealed targets have been considered in this chapter. Some parameters are within the control of the designer and operator, and these relate to frequency of operation, antenna radiation patterns, polarization, and whether the system operates as radar or as a passive system. The factors that can be exploited as means of detection, recognition, and identification relate to specific indicators of the target, whether this is associated with explosives (absorption or reflection spectra) or with humans (Doppler or micro-Doppler signatures). The difficulty of detection is increased by the intervening path losses, whether they are due to atmospheric absorbers for standoff systems or clothes or earth materials for close-in detection systems. The effect of clutter is to make detection of the target more difficult and is particularly noticeable for radar systems working through the rubble of earthquakes. As control of the spatial radiation pattern is critically important, the next chapter considers the design issues associated with antenna design.

CHAPTER 3

Antennas

3.1 INTRODUCTION

This chapter provides an introduction to the key parameters and physics of antennas and is intended as a primer for readers unfamiliar with the principles and design of antennas. Further information on antennas is found in the literature [96–107], which provides a variety of treatments and useful further reading and references. The author is grateful for the permission of the IET to use copyright material from Danielss [2] for this chapter.

The definitions relating to antennas are taken from IEEE Standard Definitions of Terms for Antennas, IEEE STD 145-1983, which defines an antenna as that part of a transmitting or receiving system designed to radiate or to receive electromagnetic waves.

An antenna is a device for coupling energy from a source of RF energy into a transmitting medium, which is normally air, but in the case of concealed targets hidden under clothing, in building structures, or buried in the ground it can be considered a lossy dielectric. An antenna is also a device that can absorb electromagnetic energy emitted in free space and efficiently deliver it to a connected receiver. When reciprocity applies, then the antenna functions equally when used as part of a receiving or a transmitting system. An antenna can be used to transmit energy, receive energy, or both depending on the design of the radar or RF system.

The angular response of an antenna can vary depending on the requirements. Antennas may be omnidirectional or directional. An omnidirectional antenna has an essentially nondirectional pattern in a given plane of the antenna and a directional pattern in any orthogonal plane. While an antenna equally sensitive in radiating to, or receiving from, all angles in space (or *isotropic*), as shown in Figure 3.1, is a practical impossibility, the concept of an

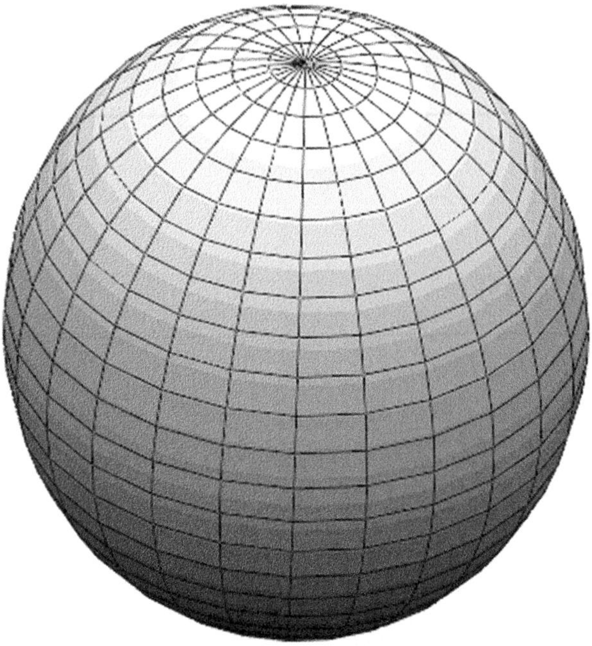

FIGURE 3.1 Isotropic surface radiation pattern.

omnidirectional or isotropic antenna, which is one that radiates energy uniformly in all directions, is a useful, if fictitious, concept to enable the characteristics of real antennas to be referenced.

Directional antennas radiate energy in patterns of lobes or beams that extend outward from the antenna in one direction in a particular plane for a given antenna position. The radiation pattern also contains minor lobes, otherwise known as side lobes or back lobes, but these lobes are weak and may have little effect on the main radiation pattern, although they can cause targets other than the wanted target in the main beam to be detected. In typical radar systems the main lobe may vary in half-power beam width in both azimuth and elevation.

Antennas have two important characteristics: directivity and power gain. The directivity of an antenna refers to the degree of sharpness or directionality of its beam. If the beam is narrow in either the azimuth or elevation plane, the antenna is said to be directional in that plane. Conversely, if the beam is wide in either plane, the directionality of the antenna in that plane is low. When the directivity of an antenna is increased, this implies a directional angular response when the beam or main lobe is narrowed, less power is required to cover the same range because the power is concentrated.

The power gain of an antenna is directly related to directivity but also includes effects of losses within the antenna. Practically, it is measured by the

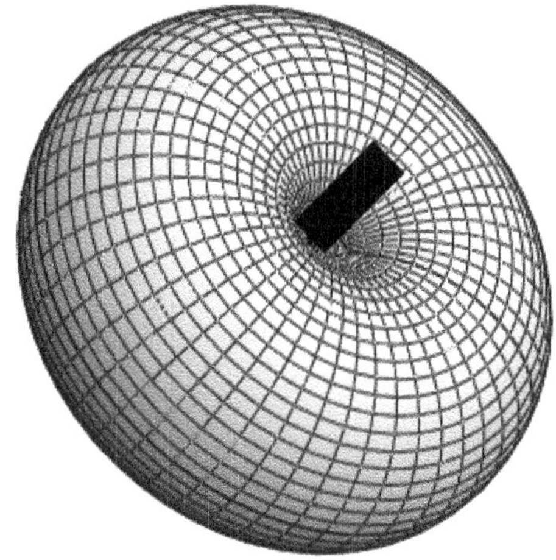

FIGURE 3.2 Radiation surface pattern of simple dipole.

ratio of the radiated power from the antenna to that of a co-polarized reference (basic) antenna, the gain of which is accurately known a priori. In order to measure power gain, both the reference antenna (usually dipoles or horns) and the antenna under test (AUT) must have been excited or fed in the same manner, and each must have radiated from the same position. A single point of measurement for the power gain ratio must lie within the radiation field of each antenna. An antenna with high directivity usually has a high power gain, and vice versa. The radiation pattern of a typical reference antenna such as a dipole is shown in Figure 3.2 where the dipole lies in the vertical plane. This pattern is omnidirectional in azimuth and the polarization is vertical.

3.2 ANTENNA PARAMETERS

This section describes the basic antenna parameters that need to be considered when designing and operating systems for the detection of concealed targets.

3.2.1 Antenna Directivity

The directivity of an antenna is given by the following expression:

$$D = \frac{\text{average radiation intensity}}{\text{maximum radiation intensity}} \tag{3.1}$$

where in the case of a transmitting antenna the radiation intensity $P(\theta, \phi)$ is defined as the radiated power per solid angle (watts/sterad).

Essentially, directivity is derived by normalizing the radiation pattern by the integrated total power and leads to

$$D(\theta, \phi) = 10 \log\left[\frac{4\pi P(\theta, \phi)}{\iint P_{in}(\theta, \phi) \sin\theta \, d\theta \, d\phi}\right] \tag{3.2}$$

It can be proved that for the case of antennas with a well-developed single main beam the following approximation is valid:

$$G_D = \frac{4\pi}{\theta_{el}\phi_{az}} \tag{3.3}$$

where θ_{el} is the elevation half-power beam width in radians and ϕ_{az} is the azimuth half-power beam width in radians.

The half-power beam width (HPBW) of an antenna is an expression, in degrees or radians, of the width of the radiated beam between the half-power or 3 dB points (down from the peak of the beam). Many antennas will exhibit one HPBW in azimuth and a different HPBW in elevation written as $HPBW_A$ and $HPBW_E$. An antenna described as omnidirectional will have equal coverage in all directions. A typical wideband omnidirectional (in azimuth) antenna will have a $HPBW_A$ of 360° and $HPBW_E = 50°$. The directivity of an antenna is generally combined with efficiency and expressed as gain, as described in the next section.

3.2.2 Antenna Gain

The directivity of an antenna is generally combined with the efficiency and expressed as gain (G). Usually, the gain of an antenna is expressed in decibels, 10 $\log_{10}$ (numerical gain), above the signal from an equal magnitude accepted power fictitious reference isotropic radiator. The gain in decibels for this reason is designated dBi.

The gain expressed for a lossless antenna is generally the maximum or peak gain. The power gain of an antenna is defined as G = maximum radiation intensity/radiation intensity from an ideal isotropic antenna. It is assumed that the reference isotropic antenna has accepted the same power P_{in} as the antenna under discussion It is easy to prove that the gain is related to directivity by the following simple relation:

$$G = \eta_r D \tag{3.4}$$

where D = directivity
η_r = *radiation* efficiency of the antenna

Note that the gain of an antenna includes contribution due to resistive losses within the antenna. A related concept to power gain is the effective aperture A_{eff} of the antenna that more appropriately characterizes the antenna performance when used to receive radiation. In particular, this can be seen as a fictitious planar aperture that can capture the same power as the one actually received by the antenna. For true aperture antennas (such as reflectors, horns lenses, etc.) the effective aperture relates to the real projected aperture A by the simple relation:

$$A_{eff} = \eta_a A \tag{3.5}$$

With η_a the antenna efficiency that combines factors about how efficiently the aperture is utilized as well as what ohmic loses are incurred.

Gain and effective aperture are related by the following basic relation:

$$G = \frac{4\pi A_{eff}}{\lambda^2} \tag{3.6}$$

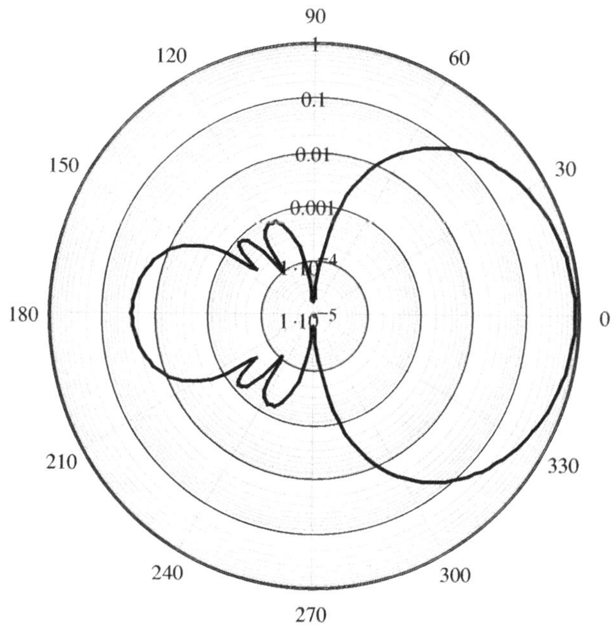

FIGURE 3.3 Typical far-field radiation pattern of an antenna showing main lobe, side lobes, and back lobes.

The gain can also be expressed as

$$G \approx \frac{4\pi\eta_r}{\theta_{\mathrm{EL}}\theta_{\mathrm{AZ}}} \tag{3.7}$$

where the elevation and azimuth 3-dB angles are as before. A typical example of the far-field radiation pattern is shown in Figure 3.3.

3.2.3 Antenna Efficiency

The radiation efficiency of an antenna is the ratio of the total power radiated by an antenna to the net power accepted by the antenna from the connected transmitter. This quantity relates to the fact that all antennas suffer from losses. Antenna efficiency on the other hand expresses how efficiently the antenna aperture is utilized to form a beam. This combines both ohmic losses as well as illumination characteristics of the aperture fields distribution. A simple horn antenna, for example, will not be as efficient as a perfect aperture of the same size because of tapered amplitude and phase field distribution. The total efficiency of an antenna combines impedance match with other factors such as aperture and radiation efficiency to give the overall radiated signal for a given input. The best and most widely used expression of this efficiency is to combine overall efficiency with directivity (of the antenna) and express the efficiency times directivity as the true gain.

3.2.4 Side Lobes and Back Lobes

The side lobes are particular features of the radiation pattern topology representing local maxima toward directions other than the main beam. High side lobes can cause unwanted interference to/from other systems, hence the maximum sidelobe level tolerated is often dictated by the regulatory bodies. For antennas producing a single forward beam the front-to-back ratio (F/B) is the difference between the peak gain of the antenna and the radiation in the back of the antenna (often 180° from the peak of the beam). In the case of bistatic antenna arrangements, a low level of side lobes and back lobes is important to reduce the cross-coupled energy that could saturate the receiver and reduce reflections from structures behind the antenna.

3.2.5 Bandwidth

An antenna will radiate energy over a defined bandwidth. This is sometimes defined as that bandwidth over which the input voltage standing-wave ratio

(VSWR) is 2 : 1. However, in the field of broadband antennas, other VSWR limits may be used. If a limit, say 1 : S is adopted, then the fractional bandwidth of the antenna is defined by

$$B = 2\left(\frac{F_h - F_l}{F_h + F_l}\right) \tag{3.8}$$

where F_h = highest frequency limit for the 1 : S VSWR performance
F_l = lowest frequency limit for the 1 : S VSWR performance

An ultra-wideband antenna has a bandwidth equal or greater than 25% of the fractional bandwidth about the carrier frequency. Most GPR systems need to use antennas that have at a minimum octave bandwidth and often up to decade bandwidth.

3.2.6 Polarization—Linear, Elliptical, and Circular

Polarization has been discussed in detail earlier and is summarized as follows. In a specified direction from an antenna and at a point in its far field, the polarization of an antenna is the polarization of the local plane wave, which is used to represent the radiated wave at that point. The polarization of an antenna is often viewed as the locus of the tip of the transmitted (or received) electric field (E plane). The optimum polarization for a system depends on the polarization of other antennas in the system. An infinite number of polarizations exist, being described as linear, elliptical, and circular. For a linear antenna three possibilities generally seen are vertical, horizontal, and slant linear. It is important to match linear polarizations for transmit-and-receive antennas. A linear polarization mismatch can result in severe loss and practically figures of 20-dB loss are not uncommon for severe cases.

Circular polarization is generally given as right-hand circular polarization (RHCP) or left-hand circular polarization (LHCP). The cross polarization of an antenna is defined as the polarization orthogonal to a specified reference polarization.

3.2.7 Antenna Phase Center

The phase center of an antenna is considered to be that point where the source of an expanding spherical shell of radiation is situated. The concept of a phase center is somewhat artificial, as all radiation has to originate from an oscillating current source that occupies some physical space rather than a point. As noted earlier the isotropic radiator is a useful fiction, but a real source of uniform radiation in all directions is difficult to achieve.

The importance of the concept of antenna phase center relates to the characteristics of the radiated field. For example, with an antenna such as a log periodic or spiral, the physical position of the source of radiation at a particular frequency will vary along either the length of the antenna or the position across the antenna. With horn antennas the phase center will depend on the aperture distribution and the taper of the horn, and the resultant far-field pattern will be affected by the variation of phase center.

For broadband signals the phase centre movements may present a distortion of the received waveform, and as such one either has to use antennas with well-fixed phase centre or prepare to compensate by processing.

3.2.8 Antenna Patterns

Antenna patterns show in a graphical manner the directivity or gain of the antenna versus angle. Generally, two patterns are provided in orthogonal directions and are called the principal plane cuts. Typical examples are the azimuth and elevation plot of antenna gain. These patterns will show not only the gain and beam width for the main beam of the antenna but also side lobes and the back lobe of the antenna. Such plots can be in polar form as already shown in Figure 3.3 or Cartesian form as shown in Figure 3.4.

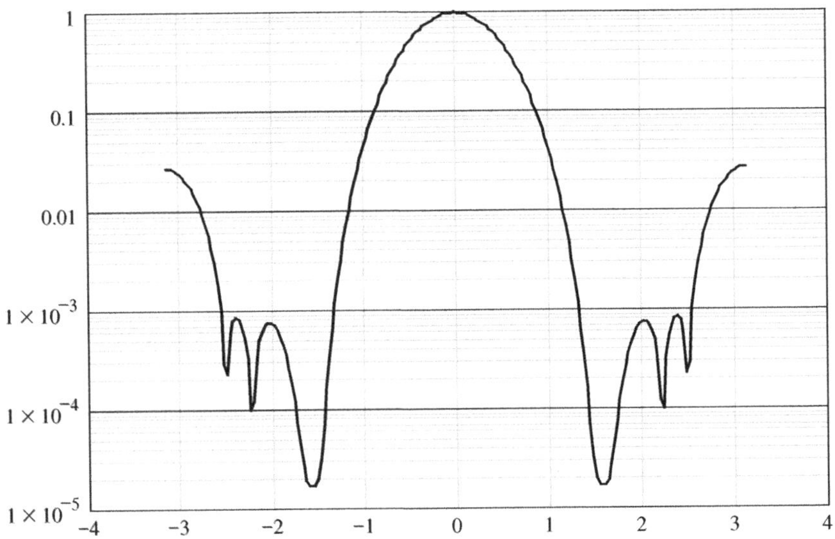

FIGURE 3.4 Typical linear plot of antenna radiation pattern in decibels versus radians.

3.2.9 Time Side Lobes and Ring-down

Antennas that have to work at very short ranges with embedded electromagnetic energy traveling times of some tens of nanoseconds have stringent requirements in terms of self-generated clutter. Therefore, the rate of decay of energy stored within the antenna is a key parameter in defining the inherent self-clutter of the complete radar system. The requirement covers operation in both the time domain and frequency domain; however, it is easiest to understand the issues when described in the time domain for wideband systems involving pulse-type signals. The time side lobes will obscure targets that are close in range to the target of interest, in other words the resolution of the radar can become degraded if the impulse response of the antenna is significantly extended. Two examples are shown of the effect of time side lobes. The linear amplitude of the time-domain waveform is shown in Figure 3.5.

Although the linear plot suggests that the waveform is marginally acceptable, the graph in decibels in Figure 3.6 shows that the time side lobes would limit this particular radar's sensitivity to a dynamic range of 30 dB at 5 ns, assuming a signal-to-side lobe ratio of 10 dB. The result is that targets whose amplitude is less than the time sidelobe level would not be able to be detected unless some form of deconvolution was employed.

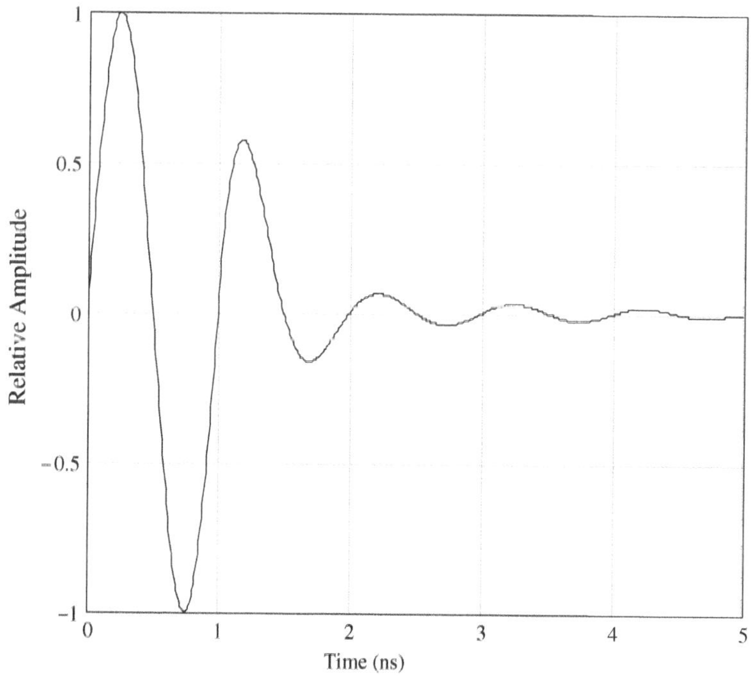

FIGURE 3.5 Linear amplitude of radiated impulse with time side lobes.

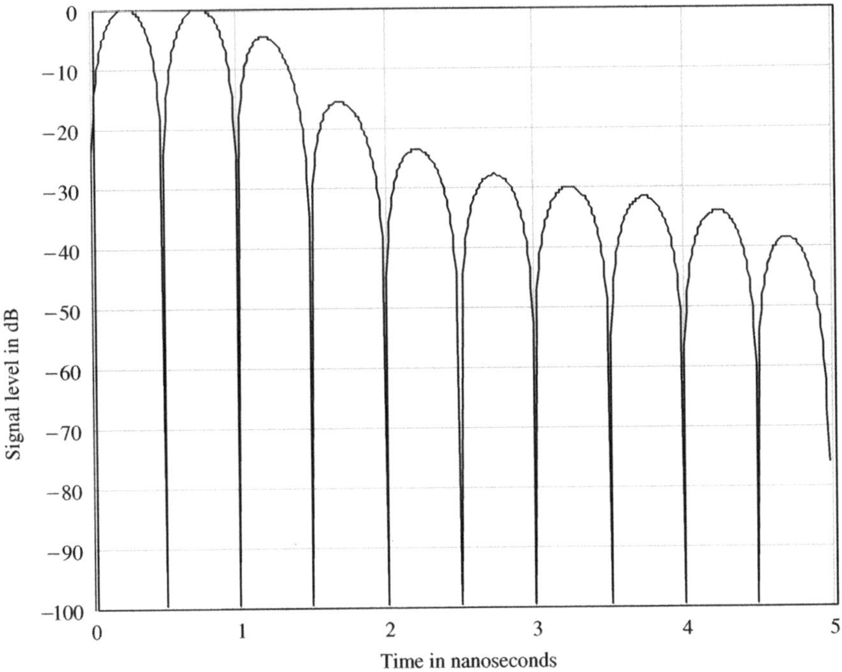

FIGURE 3.6 Amplitude in decibels of radiated impulse in Figure 3.5.

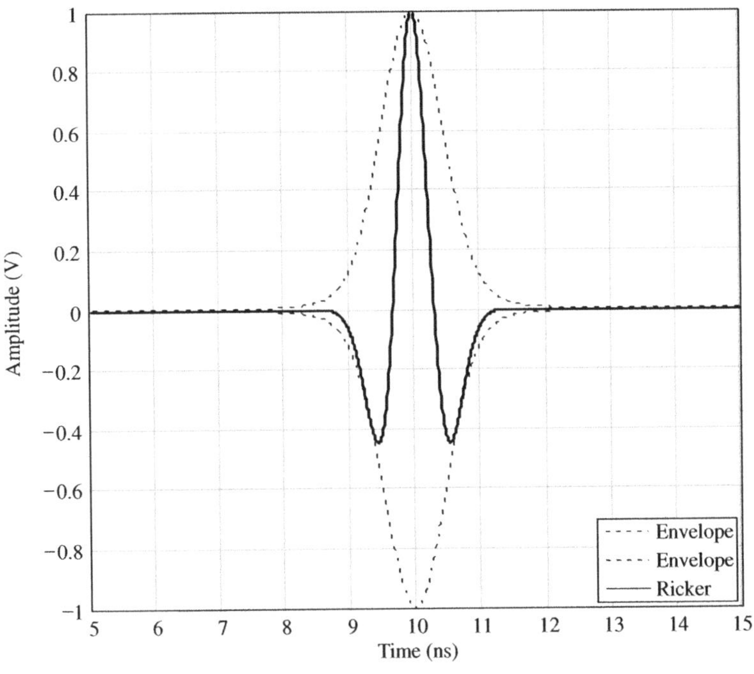

FIGURE 3.7 Ricker wavelet in the time domain showing envelope of the function.

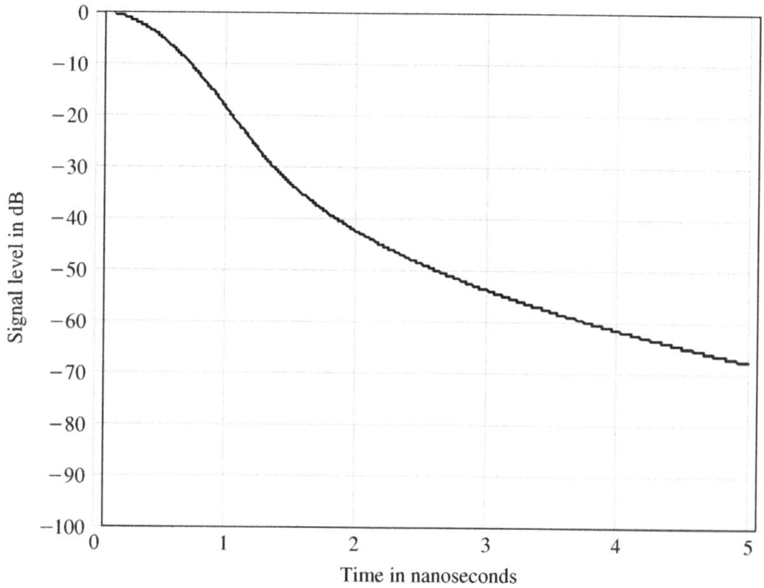

FIGURE 3.8 Amplitude in decibels of the envelope of the Ricker wavelet.

In time-domain radar systems the ideal radiated waveform is the Ricker wavelet, shown in Figure 3.7, which has a much better rate of decay, as shown in Figure 3.8, and thus would enable a radar to achieve a better range resolution compared with a signal with a lower rate of time decay. Control of the time side lobes in any type of radar is a fundamental issue for system performance.

3.2.10 Antenna Footprint

The azimuth and elevation resolution of an antenna is defined by the characteristics of the antenna and the signal processing employed. In general, radar systems (apart from SAR), require a high-gain antenna to achieve an acceptable plan resolution. This necessitates a sufficiently large aperture at the lowest frequency to be transmitted. To achieve small antenna dimensions and high gain, therefore, requires the use of a high carrier frequency, which may not penetrate the material to sufficient depth. When selecting equipment for a particular application, it is necessary to compromise between plan resolution, size of antenna, the scope for signal processing, and the ability to penetrate the material. Plan resolution improves as attenuation increases, provided that there is sufficient signal to discriminate under the prevailing clutter conditions. In low attenuation media the resolution obtained by the horizontal scanning technique is degraded, but only under these conditions do synthetic aperture techniques increase the plan resolution. Essentially, the ground attenuation has the effect of placing a "window" across the SAR aperture and the higher the attenuation the more severe the window. Hence in high-attenuation soils SAR techniques may not provide any useful improvement to GPR systems. SAR techniques have been applied to GPR but very often in dry soils with low attenuation.

The plan resolution of a radar system is important when localized targets are sought and when there is a need to distinguish between more than one target at the same depth. When the requirement is for location accuracy, which is primarily a topographic surveying function, the system requirement is less demanding.

The effect of the radiation footprint on the ground can be seen from Figure 3.9 where the distance between the radiating source and the ground surface has been increased from 0.1 to 0.5 m (left to right). The ground area is 2 m by 2 m, and it can be seen that the width of the 3-dB footprint increases considerably as the source is raised from the ground. The effect of this on the image resolution is also considerable as the convolution of the antenna pattern with the target causes a blurring of the target image, as shown in Figure 3.10. The simulation shows the blurring effect on a number of targets of different size and shape and is essentially the convolution of the antenna footprint with the point spread function of the targets.

Unlike conventional radars, which generally use a single antenna, most GPR systems use separate transmit-and-receive antennas in what has been termed a bistatic mode, which then adds a further convolution to the antenna footprint.

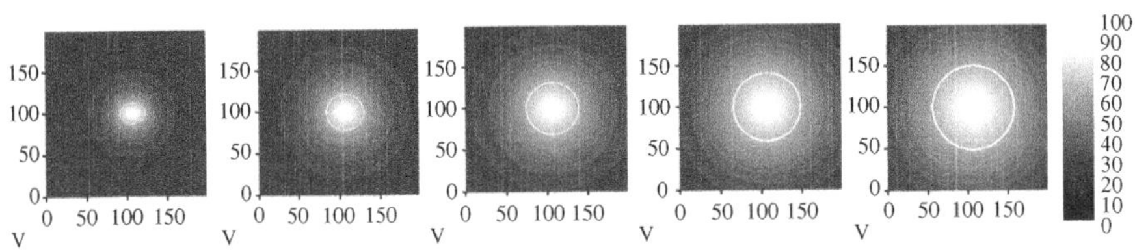

FIGURE 3.9 Radiation footprint on the ground from an isotropic source.

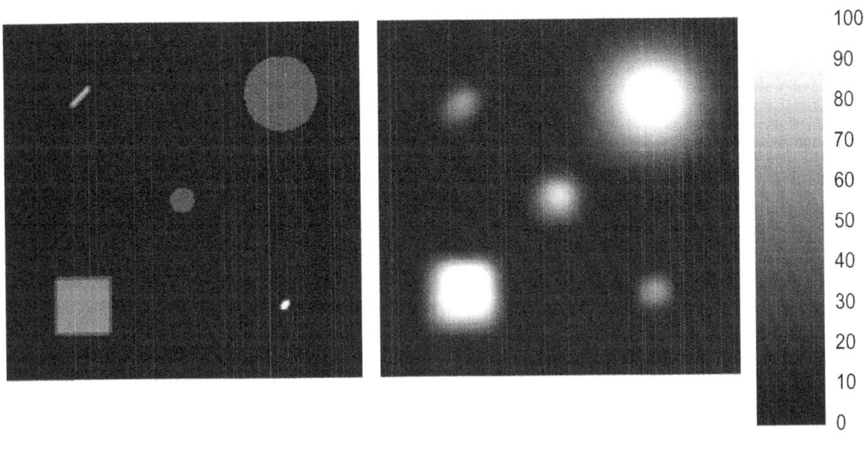

FIGURE 3.10 Effect of convolution of antenna footprint on radar image.

Note that as the antenna configuration for GPR systems is normally mobile, the term bistatic is not really relevant because the antennas are mobile and their separation is such that the bistatic angle is close to zero.

3.3 APERTURE ANTENNAS

Many radar systems use an aperture antenna to provide illumination of the target. Such apertures are often formed from waveguide horns, lenses, and reflectors. A typical waveguide horn is shown in Figure 3.11. The E field is usually in the vertical direction and the H field in the horizontal direction.

The far-field patterns of antenna apertures may be calculated by a number of analytic techniques, and a straightforward approach is to compute the Fourier transform of the aperture distribution function. Ramsay [108] notes that if the aperture distribution is unipolarized along the x direction and given by $f(x, y)$ where x and y are the coordinates, then the E field is given by

$$E(\theta, \phi) = \iint f(x, y) e^{j \sin[\theta x \cos(\phi) + y \sin(\phi)]} dx\, dy \tag{3.9}$$

There are cases where the distribution function can be separated into separate azimuth and elevation patterns by treating the antenna as a line source in x and y planes independently. The phase distribution across the aperture needs to be taken into account as this causes increased sidelobe levels but can be corrected by means of a suitable lens. Amplitude tapering across the aperture also improves the sidelobe level in the far field. As described earlier, it is important to account for the near-field–far-field boundary, as this will influence the illumination of the target as can be seen in Figure 3.12, where the illuminated field only spreads out after the near field - far field boundary. Although a large aperture will provide a higher directivity in the far field, the

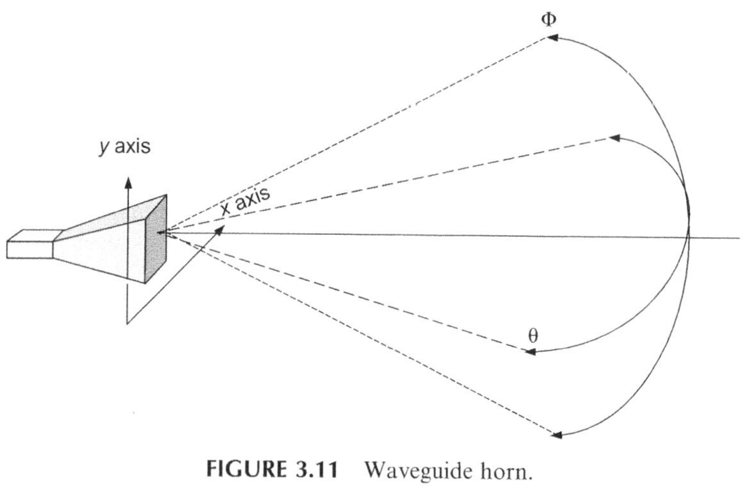

FIGURE 3.11 Waveguide horn.

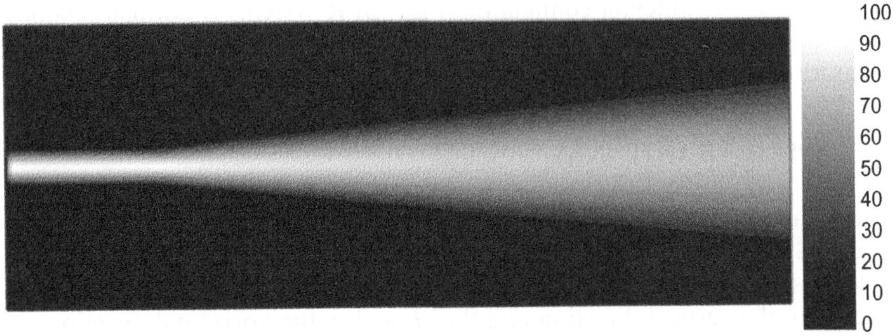

FIGURE 3.12 Representation of near–far field boundary showing divergence in the far field.

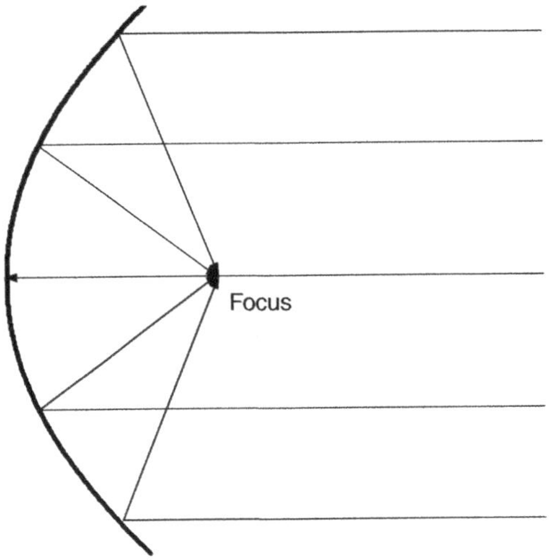

FIGURE 3.13 Parabolic reflector antenna.

width of the near-field pattern is set by the size of the aperture. A simplified representation of the near–far field boundary as a function of range can be shown using a ray diagram of an antenna pattern, and it can be seen that the antenna aperture always defines the extent of the near-field illumination.

At long wavelengths or for high gains, the length of a horn becomes excessive so that a reflector arrangement is employed. The reflector shape that can focus plane waves onto a single point must preserve the phases of all waves reaching the focal point. Thus, the total path lengths traversed by all portions of the reflected wave must be the same, and this requirement determines the shape of the desired reflecting surface. Clearly, the surface must be rotationally symmetric about its axis as is shown in Figure 3.13. All the incoming rays are reflected and pass through the focus or are radiated at the focus and then formed into a collimated beam.

The literature is well provided with information on the design and performance of reflector antennas, and the reader is referred to the introduction for information.

3.4 ANTENNAS FOR PROXIMAL OPERATION

3.4.1 Introduction

Systems for the detection of concealed targets that are operated very close to the target can be termed a proximal mode to distinguish from systems that operate standoff distances mainly in the far field of the antenna. Therefore, it is useful to describe the characteristics of antennas that are operated very close to a dielectric.

Where an antenna is closely coupled to a lossy dielectric, the design must consider additional requirements to those for antennas operating in free space. The type of radar system must also be considered in deciding the specification of the antenna for the following reasons. When the radar system is a time-domain system that applies an impulse to the antenna, there is a requirement for linear phase response, and this means that only a limited number of types of antenna can be used unless the receiver uses a matched filter to deconvolve the effect of the frequency-dependent radiation characteristics of the antenna. When the radar system is frequency modulated or synthesized, the requirement for linear phase response from the antenna can be relaxed, and log periodic, horn, or spiral antennas can be used as their complex frequency response can be corrected if necessary by system calibration. The overall configuration is further complicated by the use of separate transmit-and-receive antennas, which causes a convolution of the separate radiation patterns to form a composite pattern.

The use of separate transmit-and-receive antennas is dictated by the difficulty associated with operation with a single antenna, which would require an ultra-fast transmit–receive switch or a wideband circulator. As it is not yet possible to obtain commercially available and low-cost ultra-fast transmit–receive switches to operate in the subnanosecond region with sufficiently low levels of isolation between either transmit-and-receive ports or breakthrough from the control signals, most proximal radar systems use separate antennas for transmission and reception in order to protect the receiver from the high level of transmitted signal.

Therefore, the cross-coupling level between the transmit-and-receive antenna is a critical parameter in the design of antennas for such radars, and satisfactory levels are often achieved by empirical design methods. Typically, a parallel dipole arrangement achieves a mean isolation in the region of −50 dB [using additional means of isolation such as baffles and radar-absorbing material (RAM) between them], whereas a crossed dipole arrangement can reduce levels of cross coupling from −60 to −70 dB. For the crossed dipole arrangement such levels are highly dependent on the standard of mechanical construction, and a high degree of orthogonality is necessary. The crossed dipole is sensitive to variations in antenna-to-surface spacing, and it is important to maintain the plane of the antenna parallel with the plane of material surface to avoid degrading the isolation.

3.4.2 Coupling Energy into the Ground or a Dielectric

When the antenna is closely coupled into a dielectric, the radiation characteristics of the antenna are affected to a considerable extent, and this presents the antenna designer with significant restrictions on the types of antennas that can be used. The propagation path consists in general of a lossy, inhomogeneous dielectric, which, in addition to being occasionally anisotropic, exhibits a frequency-dependent attenuation and hence acts as a low-pass filter. The upper frequency of operation of the system, and hence the antenna, is therefore limited by the properties of the material. The need to obtain a high value of range resolution often requires the antenna to exhibit ultra-wide bandwidth and in the case of impulsive radar systems linear phase response. The requirement for wide bandwidth and the limitations in upper frequency are mutually conflicting, and hence a design compromise is adopted whereby antennas are designed to operate over some portion of the frequency range 10 MHz to 5 GHz, depending on the resolution and range specified. The requirement for portability for the operator means that it is normal to use electrically small antennas, which consequently result generally in a low gain and associated broad radiation patterns. Valle et al. [109] considered the theoretical and experimental directivity functions for antennas for ground-penetrating radar.

The classes of antennas that can be used are therefore limited, and the following factors have to be considered in the selection of a suitable design: large fractional bandwidth, low time side lobes, and in the case of separate transmit-and-receive antennas, low cross-coupling levels. The interaction of the reactive field of the antenna with the dielectric material and its effect on antenna radiation pattern characteristics must also be considered.

The coefficients of reflection and transmission need to be considered, as the radiated wave passes through the dielectric to the target. Snell's laws describe the relationships between associated angles of incidence, reflection, transmission, and refraction. For proximal operation the efficiency of the coupling process in generally high, but this is not the case for standoff radar systems since, where lossy materials are involved, complex angles of refraction may occur. Buried targets pose a difficult detection problem for standoff radars, and their performance is strongly influenced by the ground conditions. With vertical polarization at incidence angles less than the Brewster angle, transmission losses at the air–ground interface are relatively small, but at larger incidence angles than the Brewster angle the losses increase more rapidly. The dependence of transmission loss on dielectric constant and angle of incidence suggests that this should be no more than 85°. Hence to maximize the operating range the radar should be mounted as high off the ground as is possible. Thus, for a given height, the performance of the radar will be set by the relative dielectric constant of the ground. In addition to the problem of coupling energy into the ground, the effective cross section of all targets decreases when they buried.

Measurements and modeling suggest that under conditions of negligible attenuation losses, as are expected in very arid ground or for shallow burial depths, target-to-clutter ratios are expected to be degraded on burial by approximately 10 dB for standoff radar systems. Under the same conditions the cross section of nonmetallic targets is reduced by a larger factor because of reduced dielectric contrast between it and the surrounding soil, so that in wet sandy soils nonmetallic targets are more readily detected than in dry conditions.

It is important to appreciate the effect of the material in close proximity to the antenna. In general, this material, which in most cases will be soil or rocks, or indeed ice, can be regarded as a lossy dielectric, and by its consequent loading effect can play a significant role in determining the low-frequency performance of the antenna and hence surface-penetrating radar. The behavior of the antenna is intimately linked with the material and, in the case of bore-hole radars, the antenna actually radiates within a lossy dielectric, whereas in the case of the ground-penetrating radar working above the surface, the antenna will radiate from air into a very small section of air and then into a lossy half-space formed by the material.

The behavior of antennas both within lossy dielectrics and over lossy dielectrics has been investigated by Junkin and Anderson [110], Brewitt-Taylor et al. [111], Burke et al. [112], and Rutledge and Muha [113] and is well reported. The propagation of electromagnetic pulses in a homogeneous conducting earth has been modeled by Wait [114] and King and Nu [115], and the dispersion of rectangular source pulses suggests that the time-domain characteristics of the received pulse could be used as an indication of distance.

The interaction between the antenna and the lossy dielectric half-space is also significant as this may cause modification of the antenna radiation characteristics both spatially and temporally and should also be taken into account in the system design. In the case of an antenna placed on an interface, the two most important factors are the current distribution and the radiation pattern. At the interface currents on the antenna propagate at a velocity, which is intermediate between that in free space and that in the dielectric. In general, the velocity is retarded by the factor.

The net result is that evanescent waves are excited in air, whereas in the dielectric the energy is concentrated and preferentially induced by a factor of $n^{3:1}$. The calculated far-field power density patterns in both air and dielectric (Table 3.1) are given by Rutledge [113] and these are plotted for a relative dielectric constant of 6 and 10 in Figures 3.14 and 3.15 respectively.

TABLE 3.1 Power Density Patterns in Air and Dielectric

	Power	
Plane	$S(\theta_\alpha)$ Radiation Pattern in Air	$S(\theta_\alpha)$ Radiation Pattern in Dielectric
H	$\alpha\left(\frac{\cos\theta_a}{\cos\theta_a+\eta\cos\theta_d}\right)^2$	$\alpha\eta\left(\frac{\eta\cos\theta_d}{\cos\theta_a+\eta\cos\theta_d}\right)^2$
E	$\alpha\left(\frac{\cos\theta_a\cos\theta_d}{\eta\cos\theta_a+\cos\theta_d}\right)^2$	$\alpha\eta\left(\frac{\eta\cos\theta_a\cos\theta_d}{\eta\cos\theta_a+\cos\theta_d}\right)^2$

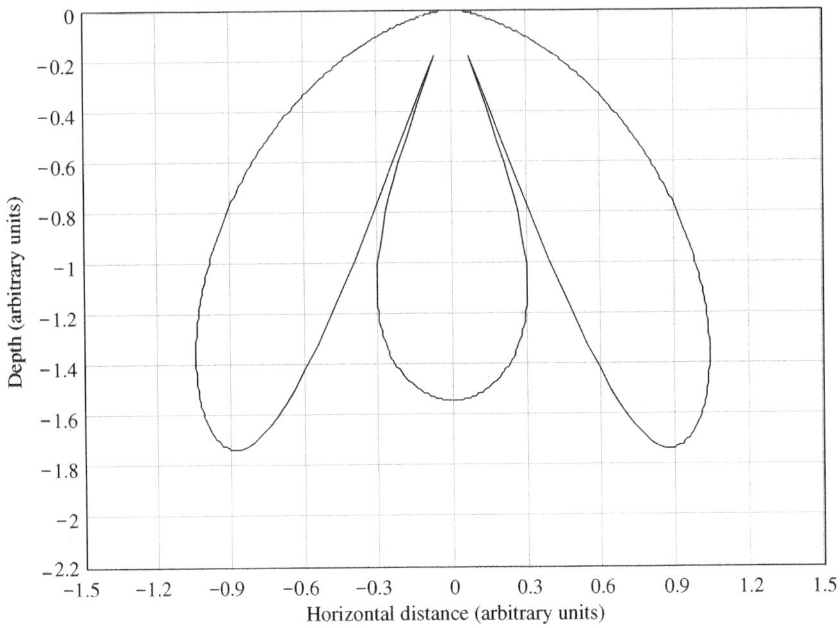

FIGURE 3.14 *E*-plane plot of far-field power density of a current element radiating into a dielectric.

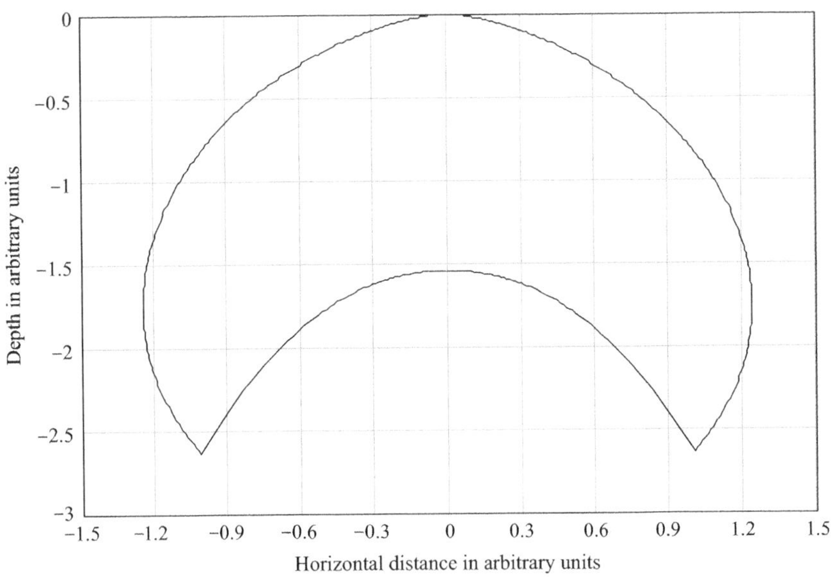

FIGURE 3.15 *H*-plane plot of far-field power density of a current element radiating into a dielectric.

The above expressions assume that the current source contacts the dielectric, whereas a more general condition is when the antenna is just above the dielectric. The side lobes in the pattern are a direct result of reactive field coupling. A significant practical problem for many applications is the need to maintain sufficient spacing to avoid mechanical damage to the antenna. It can therefore be appreciated that the effect of changes in distance between the antenna and half-space cause significant variation in the resultant radiation patterns in the dielectric. When the source interface space is increased, the antenna field patterns are modified by a reduction in the effect of the reactive field.

3.5 LINEAR PHASE ANTENNAS

Antennas used in time-domain systems need to have a linear phase frequency response. This is to ensure that the transmitted or received waveform remains coherent and is not dispersed. Examples of nondispersive antennas are the TEM horn, the bicone, the bow tie, the resistive, lumped-element loaded antenna, or the resistive, continuously loaded antenna. A typical antenna used in an impulse radar system would be required to operate over a frequency range of a minimum of an octave and ideally at least a decade, for example, 100 MHz to 1 GHz. The input voltage driving function to the terminals of the antenna in impulse radar is typically a Gaussian pulse, and this requires the impulse response of the antenna to be extremely short. The main reason for requiring the impulse response to be short is that it is important that the antenna does not distort the input function and generate time side lobes. These time side lobes would obscure targets that are close in range to the target of interest; in other words the resolution of the radar can become degraded if the impulse response of the antenna is significantly extended. All of the antennas used to date have a limited low-frequency performance unless compensated and hence act as high-pass filters; thus the current input to the antenna terminals is radiated as a differentiated version of the input function.

3.5.1 Dipoles

Element antennas such as monopoles, dipoles, conical antennas, and bow-tie antennas have been widely used for surface-penetrating radar applications. Generally, they are characterized by linear polarization, low directivity, and relatively limited bandwidth, unless either end loading or distributed loading techniques are employed in which case bandwidth is increased at the expense of radiation efficiency. Various arrangements of the element antenna have been used such as the parallel dipole and the crossed dipole, which is an arrangement that provides high isolation and detection of the cross-polar signal from linear reflectors.

It is useful to consider those characteristics of a simple normally conductive dipole antenna that affect the radiation response to an impulse applied to the

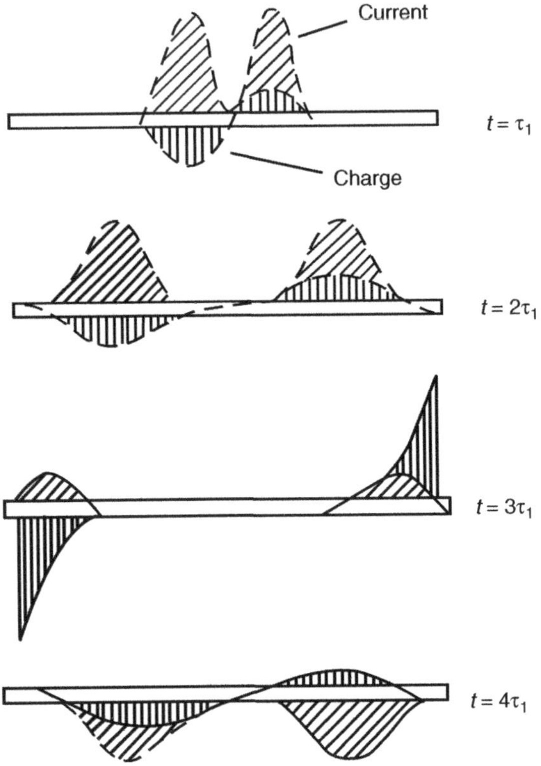

FIGURE 3.16 Current and charge distribution on a conducting dipole antenna due to an applied impulse.

antenna feed terminals. As shown in Figure 3.16 two current and charge impulses will travel along the antenna elements until they reach each end. At the end of the antenna the charge impulse increases while the current collapses. The charge at the end of the antenna gives rise to a reflected wave carried by a current traveling back to the antenna feed terminals. This process continues for a length of time defined by the ohmic losses within the antenna elements. As far as the radiation field is concerned, the relevant parameters, electric field strength, displacement current, and energy flow can be derived from consideration of Maxwell's equations.

The electric field component E_z is given by Kappen and Monich [116] as

$$E_z = -\frac{1}{4\pi\varepsilon_0} \int_{l_1}^{l_2} \left\{ \frac{1}{c} \frac{dI}{dt} + \frac{\partial q}{\partial z'} \right\} \frac{1}{r} dz' \qquad (3.10)$$

and must equal zero at the surface of the antenna. This condition can only be satisfied at certain points along the element and implies that for a lossless

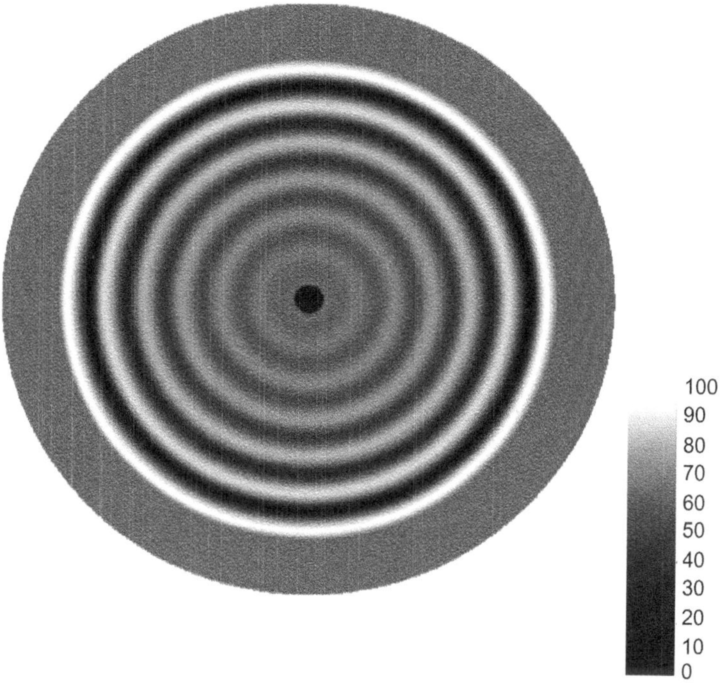

FIGURE 3.17 Radiated field pattern at the equatorial plane of a conducting dipole element due to an applied impulse.

antenna there is no radiation of energy from the impulse along the element. The radiated field is therefore caused by discontinuities, that is, the feed point and end point are the prime sources of radiation. As would be expected, the time sequence of the radiated field is extended and can be visualized when viewed from above the axis of the dipole, as shown in Figure 3.17 by the spreading of the radiation in the plane normal to the axis. For this reason the unloaded dipole is unsuitable for use with time-domain radar systems.

3.5.2 Loaded Antennas

As it is required to radiate only a single impulse in a time-domain radar system, it is important to eliminate either the reflection discontinuities from the far end of the antenna by end loading or reduce the amplitude of the charge and current reaching the far end. The latter can be achieved either by resistively coating the antenna or by constructing the antenna from a material such as Nichrome, which has a defined loss per unit area. In this case the antenna radiates in a completely different way as the applied charge becomes spread over the entire element length, and hence the centers of radiation are distributed along the length of the antenna.

FIGURE 3.18 Radiated field pattern at the equatorial plane of a resistively loaded dipole element due to an applied impulse.

In essence the electric field E_z must now satisfy Eq. (3.11), which implies that some dispersion takes place. The electric field lines for the lossy element are now different from the lossless case and are shown in Figure 3.18. Randa et al. [117] and Esselle and Stuchly [118] provide further analysis of the radiation characteristics of a resistively coated dipole antenna.

$$E_z = -\frac{1}{4\pi\varepsilon_0}\int_{l_1}^{l_2}\left\{\frac{1}{c}\frac{dl}{dt}+\frac{\partial q}{\partial z'}\right\}\frac{1}{r}dz' = R'I \qquad (3.11)$$

Altschuler [119] placed lumped-element resistors at a distance $\lambda/4$ from the end of the antenna and produced a traveling-wave distribution of current by suitable values of resistance. The distribution of current varied almost exponentially with distance along the element. Instead of lumped-element resistors, continuously distributed constant internal impedance per unit length can be used. In a classic study Wu and King [120] treated the parameters of the antenna such as input resistance, resistivity profile, and the like. The parameters of a center-fed cylindrical antenna can be characterized by a distribution of current equivalent to a traveling wave.

The cylindrical antenna with resistive loading has been shown by Wu and King to have the following properties:

The far-field pattern of the antenna comprised both real and imaginary components, that is,

$$F_m = \sqrt{F_r^2 + F_i^2} \tag{3.12}$$

where for a quarter-wave antenna

$$F_r = \frac{1 + \cos^2\theta - 2\cos\theta\left(\frac{\pi}{2}\cos\ \theta\right)}{\left(\frac{\pi}{2}\right)\sin^3\theta} \tag{3.13}$$

$$= \frac{-\frac{\pi}{2}\sin^2\theta + \left(1 + \cos^2\theta\right)\cos\theta\left(\frac{\pi}{2}\cos\ \theta\right)}{\left(\frac{\pi}{2}\right)\sin^3\theta} \tag{3.14}$$

The efficiency of the antenna is given by

$$\eta = \frac{P_r}{P_r + P_a} \tag{3.15}$$

where P_r = radiated power
P_a = absorbed power

For a resistively loaded antenna of the Wu–King type [120], the efficiency is approximately 10% but rises to a maximum of 40% for antenna lengths of 40λ.

The resistivity taper profile for a cylindrical monopole has the form given by Rao [121]:

$$R(z) = \frac{R_0}{1 - {}^{z}/_{H}} \tag{3.16}$$

where R_0 = resistivity at the drive point of the element
H = element length
Z = distance along the antenna

A graph of a resistivity versus length for a 200-mm element is shown in Figure 3.19. The overall efficiency of this type of antenna can be improved by reducing the value of R_0 and an increase of from 12 to 28% by reduction of R_0 to 0.3 R_0 was shown by Rao [121].

Further improvement in bandwidth can be gained by matching the antenna with a compensation network, and Esselle and Stuchly [118] developed a field probe with a bandwidth of 20 MHz to 10 GHz. Obviously, the use of a compensation network further reduces efficiency, but with a high impedance receiver probe, a frequency range of 10 MHz to 5 GHz was achieved. Nishioka

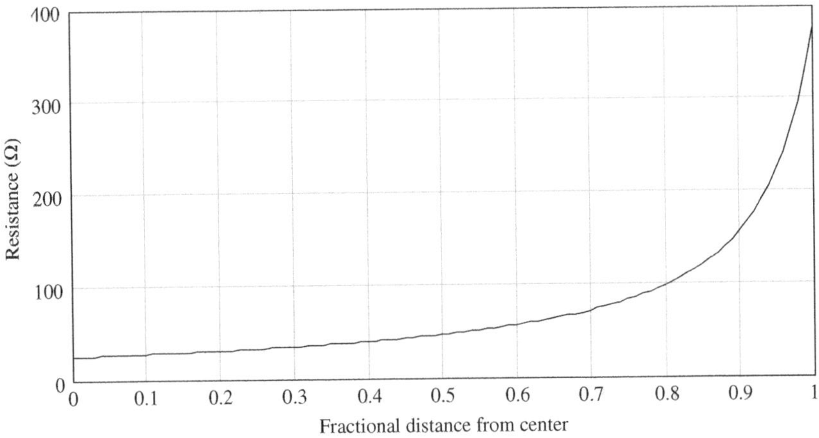

FIGURE 3.19 Resistivity profile of a 200-mm element.

et al. [122] carried out an FDTD analysis of resistor-loaded bow-tie (triangular shaped) antennas covered with ferrite-coated conducting cavity.

Resistively loaded dipoles have been used as electric field probes for EMC measurement applications, and although the frequencies of operation are well in excess of that used for surface-penetrating radar applications, it is useful to consider the general approach adopted by Maloney and Smith [123].

Randa et al. [124] developed loaded 8-mm dipoles to measure frequencies up to 18 GHz and subsequently 4-mm dipoles to measure over the frequency range 1 MHz to 40 GHz with an error of ± 4 dB. The transfer function of this antenna is on the order of -50 dB, which illustrates the penalty, which is paid for ultra-wideband width operation using loaded antennas.

A design that offers improved efficiency over the continuously loaded resistive antenna is based on a pair of segmented blade antennas arranged in a butterfly configuration and fed in phase. Each blade consists of a series of concentric conducting rings connected together by chip resistors. Radial cuts are used to reduce transverse currents. The efficiency of this class of antenna is higher than the continuously loaded dipole without serious degradation of the time-domain response. Morrow et al. [125] developed a rolled-edge ultra-wideband dipole antenna for GPR application.

3.5.3 BiConical Antennas

The biconical antenna is simple enough to evaluate to generate an approximate solution to the wave equation and has been one of the most important in yielding useful results for the antenna impedance problem. Much of this is available in the literature but Schelkunoff [104] gave particular attention to the work. He concluded among other things that the fatter the antenna the more broadband its properties. The useful feature of the biconical antenna is that it can be developed into either a TEM horn by using a sector or a Bowtie antenna if treated as a two-dimensional structure as shown in Figures 3.20 to 3.22.

3.5.4 Bow-Tie Antennas

The triangular bow-tie antenna has been widely used in commercial surface-penetrating radar systems. A triangular bow-tie dipole of 35 cm in length with a 60° flare angle can provide useful performance over an octave bandwidth of 0.5 to 1 GHz with a return loss of better than 10 dB, as shown by Brown and Woodward [126]. Evidently, without some form of end loading, such an

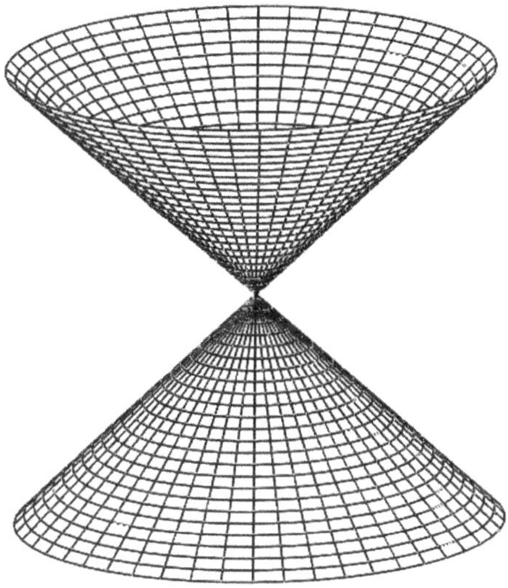

FIGURE 3.20 Basic biconical antenna.

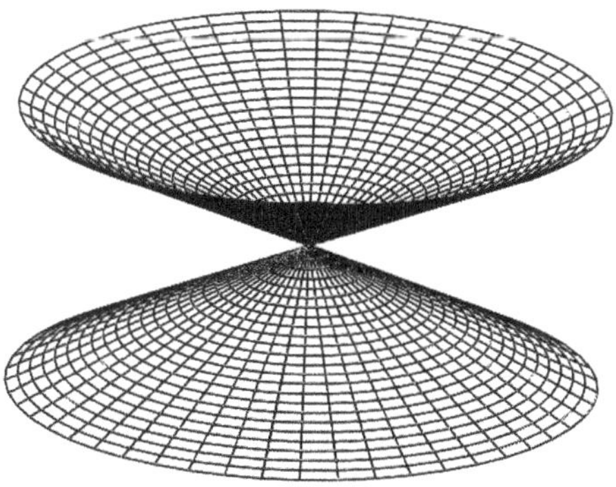

FIGURE 3.21 Biconical antenna with small angle.

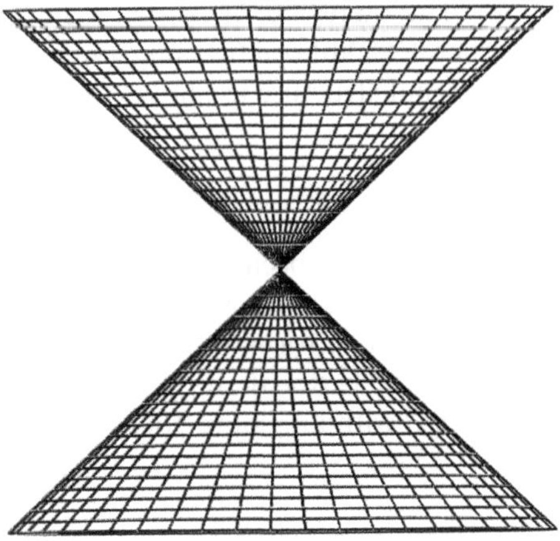

FIGURE 3.22 Biconical antenna realized as planar structure.

antenna would not be immediately suitable for use with impulse radar systems, and the triangular antenna normally uses end loading to reduce the ringing that would normally occur in an unloaded triangular plate antenna. The technique can also be used with a folded dipole, and the use of terminating loads results in a transient response equivalent to one and a half cycles as described by Young [127]. Other design options are described by Chen et al. [128] who describes as UWB full-polarimetric horn-fed bow-tie GPR antenna for buried unexploded ordnance (UXO) discrimination. Lestari et al. [129] describe improvements to the design of bow-tie antennas for pulse radiation.

3.5.5 Dielectric Road Antennas

An outline of the dielectric rod antenna is shown in Figure 3.23 with the feed section and radiating section shown. Typical dielectric values are in the region 2–6 and dielectric rod antennas have been investigated by Mueller and Tyrrell [130], Chen et al. [131, 132], chen [133], Diaz et al. [134], Yarovoy et al. [135], as well as Venkatarayalu et al. [136]. Adamiuk et al. [137] developed a 35-mm diameter dual-polarized dielectric rod antenna for operation in the 4 to 10-GHz band, with a gain of 8 dBi and cross-polar ratio better than −15 dB. Further details are provided by Danielss [2] in Section 5.9.1, which describes a Polyrod

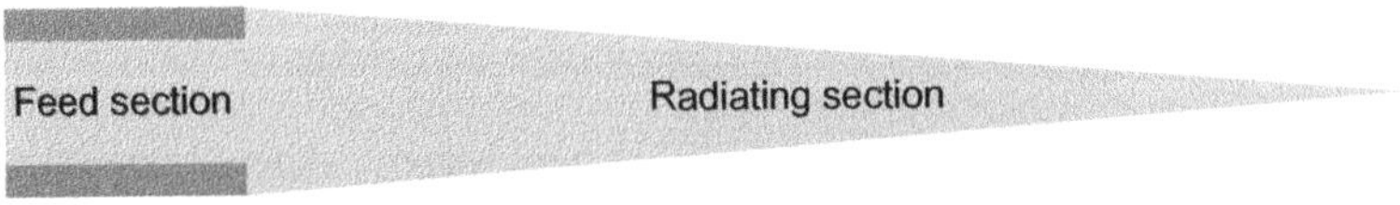

FIGURE 3.23 Dielectric rod antenna.

antenna design. The tip section requires some care in design to provide the wanted radiation pattern.

3.5.6 TEM Horn Antennas

The TEM horn antenna is important in short-range radar because of its time-domain characteristics as it is capable of supporting a forward-traveling TEM wave with little dispersion. In general, such antennas consist of a pair of conductors either flat, cylindrical, or conical in cross section, forming a V structure in which radiation propagates along the axis of the V structure as shown in Figure 3.24.

This type of antenna has directivity on the order of 10 to 15 dB; hence useful gain can still be obtained even with a terminating loss on the order of 3 to 5 dB, when loading resistors are placed either at the tips or along the length of the antenna.

The travelling-wave current on one of the cylindrical elements of a V antenna is given by Izuka [138] and is also discussed by King.

$$I_t = I_a e^{-j\beta z} \tag{3.17}$$

Hence, the azimuthal radiation field E is given by

$$E = \frac{j\omega\mu e^{-j\beta . r}}{4\pi} \int_0^l I_0 e^{-j\beta z(1-\cos\theta)} \sin\theta \, dz \tag{3.18}$$

where R = loading resistance
l = length of the element
z = distance from radiating source
0 = angle in H plane

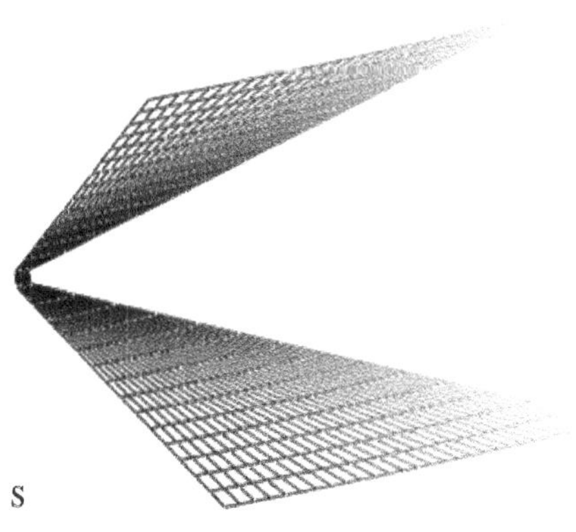

FIGURE 3.24 Basic TEM horn antenna.

This simplifies to

$$E_T(\theta) = \frac{1 - \exp[-j\beta(l_0 - l_1)(1 - \cos\theta)]}{1 - \cos\theta} \sin\theta \tag{3.18a}$$

where l_0 = element length and l_1 is the radiating start point.

However, a standing wave also exists caused by the resistive termination at the end of the antenna, and the contribution from this is given by

$$E_S(\theta) = j\frac{\exp[-j\beta(l_0 - l_1)(\cos\theta)]}{\sin\theta} \times \left\{\cos\left(\frac{\pi}{2}\cos\theta\right) + j\left[\sin\left(\frac{\pi}{2}\cos\theta\right) - \cos\theta\right]\right\} \tag{3.19}$$

The resultant field from one element can be derived from the sum of the contribution

$$E'(\theta) = E_T(\theta) + E_S(\theta) \tag{3.20}$$

and hence the field from both elements is

$$E = E'_U(\theta) + E_L(\theta) \tag{3.21}$$

where the subscript U denotes the upper element and L denotes the lower element.

The antenna will in fact radiate an impulse, which is extended in time as a consequence of the geometry of the antenna. Theodorou et al. [139] gives the pulse extension on bore sight by

$$t = \frac{L}{v}(1 - \cos\alpha) \tag{3.22}$$

where α is the half-angle between the elements and L is the element length and v is the phase velocity of waves along the antenna.

Evidently, a small flare angle and short element length help in reducing pulse extension. The electric field on bore sight is related to the time derivative of input current and on bore sight is given by

$$E = \frac{-\mu_0 L \sin\alpha}{2\pi r} \frac{\partial I_1(t - r/u)}{\partial t} \tag{3.23}$$

The impedance of the antenna should vary in such a way that the derivative of impedance at the feed and end parts is a minimum and along the antenna is low.

Typically, the characteristic impedance is given as a function of distance x as

$$Z_0(x) = Z_x \exp(-K_1 \cos K_2 x) \tag{3.24}$$

Hence,

$$\frac{dZ_0(x)}{dx} \to 0 \quad \text{for} \quad K_2 x = 0 \vee \pi \tag{3.25}$$

Usually, the feed impedance is on the order of 50 Ω and the end impedance is desired to be equal to that of free space (377 Ω). However, there is usually a difference between the transmission line wave impedance characteristic and that of wave in free space, and a design to meet given criteria in terms of return loss must take this effect into account.

Graphs of both typical antenna impedance and rate of change of impedance as a function of length are shown in Figures 3.25 and 3.26. Using this characteristic, a typical antenna–antenna time-domain response is shown in Figure 3.28. Improved directivity can be obtained using a V-conical antenna, as shown by Shen et al. [140]. A pair of triangular metal plates bent around a cone forms this. The antenna is characterized by two angles, the flare half-angle and

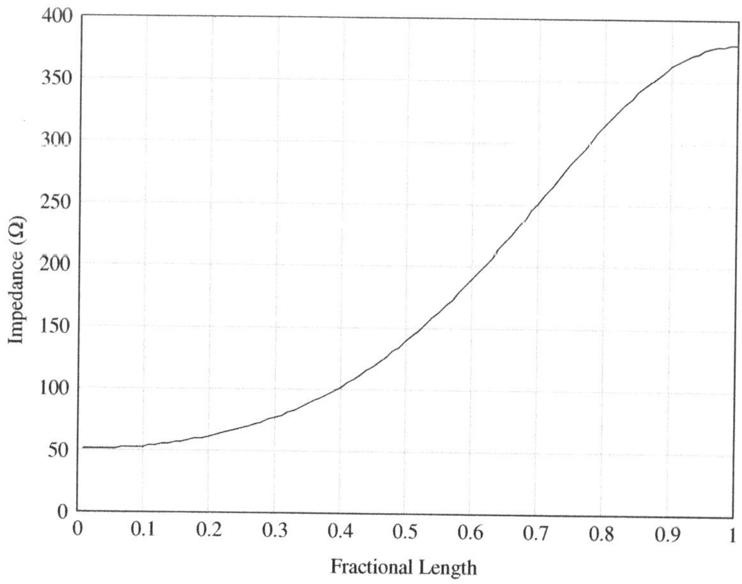

FIGURE 3.25 Characteristic impedance of a traveling-wave antenna.

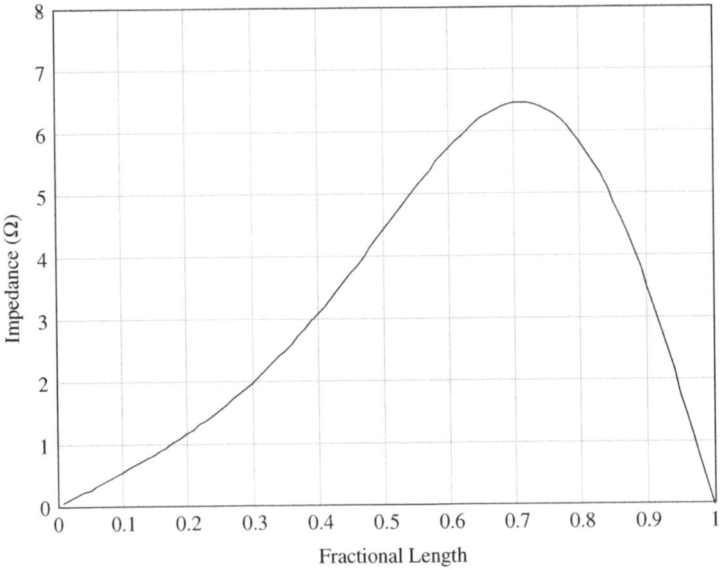

FIGURE 3.26 Rate of change of impedance of a traveling-wave antenna.

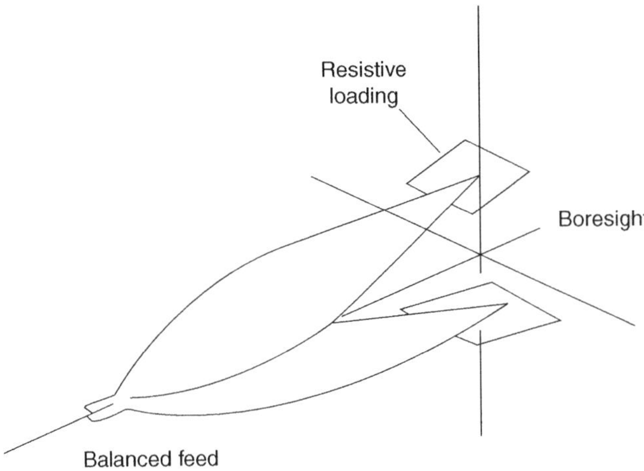

FIGURE 3.27 Generic TEM horn arrangement.

the azimuthal angle ϕ. Further developments of the generic TEM horn design as shown in Figure 3.27 from the original design first described by Wohlers [141] are to be found in studies by Daniels [142], Evans and Kong [143], Reader et al. [144]. and Foster and Tun [145].

A further development of the TEM horn is given by Martel et al. [146]. The antenna is composed of a set of spread "fingers" forming the shape of a horn, as shown in Figure 3.29 and 3.30. Each finger is a wire with a diameter of 1 mm

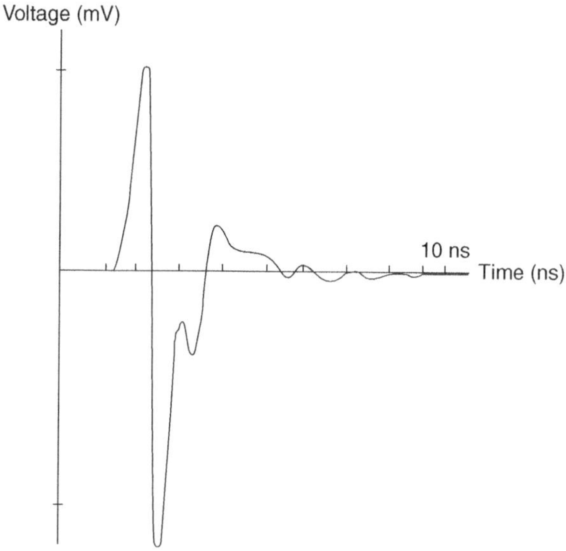

FIGURE 3.28 Time-domain response of a pair of traveling-wave antennas used in a face-to-face configuration.

and is resistively loaded at different locations along the length of the antenna. The feed component for the antenna is comprised of a 50 Ω coax, feeding a tapered parallel-plate waveguide with a width of 30 mm and a height of 7 mm. A taper along the width of the top parallel plate is used as a transition to transform a 50 Ω unbalanced line into a 50 Ω balanced line.

Optimization was undertaken on the geometry of the antenna and the values and positions of the loaded resistances. The fixed parameters were the dimensions at the antenna feed point and the length of the antenna, which for practical purposes was set to 35 cm. The height and width of the horn aperture were the variable parameters of the geometry.

An investigation on the optimum number of resistances was also undertaken. While the efficiency is not a critical issue, a compromise had to be found between the efficiency and the broadband response of the antenna. The usage of many resistances leads to a lower reflection coefficient but degrades the efficiency considerably. Reducing the number of resistances makes the antenna efficient but causes multiple reflections. After investigating the effect of the number of resistors on the performance of the antenna, the best compromise was to design the antenna with five resistors. The antenna has therefore been optimized with five resistors along the arms of the TEM horn in order to achieve a satisfactory VSWR over the frequency range [100,1400] MHz. The method of moments (MoM) was used in the optimization process. Within this framework the resistors were modeled as lumped loads.

An optimized geometry was determined with the following dimensions: $W = 30$ cm, $H = 10$ cm, $L = 35$ cm, where W is the aperture width of the

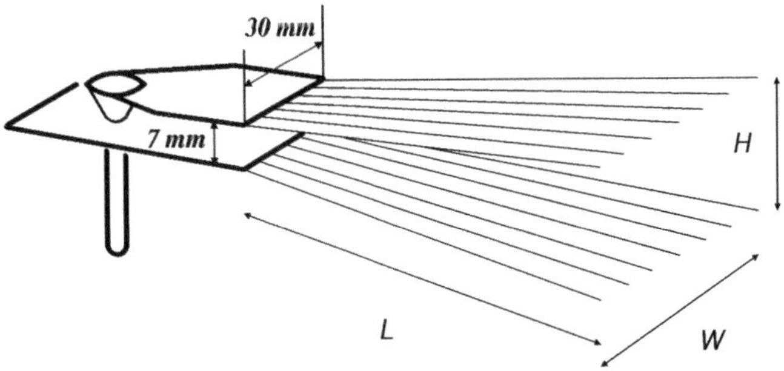

FIGURE 3.29 Antenna and feed geometry of loaded TEM horn.

FIGURE 3.30 Photograph of loaded TEM horn.

horn, H is the aperture height of the horn, and L is the length of the antenna. Figure 3.31 shows the predicted VSWR in the frequency band [0, 1400] MHz for the optimum design. The method using cascaded segments of strips agreed well with the method of Moments. The VSWR was better than 2 : 1 for frequencies between 400 and 1400 MHz. The best performance was achieved at 700 MHz.

The predicted time-domain pulse of the optimum design is shown in Figure 3.32. The shape of the time-domain antenna response is similar to a second-derivative Gaussian signal. It can be seen that most of the internal reflections have been suppressed. The rate of decrease for the unwanted ringing is better than 9 dB/ns. Other developments of the loaded V antenna are discussed by Montoya and smith [147], who describe resistively loaded Vee antennas for short-pulse ground-penetrating radar.

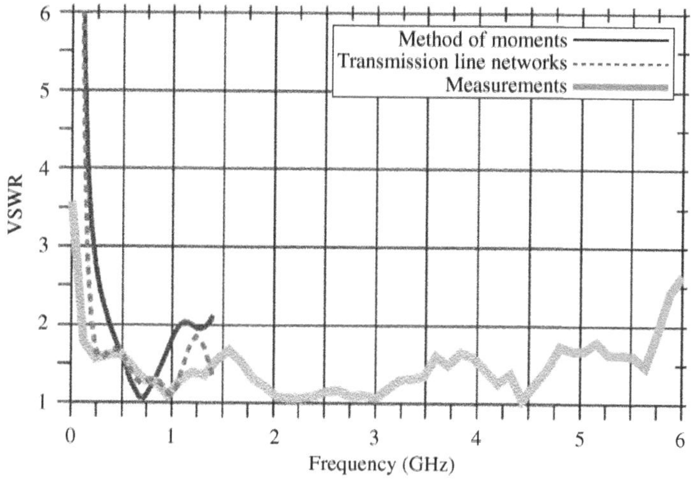

FIGURE 3.31 VSWR of loaded TEM horn antenna.

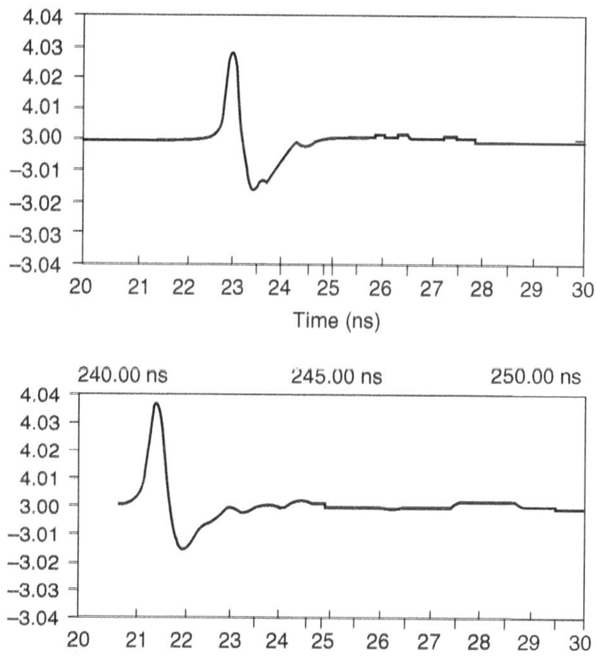

FIGURE 3.32 Predicted and actual pulse responses of TEM horn antenna.

3.5.7 Impulse Radiating Antennas

Impulse radiating antennas (IRAs) are a class of focused aperture ultra-wideband antennas designed to radiate extremely short electromagnetic pulses when fed by a fast rise time step waveform. IRAs have found applications such as foliage penetration, hostile target identification, buried object detection, and broadband jamming. A typical IRA is a reflector antenna fed by a pair of conical transmission lines connected in parallel at the focal point. A large (several meters) diameter reflector antenna is suitable for generating impulsive radiated field amplitude of between 4 and 5 kV m^{-1} with pulse duration of around 200 ps at a range of about 300 m. The spectrum of the radiated field extends from about 50 MHz to a several gigahertz.

Impulse radiating antennas are composed of a conical, nondispersive TEM feed structure that is fed at the focal point of a reflector, although sometimes a lens may be used. The radiated waveform is the time derivative of that applied to the feed because of diffraction effects. The reflector changes the TEM mode wave on the feed line into a plane wave, and the gain of the antenna results in a very narrow beam of a degree or less and the general layout is shown in Figure 3.33.

Much work has been carried out on the fundamental design of this type of antenna by Baum and co-workers [148–150], Farr and co-workers [151–156], Buchenauer et al. [157], and Tyo [158]. For an IRA to radiate well-collimated impulses, the feed point must be at the true focal point of either the reflector or the lens. Practical limitations result in defocusing, which has the adverse effect of distorting the radiated impulse and reducing the gain of the antenna. In addition to its use as a single-element antenna, the IRA can be used in arrays. Tyo [158] describes the development of a number of IRAs for use in various ultra-wideband (UWB) and UWB high-power microwave (HPM) applications. Many of these antennas are designed to behave as differentiating antennas excited by a fast-rising step waveform, thereby resulting in a large-amplitude, narrow pulse in the far field. In order to mitigate some of the limitations of reflector IRAs and lens IRAs, a new class of IRAs—the array IRA—has been proposed and examined in a number of studies. The principal benefit that is

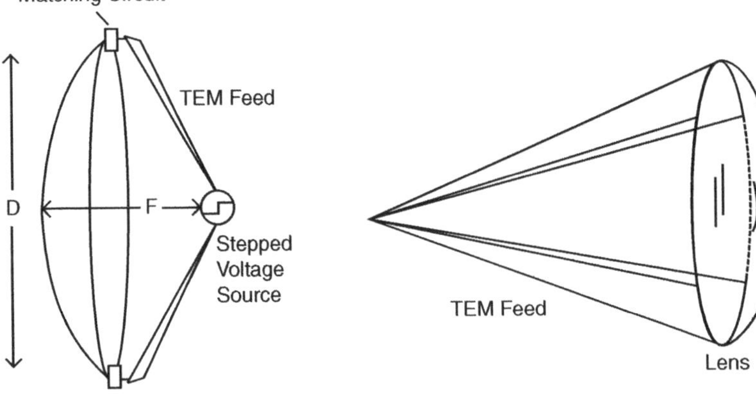

FIGURE 3.33 Generic constructions of impulse radiating antennas.

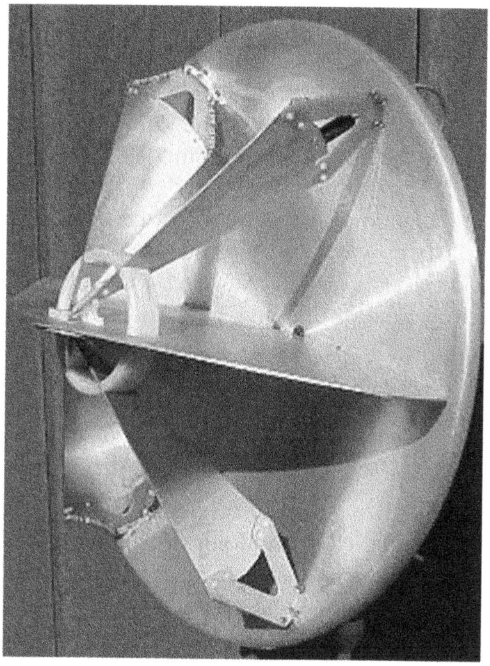

FIGURE 3.34 Impulse radiating antenna (reproduced with kind permission ©Farr Research Inc.).

realizable with an array IRA is the potential for electronic beam steering. Second, array IRAs will allow for far-field power combination from solid-state sources that are currently under development. In addition to these primary benefits, array IRAs provide a reduction in volume and the related mass of the antennas. It is important to note, however, that these benefits come at the cost of the added complexity associated with the precise timing of the array elements that is necessary for beam forming and far-field power combination. Further information on antenna design can be found at http://www.farr-research.com/ and an example of an IRA design is shown in Figure 3.34.

3.6 NONLINEAR PHASE ANTENNAS

This class of antennas has a geometry entirely defined by angles and exhibits a performance over a range of frequencies set by the overall dimensions of the structure. Typical examples are the log periodic, equiangular spiral, and conical spiral. Log periodic structures can also provide broadband performance but are not completely defined in terms of angles.

Examples of dispersive antennas that have been used in surface penetrating radar are the exponential spiral, the Archimedean spiral, the logarithmic planar antenna, the Vivaldi antenna, and the exponential horn.

3.6.1 Vivaldi Antennas

This is basically a slot-line antenna. In theory, the bandwidth of the Vivaldi antenna is infinite. In practice, the main bandwidth limitations are the aperture size, which defines the low-frequency limit, and the slot-line to microstrip transition, which defines the high-frequency limit. The Vivaldi antenna originated by Gibson [159] also falls into the class of a periodic continuously scaled antenna structure and within the limiting size of the structure has unlimited instantaneous frequency bandwidth. It provides end fire radiation and linear polarization and can be designed to provide a constant gain–frequency performance.

The Vivaldi antenna consists of a diverging slot-form guiding conductor pair as shown in Figure 3.35. The curve of one of the guiding structures follows the equation

$$z = Ae^{kx} \tag{3.26}$$

Radiation is produced by a nonresonant traveling wave mechanism by waves traveling down a curved path along the antenna. When the conductor separation is small, the traveling-wave energy is closely coupled to the conductor but becomes less so as the conductor separation increases. The Vivaldi antenna provides gain when the phase velocity of the traveling wave on the conductors is equal or greater than that in the surrounding medium. Typical radiation patterns for an elemental Vivaldi are shown in Figure 3.36 for side and top views, respectively.

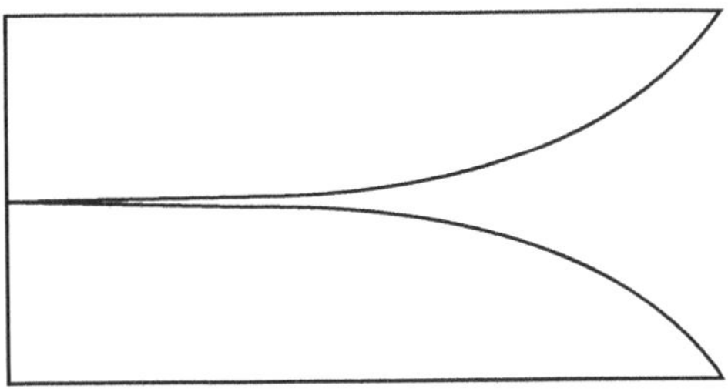

FIGURE 3.35 Vivaldi antenna.

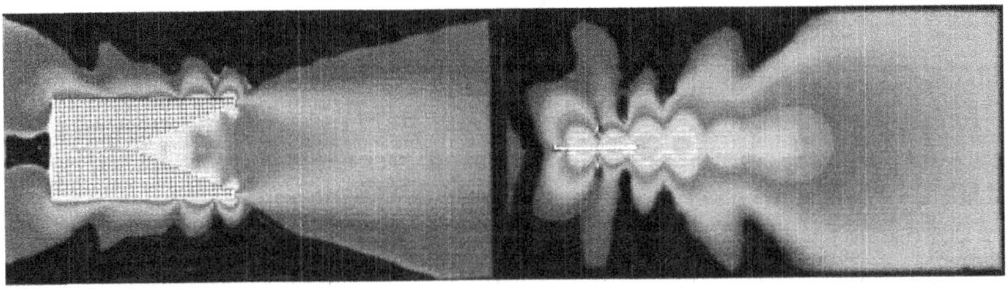

FIGURE 3.36 Calculated radiation patterns of elemental Vivaldi antenna (with kind permission ERA Technology).

The dimensions of the conductor separation, being a half wavelength, define the lower cutoff frequency and the gain is proportional to overall length. The impulse response of the antenna is extended due to the nonstationary phase center but can, of course, be corrected by the use of a matched filter. Note that the sidelobe and back-lobe radiation is significant unless suitable absorbers and screening are used.

3.6.2 Equiangular Antennas

Various developments of the spiral antenna or conical spiral antenna have been carried out by Miller and Landt [160], Pastol et al. [161], Dyson [162], Morgan [163], Kooy [164], Deschamps [165], Bawer and Wolfe [166], and Goldstone [167]. The impulse response of this class of antennas is extended and generally results in a "chirp" waveform if the input is an impulse. The main reason for this is that the high frequencies are radiated in time before the low frequencies as a result of the time taken for the currents to travel through the antenna structure and reach a zone in which radiation can take place. Phelan et al. [168] provide a near-field analysis of a wideband log-spiral antenna for 1 to 2 GHz shown in Figure 3.37, which is relevant to proximal operation.

The geometry of the equiangular spiral is defined by

$$\rho = \kappa e^{a\phi} \tag{3.27}$$

where

ρ is the radius
k is a constant
ϕ is the angle

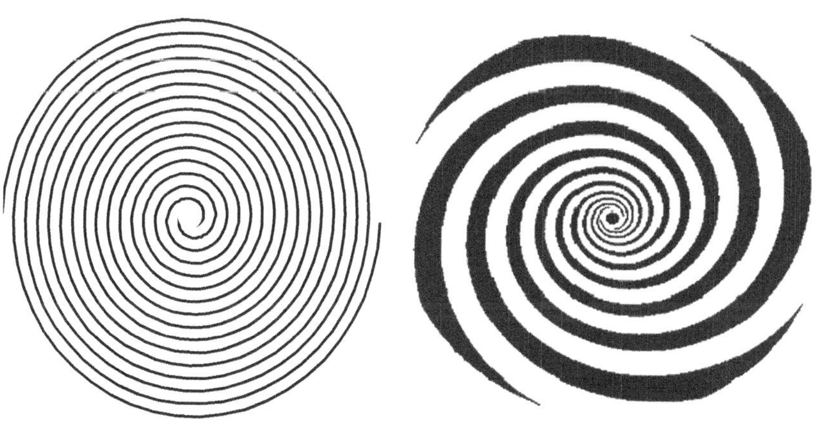

FIGURE 3.37 Archimedean spiral and equiangular planar spiral antenna.

3.6.3 Horn Antennas

Horn antennas have found most use with FMCW surface penetrating radars where the generally higher frequency of operation and relaxation of the requirement for linear phase response permit the consideration of this class of antenna. Exponentially, flared TEM horns with dielectric loading have been developed to operate over decade bandwidths [169,170]. The design of horn antennas is well covered in the literature but of particular interest is the short axial length double-ridged horn as shown in Figure 3.38.

This design can provide useful gain over a decade bandwidth using a logarithmic characteristic curve for the ridges. Typically, the short axial length horn provides a VSWR of better than 2 : 1 and a gain of 10 dB over a frequency band of 0.2 to 2 GHz for an axial length of 0.76 m. The concept of the ridged horn design can be adapted to form a quad-ridged horn operating from 0.3 to 1.9 GHz. A return loss of better than 10 dB and a cross-coupling level of better than 35 dB can be obtained. The quad-ridged horn can be used to extract information on the polarization state of the reflected signal.

An FMCW radar has been developed using an offset paraboloid fed by a ridged horn as described by Sun and Rusch [172] The arrangement was designed to focus the radiation into the ground at a slant angle to reduce the level of the reflection from the ground. Care needs to be taken in such arrangements to minimize the effect of back and side lobes from the feed antenna, which can easily generate reflection from the ground surface.

FIGURE 3.38 Short axial length double-ridged horn antenna [171].

Although horn antennas have been mostly used with FMCW systems, it is possible to radiate pulses, which will need to be suitably processed for the phase delay incurred when propagating from the ridged horn.

3.7 ANTENNA ARRAYS

A single-element antenna such as a dipole has limited potential in terms of gain; thus, an array of elements is used to generate a directive pattern either in end-fire or broadside. A broadside array will produce a directive beam, which is a maximum in the directions perpendicular to the array. Depending on the relative orientation of the elements relative to the array axis (Figure 3.39), there are two types: (a) the "point to point" and the (b) the collinear type. Schelkunoff and Friis [173] note the expressions for the radiation intensity $P(\theta,\phi)$ for both types and are noted by Eqs. (3.8a) and (3.8b), assuming all elements are equally excited in amplitude and phase (uniform array):

$$P(\theta, \phi) = 15\pi\left(\frac{I\,\Delta l}{\lambda}\right)^2 \frac{\sin^2\left[\frac{N\pi d}{\lambda}\cos(\theta)\right]}{\sin^2\left[\frac{\pi d}{\lambda}\cos(\theta)\right]}(1 - \sin^2\theta\cos^2\phi) \qquad (3.28a)$$

$$P(\theta, \phi) = 15\pi\left(\frac{I\,\Delta l}{\lambda}\right)^2 \frac{\sin^2\left[\frac{N\pi d}{\lambda}\cos(\theta)\right]}{\sin^2\left[\frac{\pi d}{\lambda}\cos(\theta)\right]} \qquad (3.28b)$$

where θ = angle in degrees
λ = wavelength in meters
N = number of radiating elements
ϕ = phase delay between elements
d = element spacing
$I\,\Delta l$ = (current) × (incremental array element length)

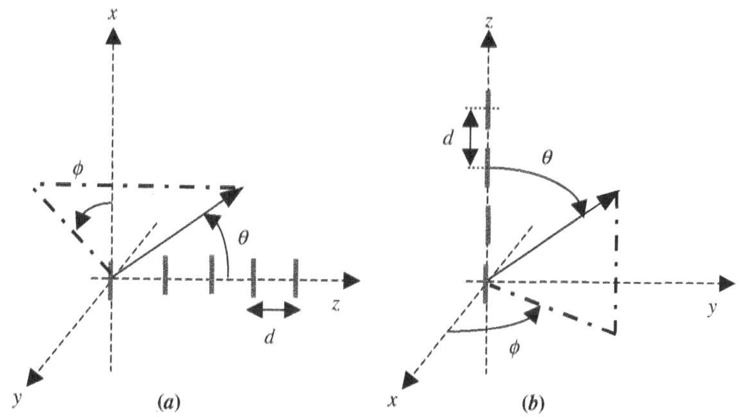

FIGURE 3.39 General arrangement of array elements: (a) point to point and (b) collinear.

A plot of the radiated power in decibels is shown in Figure 3.40, and the symmetrical nature of the radiation pattern can be seen from the polar plot. The design of antenna arrays is well covered in the literature as described in section 3.1. The element spacing has a large influence on the array factor and heavy influence on the upper limit of the antenna bandwidth. A larger element spacing results is a higher directivity. However, the element spacing is generally kept smaller than $\lambda/2$ for omnidirectional elements to avoid the occurrence of grating lobes. If grating lobes do occur, then the directivity of the array is drastically reduced.

A comparison of the radiated power at a plane of distance of 20λ from arrays with 5 elements and 11 elements and an element spacing of λ is shown in Figure 3.41. Note the impact of increasing the number of elements on the directivity of the antenna as well as the reduction in side-lobe levels. Most conventional arrays of elements use a corporate feed system to provide a single input to or output from the array of elements. The beam forming is achieved by correct phasing of the elements, which is usually fixed. More sophisticated control techniques rely on adjusting the differential phasing of the elements, which enables beam steering or even null steering. Suitable electronically controlled phase shifters or electronically controlled time delays can be used to achieve control. Tyo et al. [174] give details of beam forming in the time domain.

Array antennas are often used for ground-penetrating radar applications such as road survey and mine detection, as they increase the rate of survey of

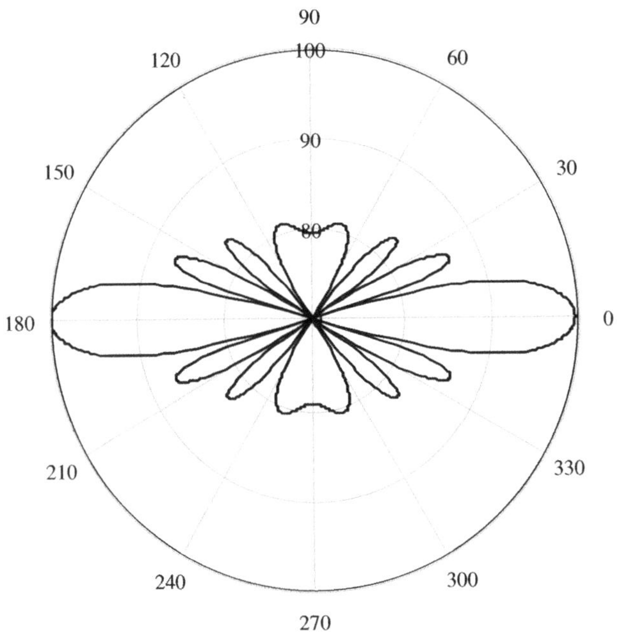

FIGURE 3.40 Radiation pattern in decibels from a 10-element broadside array.

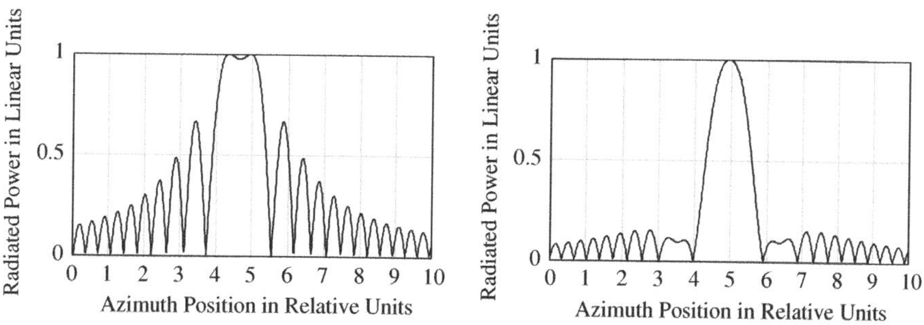

FIGURE 3.41 Comparison of the radiated power from a 5-element array (lhs) and 11-element array (rhs).

areas of the ground. There are two approaches to array design. The first is simply to take a number of single elements, slave them with a master controller, and process the data from each individual element. The key issues for the design of such arrays lie in the channel-to-channel performance and tracking over the desired operational environmental range. For time-domain systems the array must maintain calibration of both start time and time linearity for all channels to within demanding limits.

The other approach is to design the system as an integral array design and exploit the increased capability offered by combining multiple looks and SAR processing. Consideration as to whether the antenna/system should be proximal downward look or forward look is important. For radar systems being used for penetrating the ground, it is important that the reflection from the ground surface is reduced. Close coupling of the antenna array to the ground surface is one method.

When the array is operated proximally, the antenna element spacing needs to be adequate to provide proper resolution of small targets. This is closely related to the density of the elements of the antenna array. An alternative approach for standoff arrays relies on coherent subtraction, but this often means that the ground topography must be relatively smooth. When removal of the surface reflection is not adequate, then antennas operating off-normal incidence can be used. However, the effect of the grazing angle limits the potential for full 3D imaging because of the refractive index of the ground compressing the formation of the beam within the soil.

Other array designs use a two-dimensional array of precisely timed transmitters and receivers to actively image the area under the array. This provides the advantages of multiple transmitters to illuminate the scanned area with broadband energy, multiple receivers to detect back-scattered energy, and provides additional clutter rejection through time-domain synchronization of multiple impulse radars. By independently controlling the firing and gating of each transmitter and receiver, the system focuses radar energy to a point, then moves that point to scan the area under the array and the principle is shown in Figure 3.42. If the sequence of impulses is controlled in time by means

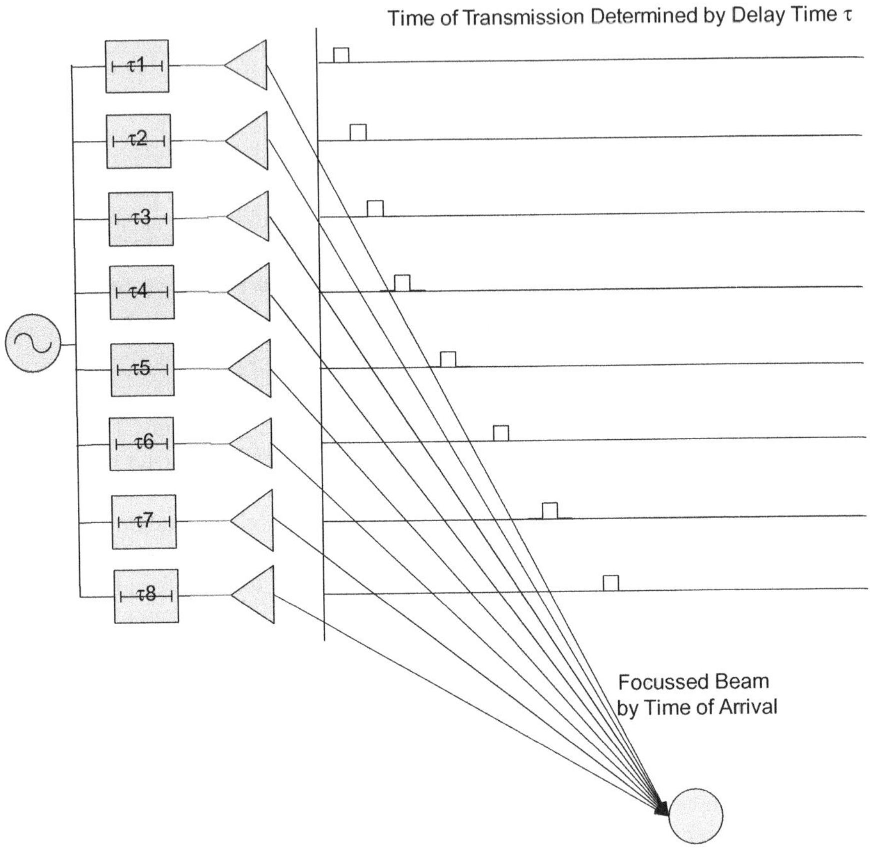

FIGURE 3.42 Beam steering by means of time delays to radiating elements.

of a differential time delay between each element, the beam position can be steered, although the interelement time delay is limited to a maximum equivalent to the distance between each element.

In terms of GPR systems, arrays of antennas are an obvious method of increasing the rate of survey of areas of the ground and have been designed and built for road survey and mine detection. Strictly speaking, these are not arrays in the sense the conventional antenna designer considers an array, but they are considered an array by the radar system designer. Some examples of antenna configurations are given below. There are two approaches to array design. The first is simply to take a number of single-channel radars, slave them with a master logger-and arrange for the data to be appropriately logged. The other approach is to design the system as an integral array design and exploit the increased capability offered by combining multiple looks and SAR processing. Consideration as to whether the antenna/system should be downward look or forward look is important. With all array systems, it is important that the surface clutter is properly removed. Close coupling of the antenna to the ground surface is one method. An alternative relies on coherent subtraction, but this often means that the ground topography must be relatively smooth. If removal of the surface clutter is not easy, then antennas operating off-normal incidence can be used. However, this in turn brings other problems, and antenna near-field effects must be accounted for, and the effect of grazing angle can be a limitation as it limits the potential for full 3D imaging because of the refractive index of the ground compressing the beam within the soil.

All array systems are geared to generating an image of the buried targets and for that accurate positioning of the array elements is crucial. This can be achieved with differential GPS (DGPS) systems coupled with the inertial navigation systems (INS)

The key issues for the design of multielement GPR systems lie in the channel-to-channel performance and tracking over the desired operational environmental range. A 32-channel GPR system, for example, must maintain calibration of both start time and time linearity for all channels to within demanding limits. In addition, the relative gain and, if used, time-varying gain profile, must also match to within close tolerances. When the antenna array is spaced off the ground, there may also be the need to compensate for variations in surface topography. A further aspect to be considered is the antenna element spacing. This needs to be adequate to provide proper resolution of the wanted target, and it can be shown that the probability of detection with respect to small targets is closely related to the density of the elements of the antenna array.

However, the ability to carry out beam forming by means of inverse synthetic aperture processing could be potentially valuable in many applications. Rutledge and Muha [175] consider the general situation of imaging antenna arrays while Anders [176] considers the specific problem of wideband beam patterns from sparse arrays. Other array options are found in standoff detection systems where digital beam-forming techniques are applied to optimize antenna patterns. The antenna beam bore sight can be simply steered by varying the relative phase of the elements so that bore sight is then steered according to the amount of phase shift. In an electronically steered array, programmable electronic phase shifters are used at each element in the array. The antenna is steered by programming the required phase shift value for each element. To form a beam both the relative amplitude and phase of the signal path to each antenna element must be controlled.

Combined amplitude and phase control can be used to adjust sidelobe levels and steer nulls better than can be achieved by phase control alone. The combined relative amplitude and phase shift for each antenna is termed a complex weight. Either analog amplitude and phase shift control can be used or digital beam forming is achieved with dedicated digital signal processing (DSP) integrated circuits, and a basic system diagram for an eight-element receiver array system is shown in Figure 3.43. The RF is down-converted to a frequency suitable for

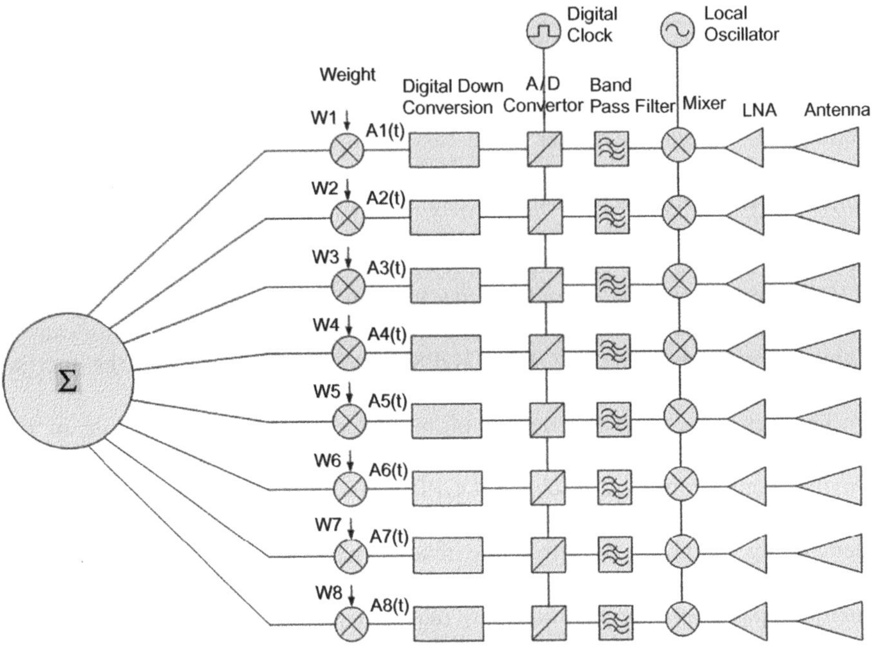

FIGURE 3.43 Digital beam former.

direct analog/digital (A/D) conversion, and then the output from each element is suitably weighted before being integrated to generate a formed beam.

3.8 SUMMARY

The antennas used for radar systems for the detection of concealed targets are defined by the frequency of operation as well as the requirements for portability and design architecture. When the radar system is operated in standoff mode, then conventional antenna designs can be employed provided the antenna is working in its far-field mode. The designs are well covered in the literature, although impulse radiating antennas tend to be less familiar.

For close-in detection or proximal operation the dielectric has a significant influence on the propagation characteristics of the antenna and must be fully considered in the design process. When the dielectric is lossy, the frequency of operation tends to be a few gigahertz, and hence the antennas are usually electrically small and consequently exhibit low gain. However, the bandwidth of the antennas is very much greater than that normally used in conventional radar systems, and surface-penetrating radars generally demonstrate very high range resolution.

All the classes of antenna discussed can be used in synthesized, FMCW, or noise-modulated radars. For time-domain systems the resistively loaded dipole, bow-tie, and TEM traveling-wave antenna are used because of the need for a linear phase response. When matched filtering can be incorporated in impulse radars, then either horn or frequency-independent antennas can also be considered.

For many basic structures such as dipoles, bow-tie, and traveling-wave antennas attention must also be given to the means by which the antenna is fed from the transmitter. Generally, an antenna is a balanced structure, but where cables are used to connect the antenna to the transmitter or receiver, some means is needed of transforming from the unbalanced configuration of the feed cable to the balanced structure of antenna. On this frequency range, baluns are generally commercially available or, alternatively, purpose-designed units can be constructed.

CHAPTER 4

Nuclear Quadrupole Resonance

4.1 INTRODUCTION

The author is extremely grateful for the invaluable assistance of Professor John Smith, Emeritus Professor of Chemistry, Kings College, London, Dr. Tim Rayner of Rapiscan, and Nigel Hunt of ERA Technology in the preparation of this chapter. The author is grateful for the kind permission of Professor Smith to use material particularly related to the detection of narcotics.

Nuclear quadrupole resonance (NQR) or quadrupole resonance (QR) is a spectroscopic radio-frequency (RF) technique in which the received signals are caused by the interaction of the electric quadrupole moment of the quadrupolar nuclei in the sample with the electric field gradient (EFG) of their surroundings, when radiated with RF energy. The quadrupole moment, nonzero only for nuclei with spin quantum number I greater than or equal to 1, is a nuclear physics parameter describing the distribution of charge in the nucleus. Smith [177, 178] Hoult and Richards [179], Marino and Klainer [180], Grechishkin and Sinyavskii [181], and also Garroway and co-workers [182, 183] describe the principles of NQR, which is a similar technique to nuclear magnetic resonance (NMR), but differs in that quadrupole resonance takes advantage of a nonzero EFG at the nucleus to be detected. These gradients are due to the distribution of the electrical charge in its vicinity and therefore strongly depend on the chemical structure of the material. NQR does not require an external magnetic field and can be considered a form of zero magnetic field NMR spectroscopy.

Nuclear quadrupole resonance requires that the RF field must penetrate the material, and the NQR signal cannot be obtained from a completely screened target. However, in some situations, such as vehicles carrying explosives, sufficient energy can be caused to radiate inside the means of transport and permit NQR detection. When targets are metallic mines, the mass of metal will affect the tuning of the exciting coil or probe, and this may be used as a means of identifying suspect packages where the mass of metal is high but no NQR signal can be detected.

The quadrupole effect is essentially due to an interaction of the nuclear charge density with the electrostatic potential of the surrounding electron clouds. Garroway [128] notes that a moment expansion of this electrostatic interaction shows that the important coupling is between the nuclear quadrupole moment and the second derivative of the electric potential, which is the gradient of the electric field). The energy level diagram and allowed transitions for a spin-1 nucleus such as ^{14}N are shown in Figure 4.1. There are three allowed transitions in the general case, one frequency (ν_x or ν_+) being the sum of the other two (ν_y or ν_-, ν_z or ν_0), the vast majority of which lie between 0.2 and 6 MHz. These frequencies are related to quantities known as the quadrupole coupling constant (e^2qQ/h) and asymmetry parameter η by Eq. (4.1):

$$
\begin{aligned}
\nu_x &= \frac{3}{4}\frac{e^2qQ}{h}\left(1+\frac{\eta}{3}\right)\\
\nu_y &= \frac{3}{4}\frac{e^2qQ}{h}\left(1-\frac{\eta}{3}\right)\\
\nu_z &= \frac{1}{2}\frac{e^2qQ}{h}\eta
\end{aligned}
\tag{4.1}
$$

where e = charge on the electron
h = Planck's constant
q = q_{zz}, the maximum principal component of the electric field gradient tensor
Q = nuclear electric quadrupole moment
η = asymmetry parameter

Also, η is defined as the difference between the other two principal components (q_{xx} and q_{yy}) of the electric field gradient tensor divided by q; it is a positive number lying between zero and one. Spin-$\frac{3}{2}$ nuclei such as ^{35}Cl have two doubly degenerate level transitions as shown in Figure 4.2, which give rise to just one frequency equal to

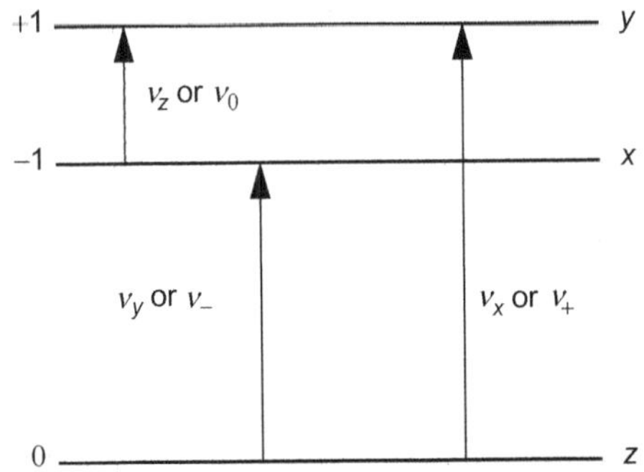

FIGURE 4.1 ^{14}N quadrupole levels and allowed transitions.

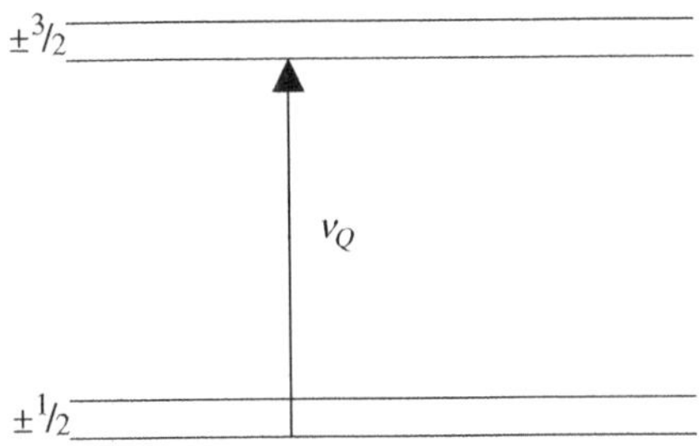

FIGURE 4.2 ^{35}Cl quadrupole levels and allowed transition.

$$\nu_Q = \frac{1}{2}\frac{e^2qQ}{h}\left(1+\frac{\eta^2}{3}\right)^{1/2} \tag{4.2}$$

The operation of an NQR detection process is that the RF magnetic field caused by a sequence of RF pulses of peak value up to a kilowatt or more in the frequency range 0.5 to 6 MHz for ^{14}N is applied to the material. The energy that is applied causes a change in the alignment of the nuclear quadrupole moment in the electric field gradient and hence a precession of the former about the latter. This is equivalent to the increase in energy from the lower to a higher energy level state. When the external oscillating electromagnetic field is removed the nuclei return to their original states, releasing very low levels of RF energy in a form characteristic and unique to the material. Hence the natural decay of the originally raised energy state back to the rest state generates a detectable RF magnetic field. Figure 4.3 show a diagrammatic representation of the approximate frequencies for particular explosives and narcotics all of which are nitrogen based. Note that the basic forms of cocaine and heroin, referred to in this figure as the hydrochloride forms, are not readily

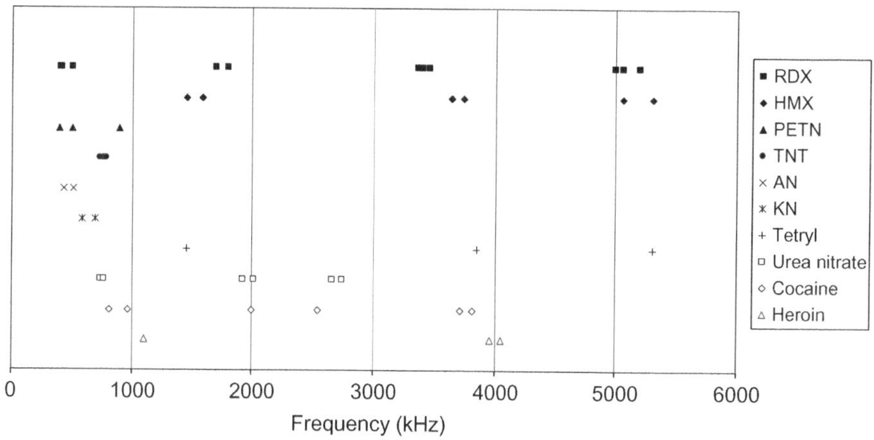

FIGURE 4.3 Approximate NQR resonant frequencies for various materials.

detectable. The RF pulse can also be considered to tip the nuclear quadrupole moments away from their equilibrium orientation in the electric field gradient of their surroundings by a flip angle α. When the applied field is removed, the altered nuclear quadrupole moments precess at their allowed frequencies, inducing an RF voltage in the same RF coil as was used to excite them.

Not all materials exhibit a NQR signal and liquid explosives or liquid narcotics are not detectable. The NQR technique identifies the RF spectral characteristics of a particular material. The signals are usually observed in material in solid form and in suspension, although material in solution cannot be detected. The material under investigation may be placed inside a coil (internal field detection) or outside a coil (external field detection). In the case of external fields and small samples of material, the reradiated signal in the order of 0.1 pT (picoteslas) depending on distance. When a coil contains a sample, the field can be considered to be mainly internal to the coil, and it is easier to generate larger fields in the range greater than 100 pT.

4.2 PULSE SEQUENCES

To obtain the greatest received signal-to-noise ratio, a range of excitation sequences is used that are in general tailored to the specific material under investigation. The basic pulse sequence is termed free induction decay (FID) and is designed to match the characteristics of the material. Other pulse sequences that are commonly used are pulsed spin locking (PSL) [or spin-locked spin echo (SLSE)] [184], steady-state free precession (SSFP), which includes phase-alternated multiple-pulse sequence (PAPS), or alternatively stochastic noise excitation.

The initial intensity of the FID is defined by the flip angle α, which is given by

$$\alpha = \gamma B_1 \tau_\omega \tag{4.3}$$

where γ = gyromagnetic ratio of the ^{14}N nucleus, equal to $1.932. \times 10^7$ rad s^{-1} T^{-1}
B_1 = amplitude of the oscillating magnetic field
τ_ω = time of applied pulse

All NQR frequencies and their associated relaxation times are characteristic of a given material and can be used to identify both the type of material and its mass as the amplitude of the received signal is proportional to the latter. The time taken to achieve detection depends not only on the signal to noise but also the relaxation time. In an FID sequence the relaxation time T_1 determines how rapidly the pulse sequence can be repeated, and the parameter T_2 restrict the maximum length of the "spin-echo" sequence used. In FID sequences T_1 limits the rate of pulse repetition because the material has to recover to its original state. The time can vary from 10 ms to many tens of seconds for particular ^{14}N-containing materials at room temperature The signal-to-noise ratio depends on the mass of material, the strength of the applied field, and the frequency of excitation. In general, the signal-to-noise ratio increases as a function of $\omega^{3/2}$ so that materials whose resonance is below 1 MHz are more difficult to detect than materials whose resonance is above 1 MHz.

Nuclear quadrupole resonance signals at these frequencies are usually detected by means of pulsed RF techniques in which the ^{14}N nuclei are subject to bursts of RF radiation at, or near to, their NQR frequency and the resulting transient signals monitored in the quiescent periods between pulses. A representation of a FID sequence (not to scale) is shown in Figure 4.4: the upper graph shows the applied RF pulse train, the middle graph shows the received RF pulse sequence, and the lower graph the rectified and detected sequence. A more detailed representation is shown in Figure 4.5.

In FID sequences both the frequency and the relevant relaxation times T_1* and T_2* are functions of temperature. Because the exact temperature of the target substance may not be known, pulse sequence parameters that are broadly effective over a band of temperature are used, typically −10 to 15°C in real-world applications. Some form of temperature monitoring is needed to ensure that the target is not at temperatures outside this temperature range. A simulation of the transmitted pulse (upper), received signal (middle), and the detected signal (lower) is shown in Figure 4.5. Note that scales are arbitrary as the figure is simply intended to illustrate the main FID signals following a single pulse.

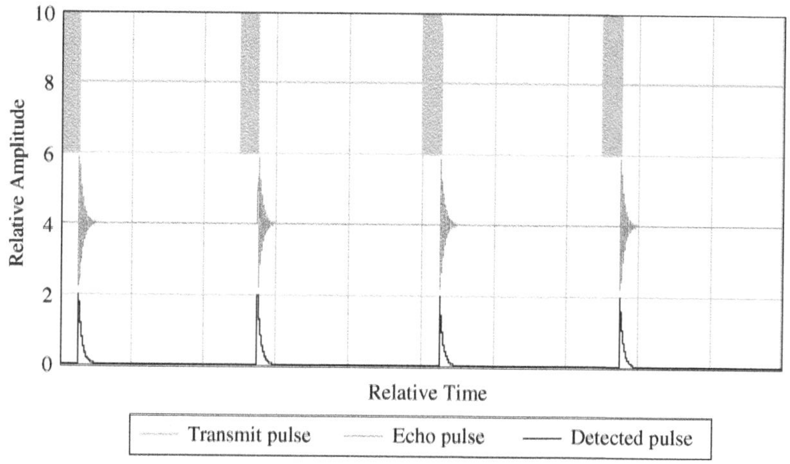

FIGURE 4.4 Idealized free induction decay (FID) sequence.

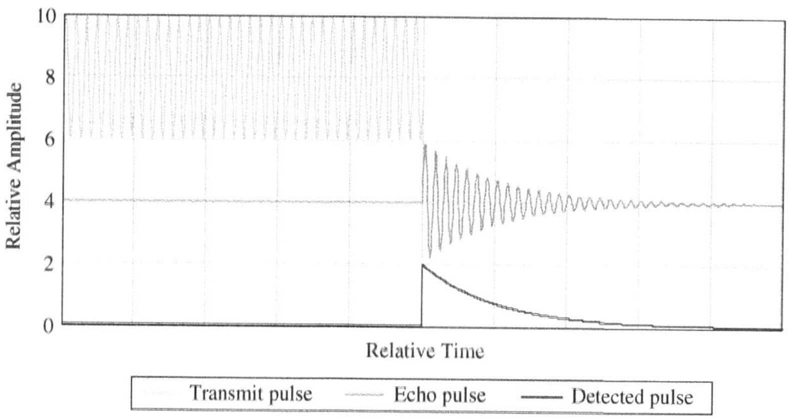

FIGURE 4.5 Simulation of basic FID response from generic material.

The key issues for detection are the signal-to-noise ratios (SNR), which can be improved by averaging at the expense of time to detect, and the effect of general electrical noise and RF interference, which is a major difficulty in the medium wave wavebands when systems are used in outside and uncontrolled environments. Other causes of interference can be due to piezoelectric responses from silica in the soil and responses from certain metals such as nickel and metal coatings. Careful design of the pulse sequences, signal processing, and hardware can alleviate these to a great extent.

While FID is the decaying signal observed immediately following a pulse, an echo is a regenerated signal with maximum intensity between RF pulses in a multiple-pulse train. For this purpose, it is customary to use extended trains of pulses accumulating the observed responses between pulses to enhance the SNR. One example is known as pulsed spin locking (PSL) and can be represented by

$$\alpha_{0^\circ} - (\tau - \alpha_{90^\circ} - \tau -)_n \tag{4.4}$$

where α, denoting flip angle, represents the pulse width, selected to optimize the signal, the subscripts denote the RF phase, τ determines the pulse spacing, which is 2τ after the first two pulses, and n determines the number of pulses in the train whose optimum value depends on the relaxation times of the material. The entire pulse sequence is repeated, often several hundred times, for further signal averaging depending on the quantity of material in the sample and in the case of remote detection, its distance from the antenna. Spurious signals, such as observed from piezoelectric materials such as sand, need to be eliminated usually by cycling the phases of the RF pulses and signals before processing, which can be in the time domain or frequency domain after Fourier transformation. Since the pulse sequence repetition time is limited by the need to wait for the nuclear spins to recover their equilibrium magnetization, and this time is determined by the spin-lattice relaxation time T_1, this is an important parameter, which needs to be measured before any measurements are undertaken. The other important relaxation times are T_2, which determines the echo decay time in a two-pulse sequence, and T_2e, which determines the usually longer multiple-echo decay time within a multiple pulse sequence, such as PSL.

Figure 4.6 shows a simulation of the decaying FID and the echo signal from a material after the last pulse of a two-pulse train with a known pulse time separation. As the pulse spacing is increased, the intensity of the echo decays with a time constant T_2, which is known as the spin–spin relaxation time.

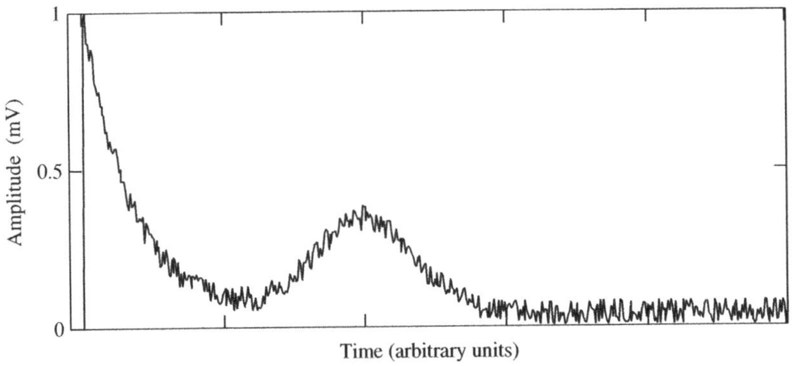

FIGURE 4.6 Simulation of echo caused by multiple pulses.

Smith [177,178] explains this process as a refocusing of the de-phased signals produced after the FID.

The refocusing effect can be repeated by further pulses provided always that $\tau < T_1$, and multiple-pulse sequences can be used to generate long echo sequences, an example of which is the strong off-resonance comb SORC where the RF is just off resonance. The pulse spin locking or spin-locked spin echo (SLSE) given in Eq. (4.4) is a typical example. Note that the first pulse has to match a flip angle of $90^{\circ}{}_{\text{eff}}$, but subsequent pulses may have other values.

A simulated train of PSL echo signals for RDX is shown in Figure 4.7, where it can be seen that echoes are refocused several times. PSL sequences are appropriate for compounds in which $T_1 >> T_2$. When $T_1 \approx T_2$, SSFP pulse sequences can be used to produce long echo trains, and these are referred to as phase-alternated multiple-pulse sequence (PAPS) and have the form as given in Eq. (4.5):

$$\alpha_{0^{\circ}} - (\tau - \alpha_{180^{\circ}} - \tau -)_n \tag{4.5}$$

The important advantage of multiple-pulse sequences is that the many echoes they generate can be summed in a very short time to produce an enhancement in the SNR that can be as large as $n^{\S}$, where n is the number of equal intensity signals accumulated. Alternatively, the long echo train can be treated as a slowly decaying signal, governed by a time constant T_{2e}, the Fourier transformation of which can give a much sharper signal of enhanced SNR.

Rudakov [185] applied a modification of the SSFP multipulse technique to a sample of nitrogen-14 NQR in the sample of RDX. Based on the use of additional blocks of the pulses and phase-cycling technique, this technique enabled the suppression of the coherent noise signals such as the magneto-acoustic and piezoelectric signals or the ringing signals from the NQR coil.

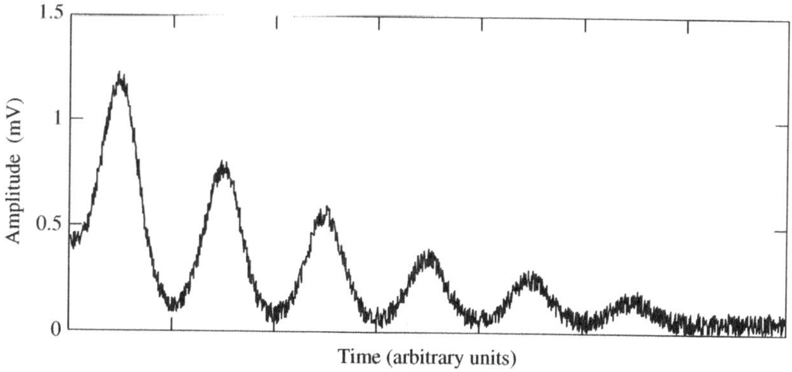

FIGURE 4.7 Pulse spin locked sequence for RDX (modeled sequence).

Garroway et al. [186] described an alternative technique based on the stochastic excitation of the sample. Stochastic excitation uses randomly phase-modulated RF pulses of low flip angle and so is a steady-state method. The general method for exploiting stochastic methods is by means of generating a sequence of random or pseudorandom pulses and cross correlating the received NQR signal with the original pulse sequence, thereby generating a free induction decay signal. Careful choice of a maximal length binary sequence and the use of Hadamard transform enables the FID signal to be recovered. The benefits claimed by Garroway et al. [187] are that the stochastic excitation generates phase variations of 0° to 10° rather than the 100° commonly used in traditional pulse sequences. This results in a reduction in peak power by a factor of 100. While the prospect of much lower power is highly attractive, the stochastic approach still needs development.

The temperature dependency of the spectral lines is also a major issue and careful selection of the least sensitive resonant frequencies limits the effect of temperature variation. Essentially, the resonant frequencies change as a function of frequency and not all at the same rate as shown in Figure 4.8 for the six $\nu+$ frequencies of monoclinic trinitro toluene (TNT) over the range of 240 to 320 K and for heroin over the range 260 to 370 K as shown in Figure 4.9. This means the temperature of the sample must be known in order to optimize the center frequency of the exciting pulse sequence. Buess et al. [188] investigated methods of compensating for temperature dependence and acoustic ringing.

Most research groups involved with this work have paid attention to this aspect of NQR detection. Ostafin and Nogaj [189] noted that the measurements of the NQR response of a commercial TNT sample manufactured in Poland

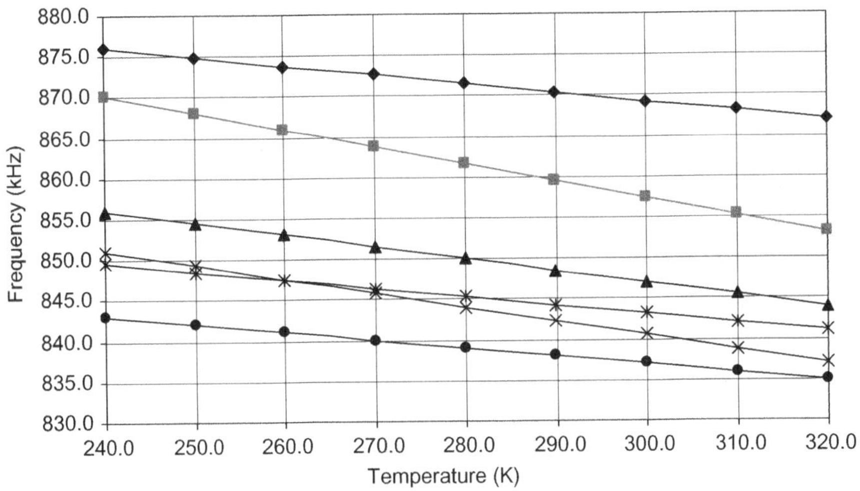

FIGURE 4.8 Temperature dependency of TNT frequencies from 240 to 320 K.

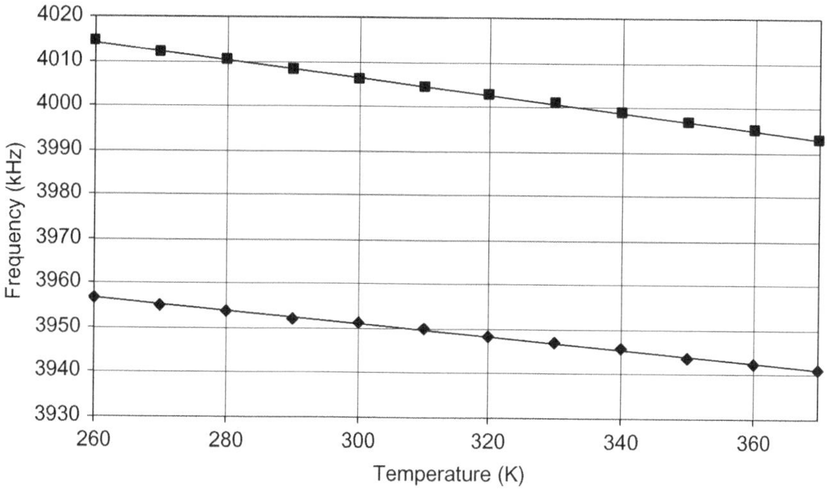

FIGURE 4.9 Temperature dependency of ν_+ and ν_- frequencies of heroin from 260 to 370 K.

were successful at 77 K but was unable to detect any response at room temperature. Jakobsen et al. [190] developed a nonlinear least squares detector that exploited the temperature dependency of the NQR frequencies as a way of enhancing the SNR and carried out numerical simulations on both synthetic and real measured data.

Equation (4.6) can be used to predict the SNR for samples inside coils:

$$\mathrm{SNR} = \frac{V_c^{1/2} Q^{1/2} A \zeta h^2 \gamma_N N_0 \nu_Q^{3/2} \mu_0^{1/2}}{30.1 kT (kTBF)^{1/2}} \tag{4.6}$$

where V_c = coil volume
Q = loaded Q factor
A = constant close to unity
ζ = filling factor of the coil
γ_N = gyromagnetic ratio for ^{14}N
N_0 = number sensitivity of the ^{14}N nuclei
ν_Q = NQR frequency
T = absolute temperature in K
B = receiver bandwidth
F = noise factor
k = Boltzmann's constant

From this expression it can be shown that for an adequate SNR a coil volume in excess of 174 litres (88 × 60 × 33 cm^3) is appropriate for smaller items of mail and 10 litres (1 × 1 × 1 m^3) for passenger baggage for explosives such as RDX. Note that Hoult and Richards [178] suggest that $\nu_q^{7/4}$ is more appropriate than $\nu_q^{3/2}$.

4.3 SYSTEM DESIGN

4.3.1 Introduction

The hardware to transmit and receive pulses can take several forms. The most sensitive detection systems are where the sample is placed inside a coil, and these designs are used to detect explosives or narcotics concealed within in parcels, mailbags, or luggage. When the material is buried or concealed under clothing, the field is generated using a planar coil that can take many forms: single turn, spiral, and the like.

A typical QR system takes the form as shown in Figure 4.10. The target is placed within an RF field, which could be inside a probe coil, which could take the form of a single or multiturn coil, and a sequence of RF pulses specifically designed to excite a response from the substance of interest is applied to the probe coil via a switch control unit. The pulse sequence is generated by the pulse sequencer and amplified by an RF power amplifier. Since it is important to achieve maximum sensitivity in the receiver, the pulses are terminated abruptly by a circuit referred to as a *Q* damper in order that the low-level response signals from the target can be detected. These are amplified by a low noise receiver and analyzed by means of a spectrometer. Overall control of the system is handled by means of a system control module, which interfaces with each of the subsystems and with the operator.

Automatic tuning will also be needed to set the probe coil to the desired frequency of operation for the substance in question. For reasons of cost this often restricts the initial range of frequencies that can be accommodated,

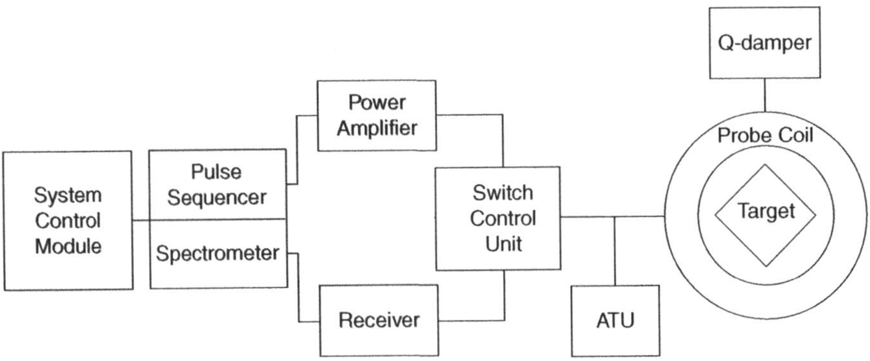

FIGURE 4.10 Generic arrangement of QR system.

although it is possible to cover a wider range of frequencies by the use of more comprehensive designs using multiple tuning networks or parallel tuned circuits. These designs are more complicated and require significant effort and attention to reduce stray capacitances and line losses compromising Q and hence performance.

4.3.2 Transmit-and-Receive Coils

The main challenge in the design of coils for NQR applications is ensuring that the very weak signal from the sample is not masked by the decaying energy from the transmit pulse. High Q factors of the coils mean both slow buildup and a long ring-down time after the transmit pulse ceases, and this energy must be reduced to very low levels, often by means of Q damping circuits, to permit detection of the signal from the sample. When the sample can be placed inside the coil, then a long solenoid can be used and the SNR is considerably improved. When the sample is external to the coil, it is a major challenge to achieve an adequate SNR. Grechishkin and Grechishkina [191] considered the optimal conditions for observations of signals in an integrated NQR detector and analyzed the coil structure in detail.

The transmit-and-receive coils used for NQR systems take many forms, and the simplest design for creating an external field is a single turn whose radiated field is given by Eq. (4.7), which describes the field on the axis of a circular coil and is shown in Figure 4.11.

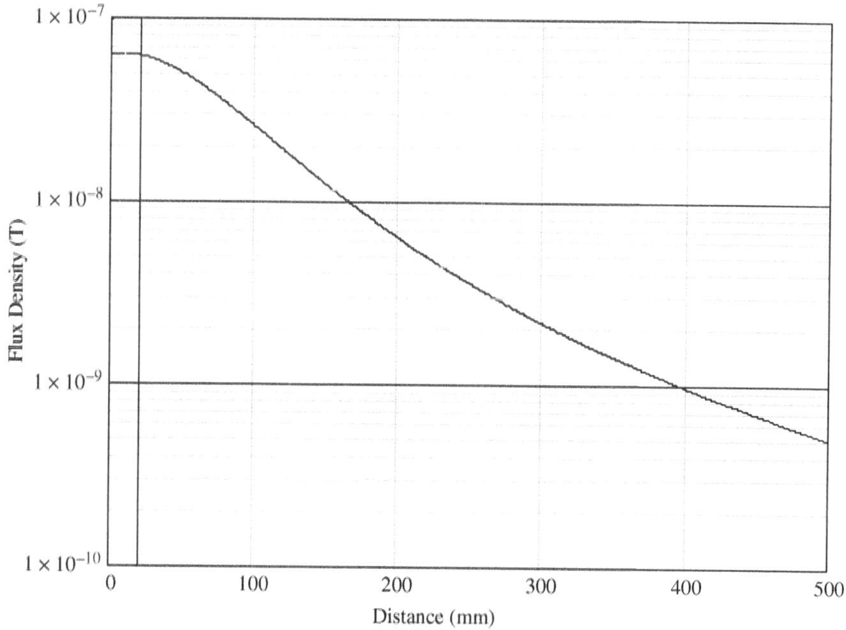

FIGURE 4.11 Magnetic flux density on the x axis of a circular coil of radius 100 units.

$$B = \frac{\mu_0 . IN}{2} \frac{r^2}{(r^2 + x^2)^{3/2}} \tag{4.7}$$

where B = magnetic flux density in teslas
I = current in amps
N = number of turns
μ_0 = absolute permeability of free space in H m^{-1}
r = coil radius
x = distance on the axis of the coil

It can be seen from Figure 4.11 that the field falls off rapidly beyond about one coil radius, and at a distance of four times the radius the field is two orders of magnitude lower than that at the center of the coil. This results in the maximum search depth for an NQR detector being largely determined by the coil size and being in the order of 5 to 20 cm for many applications. There is a limit to the radius of the coil in terms of its effective capture area, and beyond a given radius no further flux is captured. This effect is shown in Figure 4.12 where the magnetic flux from a small source on the axis of a coil has been plotted as a function of the angle between the source and the wires of the coil. The vertical axis shows magnetic field and the horizontal axis distance. As the radius of the coil is increased, the field gradually diminishes to zero at 54.7°.

The Q of NQR coils is a key design parameter, and high values are desirable to achieve good discrimination for the wanted center frequency; however, this means that only one material at a time can be interrogated with a particular tuned circuit, which is a disadvantage. Rapid methods of tuning between selected frequencies have been developed, but the high voltages and currents present during the transmit pulse require high breakdown designs, which complicates the design of the coil and associated tuning elements. In addition high values of Q also limit the rise and fall times of the applied pulses. Programmable Q switching circuits are described by Hirschfield and Klainer [192] whereby the Q is reduced during transmission, enabling the rapid rise and fall times of the transmitted pulse, kept low to keep ring-down low and then increased to obtain maximum sensitivity for the receiver. Fukushima and Roeder [193] describe the practical details of hardware, which although primarily concerned with NMR, is also highly applicable to NQR. Suits and co-workers [194, 195] investigated the practical issues involved with gradiometer coil approaches as well as the optimization of the self-shielded gradiometer. The challenge of radio-frequency interference (RFI) suppression has been addressed by many approaches, including additional coils to detect the interference and provide cancellation as described in a patent by Barrall et al. [196].

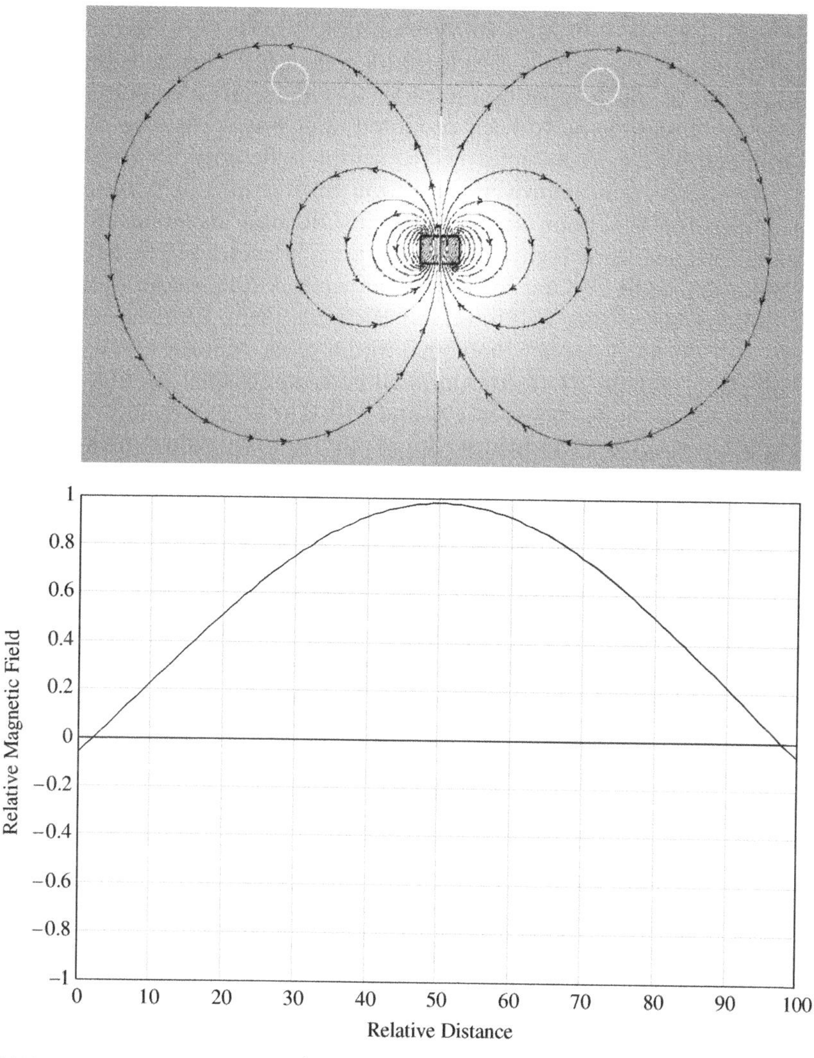

FIGURE 4.12 Magnetic field between the windings of a coil as a function of angle between the source and wire.

The signals radiated by the sample are at a very low level and correspond to flux densities in the region of picoteslas. This means that the design of the receiver is critical to achieving as high an SNR as possible. Major contributors to noise are the thermal or Johnson noise in the receive coil as well as noise in the first stage of the receive amplifier. Garroway et al. [197] calculate the mutual inductance between the sample and the receiver coil by considering the sample as modeled by an equivalent coil, and an example of this approach is shown in Figure 4.13. Note the low mutual inductance for small-diameter samples.

Garroway et al. [198] note that typical signal received pulse voltages for AP mines lie in the region of 10 to 20 nV root mean square (rms). Given that the receiver noise is in the region of 20 to 30 nV rms, it is necessary to average the signal by multiple pulse repetition in order to achieve an adequate SNR.

More exotic designs have been proposed, and one realization based on a high-temperature superconductor (HTS) sensor is described by Wilker et al. [198], who noted that improvement in the SNR could help overcome the shortcomings of the poor sensitivity of NQR and its concomitant long measurement times and high false alarm rates. Wilker [20] reasoned that as the SNR of an NQR coil improves approximately as the square root of the Q value as described earlier by Hoult and Richards [199], then the use of HTS sensors with Q values greater than 5000 would offer significant improvements in SNR given that the Q of conventional coils rarely exceeds 500 and more often provided Q values between 100 and 150. The characteristics of high Q-value sensors of narrow bandwidth require that their frequencies must be tuned to compensate for the temperature dependence of the NQR response, and its Q value must therefore be reduced to quickly dissipate stored energy. The HTS coils were made from two 0.5-mm thick, 51-mm-diameter $LaAIO_3$ or two 0.45-mm thick, 51-mm-diameter sapphire wafers coated on both sides with a 0.6-pm film of $Tl_2Ba_2CaCu_2O_8$ (TBCCO). The films typically have a critical temperature above 102 K and a surface resistance measured at 10 GHz of less than 300 μΩ on each side. For the devices on individual wafers, typical Q values were greater than 8000 for input powers of 0.1 mW. To Q damp, the HTS coil was loaded with both a resistor and a capacitor, which reduced the Q value to less than 100. Another approach to Q damping is taken by Peshkovsky

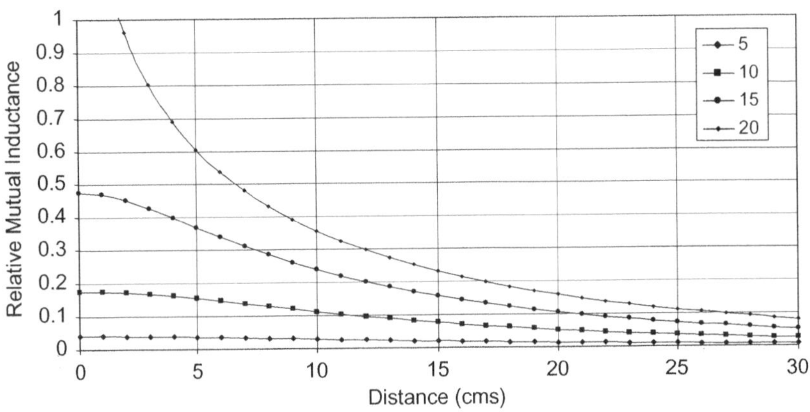

FIGURE 4.13 Relative mutual inductance between receive coil and sample for a receive coil diameter of 20 cm radius and various sample diameters in centimeters.

et al. [200] who developed a novel active ringing suppression circuit, inductively coupled to the main coil with a conveniently positioned low-inductance series transformer and achieved a signal-to-noise improvement of nearly 5.

4.3.3 Receiver and Coil Considerations

With reference to Figure 4.10 and Figure 4.14 the key elements of an NQR system can be understood. The fundamental design issues with the receiver and processor, which is essentially a spectrometer, concern the bandwidth and the SNR that can be obtained. Given the wide spread of resonant frequencies from 0.3 to 5 MHz for explosives or narcotics and hence the need for tuning over the band as well as the requirement for a high SNR, the design of the coil and receiver is nontrivial. The probe coil has to meet the conflicting requirements of high Q to achieve sensitivity and low Q to meet the need for a reasonable bandwidth within a short recovery time. The coil is normally constructed from one or more turns of copper and has to be tuned by means of high-voltage, low-loss capacitors switched in using relays. The capacitors must provide a completely linear characteristic until their rated voltage breakdown value and not exhibit any piezoelectric ringing. As the coil is coupled to both the transmitter and receiver, the latter must be protected from the transmitted high-voltage pulses either by means of isolation diodes or fast switches. The receiver preamplifier should have a low noise figure. The impedance matching of the coil to the low-noise preamplifier is critical, and Miller et al. [201] discuss the issue involved with optimizing the coupling between the probe coil and the preamplifier by mismatching their impedances. Fushikama and Roeder [193] discuss practical design issues, very relevant to the design of NQR systems, in detail in their book on NMR design techniques. Given the variety of approaches that can be taken, the designer has a range of options depending on the application. The challenge is to optimize the SNR and methods that do not require changes to the probe coil circuit such as applying a following pulse

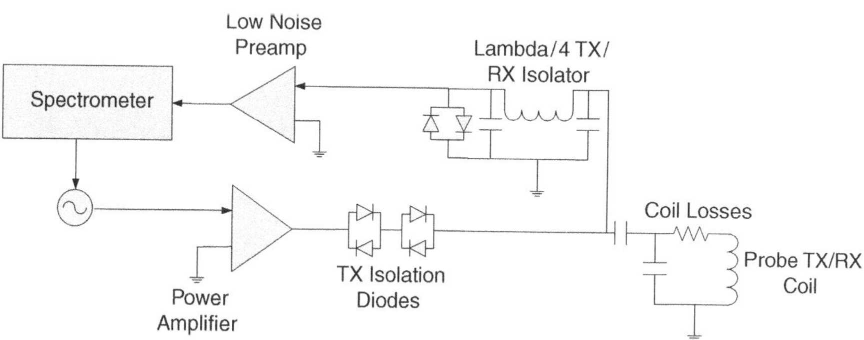

FIGURE 4.14 Typical arrangement of NQR probe coil and transmitter–receiver.

out of phase with the main transmit pulse or overcoupling the receiver and probe coil do not gain major improvements.

It was also noted by Peshkovsky et al. [200] that issues associated with the receiver protection diodes might inadvertently increase the ring-down time when Q switching is not employed. The methods of reducing interference range from using the output from a separate coil to monitor the background interference, detecting, at regular intervals, the output from the main receive coil as part of the pulse sequence with the transmitter disabled, or carrying out a spectral analysis of the interfering signal.

4.4 SIGNAL PROCESSING

The data output from a QR system takes the form of a repetitive series of transient signals of a known center frequency or range of frequencies whose short-time amplitude time envelope decays exponentially as shown in Figure 4.15.

The RF signal immediately following a pulse is known as free induction decay and is a type of short-time sample waveform. In general, this is a repetitive function, which is intrinsically formed by virtue of the physical process rather than being classified as a truncated function. Truncated time series can be analyzed and indeed recovered provided the truncation window is known. The time-domain signature of the QR response may be free induction decay or a series of echoes between pulses in a multiple-pulse sequence, each of which is a time-reversed image of the FID response, which precedes it. The echo series is multiplied by a window function that decays exponentially with time. The echo may be thought of as a repetitive noncausal function, which could be derived from the Hilbert transform of the FID response.

In essence, the signal-processing requirement is that of a time-series analysis with particular reference to the characteristics of the time series and is described by Klainer et al. [202] as well as Tantum et al. [203]. The principal characteristics of the QR signals, which can be classified as a nonstationary time series, are a usually exponential decay, the presence of multiple frequencies, a

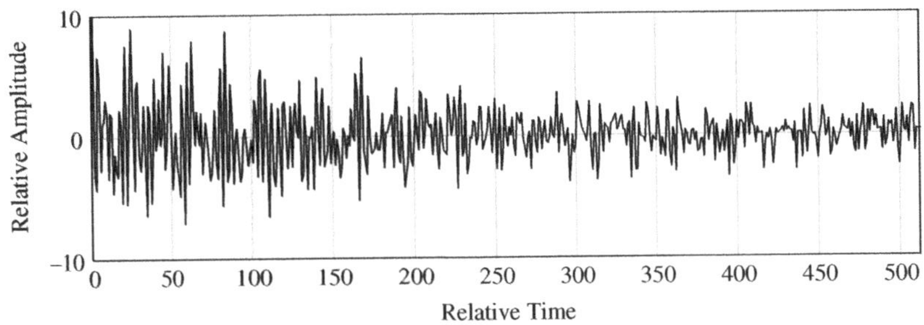

FIGURE 4.15 Simulation of the FID time-domain response of an NQR signal, with a 10-dB signal-to noise-ratio from a hypothetical material.

short-time response, an echo series, either causal or noncausal, and, most importantly, a low SNR. Normally, considerable averaging is applied to recover the weak signal from the noise, and this is achieved by repetitively subjecting the sample to a series of phase-coded, transmitted pulses. Tantum et al. [204] consider methods of signal processing for NQR application of landmine detection. Generally, it is considered that Fourier methods are most appropriate to the analysis of statistically stationary functions, whereas nonstationary functions may be better analyzed by alternative methods.

The limitations of the Fourier method of spectral estimation are that it has a frequency resolution that is approximately equal to the reciprocal of the sample time window in seconds. The action of sampling the data for a defined time causes leakage of energy from the main lobe of a spectral response into the side lobes.

The original simulated QR time series shown in Figure 4.16 comprised four exponentially damped signals of relative amplitude: 0.6, 0.8, 1, and 0.8. A comparison between the Fourier spectral analysis and the maximum entropy spectral analysis (MESA) as shown in Figure 4.17 as well as the autocorrelation method using the Yule–Walker algorithms shown in Figure 4.18 indicates that the signal-to-noise performance and estimation of the signal amplitudes of the MESA processing is the best of the methods both in terms of noise and spectral estimation. However, care in the selection of parameters for these alternative signal recovery methods is important to realize optimum spectral resolution.

Alternative methods, such as the Prony algorithm, which model functions as decaying exponentials, may therefore be appropriate. However, it is well known that early Prony algorithms are very vulnerable to noise and have met

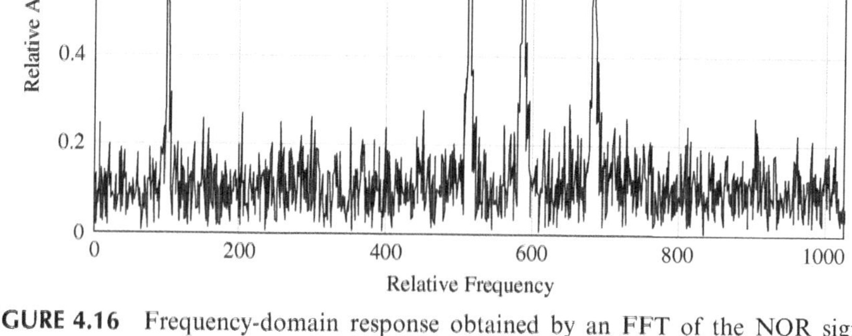

FIGURE 4.16 Frequency-domain response obtained by an FFT of the NQR signal from Figure 4.15.

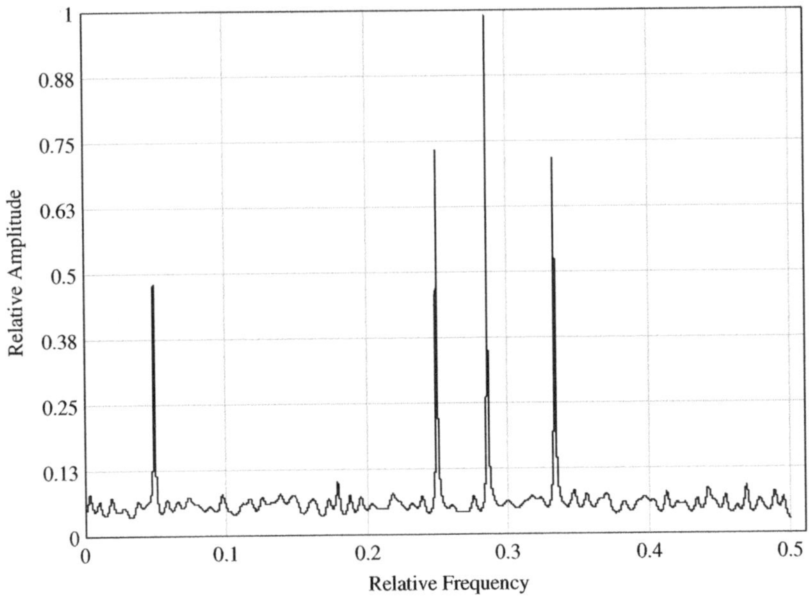

FIGURE 4.17 Frequency-domain response using the maximum entropy spectral analysis (Burg algorithm) of the NQR signal from Figure 4.15.

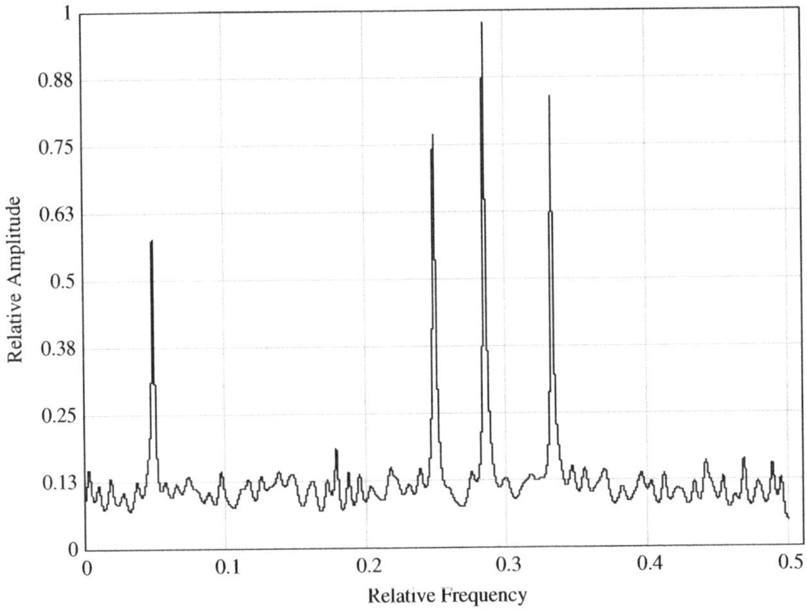

FIGURE 4.18 Frequency-domain response using the autocorrelation spectral analysis (Yule–Walker algorithm) of the NQR signal from Figure 4.15.

with mixed fortunes; more modern adaptations such as the eigenvalue realization of Prony may therefore be give better results. Wavelet transforms such as the Wigner or Short Time Fourier Transform (STFT) may be better adapted to the transient nature of the QR response, but again noise may prove a limiting factor as reported by Kercel et al. [205]. In the case of the multiple spectral lines generated by TNT, some form of template correlation may prove useful in improving the signal to noise and classification performance.

There is, therefore, the capability for improved sensitivity by selection of suitable processing methods and a short list of suitable candidates can be considered as follows:

Autoregressive spectral density estimation (AR)
Moving-average power spectral density estimation (MA)
Autoregressive–moving-average power spectral density estimation (ARMA)
Pisarenko harmonic decomposition
Prony energy spectral density estimation
Prony spectral line estimation
Maximum-likelihood method (MLM)
Maximum entropy methods (MEM)
Multiple signal classification (MUSIC)
Super-resolution methods
Wavelet transforms such as Wigner, STFT
S-plane representations
Spectral template correlation
Homomorphic deconvolution

Somasundaram et al. [206] noted that the NQR technique has been hampered by low SNR and is further aggravated by the presence of RF interference (RFI) and proposed a new NQR data model, particularly for the case where multiple-pulse sequences are used to generate trains of spin echoes. He refined maximum-likelihood algorithms, enabling them to optimally exploit the data model of the entire echo train and also incorporate knowledge of the temperature-dependent spin–echo decay time. The algorithms ensured accurate detection and robustness against residual RFI, even when the temperature of the sample was not precisely known, by exploiting the dependencies of the NQR resonant lines on temperature. As the algorithm is frequency selective, exploiting only those regions of the spectrum where the NQR signal is known to exist, further interference suppression was gained. Tan et al. [207] noted that although QR is vulnerable to RFI when the sensor is deployed in the field, adaptive RFI mitigation can remove most of the RFI and proposed a new Kalman filtering strategy to estimate and detect the QR signal in the post-mitigation signal. The results using both simulated data and experimental data show that the proposed algorithm can provide robust land-mine detection performance.

4.5 DETECTION OF EXPLOSIVES

Nearly all-solid high explosives contain nitrogen and can be detected by NQR techniques. Liquid explosives, such as ethylene glycol dinitrate (EGDN)/nitroglycerine, will not give rise to NQR signals at room temperature and hence cannot be detected. Miller and Garroway [208] describe the structure and chemical composition of solid explosives and note that ^{14}N signals have been detected in RDX, PETN, Semtex, and TNT over a range of temperatures about ambient. Garroway et al. [209] also describe the key issues in explosives detection using NQR. The full names and chemical formula of these substances

TABLE 4.1 Explosives Names and Abbreviations

Explosive	Abbreviation	Formula
1,3,3-Trinitroazetidine	TNAZ	$C_3H_4N_4O_6$
Ammonium nitrate	AN	$H_4N_2O_3$
Ammonium perchlorate		H_4NO_4Cl
Ammonium picrate	Expl D	$C_6H_6N_4O_7$
Cyclonite	RDX	$C_3H_6N_6O6$
Ethylenediamine dinitrate		$C_2H_{10}N_4O_6$
Guanidine nitrate		$CH_6N_4O_3$
Hexamethylenetriperoxide diamine	HMTD	$C_6H_{12}N_2O_6$
Hexanitrohexaazaisowurtzitane	HNIW or CL20	$C_6H_6N_{12}O_{12}$
Hydrazine nitrate		$H_5N_3O_{3\ 5}$
Lead styphnate		$C_6H_3N_3O_9P_b$
Mannitol hexanitrate		$C_6H_8N_6O_{18}$
Monomethylamine nitrate		$CH_4N_2O_3$
Nitrocellulose		$C_6H_7N_3O_{11}$
Nitroglycerin	NG	$C_3H_5N_3O_9$
Nitrotriazolone	NTO	$C_2H_2N_4O_3$
Octogen	HMX	$C_4H_8N_8O_8$
Pentaerythritol tetranitrate	PETN	$C_5H_8N_4O_{12}$
Picric acid		$C_6H_3N_3O_7$
Semtex-H	RDX-PETN	
Tetraamminecopper perchlorate		$H_{24}Cl_2CuN_4O_8$
Tetrazene		$C_2H_8N_{10}O$
Tetryl	Tetryl	$C_7H_5N_5O_8$
Triacetone triperoxide	TATP	$C_9H_{18}O_6$
Triaminoguanidine nitrate	TAGN	$CH_9N_7O_3$
Triaminotrinitrobenzene	TATB	$C_6H_6N_6O_6$
Trinitrobenzene	TNB	$C_6H_3N_3O_6$
Trinitrochlorobenzene		$C_6H2ClN_3O_6$
Trinitropyridine		$C_5H_2N_4O_6$
Trinitrotoluene	TNT	$C7H_5N_3O_6$
Urea nitrate		$CH_5N_3O_4$

TABLE 4.2 NQR Characteristics of Explosives

Explosive	Molecular Weight	ν₋ (kHz)	ν₊ (kHz)	T_1 (ms)	T_2 (ms)	T_2^* (ms)
AN	80.05	423.4	496.4	296.4 / 13, 423.4/16	8, 6	2.5
Comp B	As constituents			As constituents		
Comp C-4	As constituents			As constituents		
Detasheet	As constituents			As constituents		
HMX	296.16	3,737	5,300	≈ 1500	125	0.23
HMX	296.16	3,623	5,063	≈ 1800	159	0.14
KNO_3		559		1900	10	0.3
KNO_3		567		1000–10000	10	1.6
KNO_3		664		1000–10000	10	1.6
PETN	316.20		890	≈ 32000	50–60	0.9
RDX	222.26		5,240	12	7.1	0.74
RDX	222.26		5,192	13	8.2	1.70
RDX	222.26		5,047	13	6.8	0.83
RDX	222.26	3,458		12	5.7	0.65
RDX	222.26	3,410		11	6.2	0.92
RDX	222.26	3,359		15	6.3	0.74
Semtex-H						
Tetryl	287.15		5292	640	140	0.14
Tetryl	287.15	3829		1400	105	0.1
Tetryl	287.15	1463		2500	185	220
TNT	227.13	713.7		10		
TNT	227.13	714.3		4		
TNT	227.13	740		9		
TNT	227.13	743		4.2		
TNT	227.13	751		3		
TNT	227.13	768		16		
TNT	227.13		837	5		
TNT	227.13		842	5.2		
TNT	227.13		844	7.5		
TNT	227.13		848	28		
TNT	227.13		859	12		
TNT	227.13		870	2.91	7	1.2
Urea nitrate	123.08		2730	6.1		
Urea nitrate	123.08		2649	4.2		
Urea nitrate	123.08		2006	7		
Urea nitrate	123.08		1900	7		
Urea nitrate	123.08	750		11.1		
Urea nitrate	123.08	725		15.0		
Urea nitrate	123.08					

[a] v_+ Upper frequency
[a] v_- Lower frequency
[a] t_1 First Time Constant
[a] t_2 Second Time Constant
[a] t_3 Thirst Time Constant
[a] *Source:* Compiled from various sources including Garroway et al. [210] (with kind permission from Dr. Garroway and Prof. Smith [177, 178] (Tetryl)).

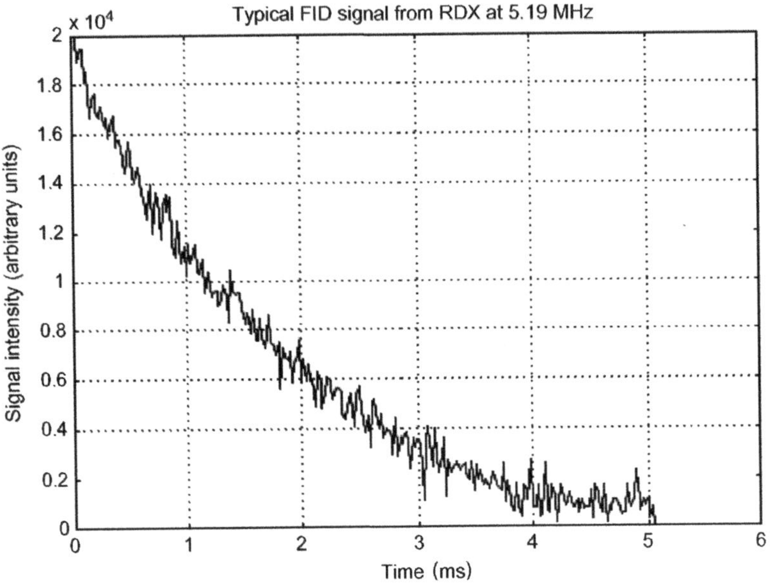

FIGURE 4.19 ^{14}N FID in RDX at 5.190 MHz (with kind permission Prof. Smith 178).

is given in Table 4.1, and their typical frequency values at room temperature are as given in Table 4.2.

In general, the detectability of RDX and Tetryl is good by NQR, but for TNT the NQR relaxation times are less favorable, and the possible presence of two crystalline polymorphs (monoclinic and orthorhombic) leads to weaker TNT signals. Two data acquisitions can be taken with a 6-s scan (using a delay of T_1 between acquisitions) for TNT with a nominal T_1 relaxation time of 6 s at 20°C, but 51 acquisitions can be obtained in a 5-min scan, giving an improvement of a factor of 5 in SNR. Hence, much effort has gone into improving NQR pulse sequences for TNT because of its low SNR. A typical FID response from TNT is shown in Figure 4.19, and the spectrum of monoclinic TNT is shown in Figure 4.20, and Hibbs [211] showed typical spectral signatures for RDX and PETN as illustrated in Figure 4.21.

Work by Mann and Hunt at ERA Technology and reported by Barras et al. [212] showed the feasibility of detecting ammonium nitrate inside vehicles. Ammonium nitrate characteristics are presented [e.g., at room temperature 423 kHz (ν_-) 496 (ν_+) kHz, temperature coefficient + 91 Hz/°C for the 423 kHz line and −300 Hz/°C for the 496-kHz line] and the signal from the ν- line at 422 kHz from a 475-kg sample of ammonium nitrate concealed inside a vehicle as shown in Figures 4.22 and 4.23.

Crowley et al. [213] describes a novel shoe scanner, as shown in Figure 4.24,developed at the GE Security San Diego Center of Excellence that uses both NQR and configuration-sensitive metal detection to identify threats

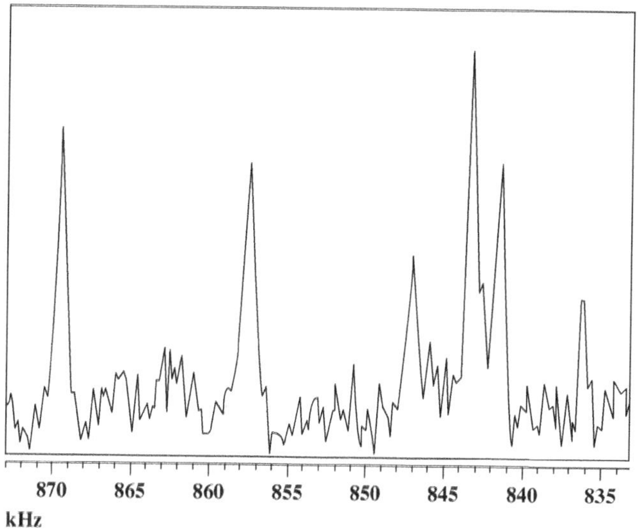

FIGURE 4.20 The 6 n^+ highest frequency lines of "creamed" monoclinic TNT (with kind permission Prof. Smith private correspondence).

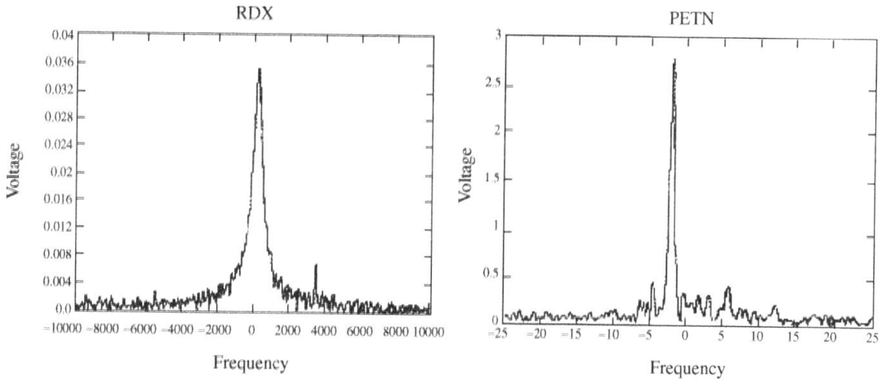

FIGURE 4.21 Typical spectral signatures for RDX and PETN (with kind permission Dr. Hibbs).

hidden in shoes. The shoe scanner was developed with an open-access chassis and scanning chamber that allows passengers to stand in the system in a natural position during the scanning process. More traditional magnetic resonance systems are closed or partially closed and cannot be used for screening personnel because the scanning chambers confine the object in question. The shoe scanner's novelty lies in a particular chassis geometry that allows both QR and metal screening.

FIGURE 4.24 Integrated shoe detector developed by GE Security (with kind permission ©GE Security and SPIE © 2007).

FIGURE 4.25 Coil with integrated preamp and battery-powered power amplifier for land-mine detection. (with kind permission ©ERA Technology).

improving the RFI rejection of the QR coil and adding degrees of freedom to the reference antennas used in adaptive RFI mitigation. New QR pulse sequences were implemented to better cancel the variable ringing encountered when operating on quartz gravel at the temperate test site.

Further technical improvements made during 2004 restored the TNT sensitivity previously traded-off for improved RFI mitigation and PER rejection. Engineering development has reduced the weight of the system by an

TABLE 4.3 Test Results (2002) Obtained by QM

Test Location	PD [90% Conf. Limit]	PFA [90% Conf. Limit]	No. of Markers
Arid test site	0.98 [0.95, 1.00]	0.04 [0.02, 0.07]	312
Temperate test site	0.98 [0.90, 1.00]	0.04 [0.02, 0.10]	134

Source: With kind permission of SPIE ©2002.

FIGURE 4.26 Quantum Magnetics concept vehicle NQR sensor (with kind permission ©GE Security).

FIGURE 4.27 U.S. Marines AMD NQR technical demonstrator coil (with kind permission ©GE Security).

order of magnitude to <65 kg, and the average power consumption by nearly 2 orders of magnitude to <100 W. The QR components for a vehicle-based technology demonstrator as shown in Figure 4.26 were planned to be developed during 2004 but subsequent developments have not been openly reported.

QM also worked on a handheld multisensor system based around the EMI/GPR detector developed for the U.S. Army and now designated the AN/PSS-14, and a photograph of this is shown in Figure 4.27 during trials.

4.7 ILLICIT DRUGS

Illicit drugs are a major problem for society, and much effort has been made to develop NQR techniques for the detection of drugs either concealed in packages or on individuals. Work at Kings College, London, has shown that ^{14}N NQR signals can be detected in several class A drugs including cocaine, heroin, and their hydrochlorides.

Haq [223] in his doctoral thesis at King's College, London, first reported double-resonance results obtained by cross-relaxation for heroin, cocaine, and cocaine hydrochloride. Later publications gave NQR results for cocaine base and the hydrochloride and heroin hydrochloride monohydrate. The cannabis group lacks a suitable quadrupolar nucleus and is not therefore eligible for detection by NQR methods.

The following sections are based on work carried out by the NQR group at King's College, London, on a pulsed Fourier transform RF spectrometer, and using a probe developed by ERA Technology as shown in Figure 4.28 but also offer comparisons with data from other laboratories. Note that the emphasis is on describing what NQR features have been found in particular substances in order to provide a guide as to what can be expected. The author is most grateful to Professor John Smith for providing the information.

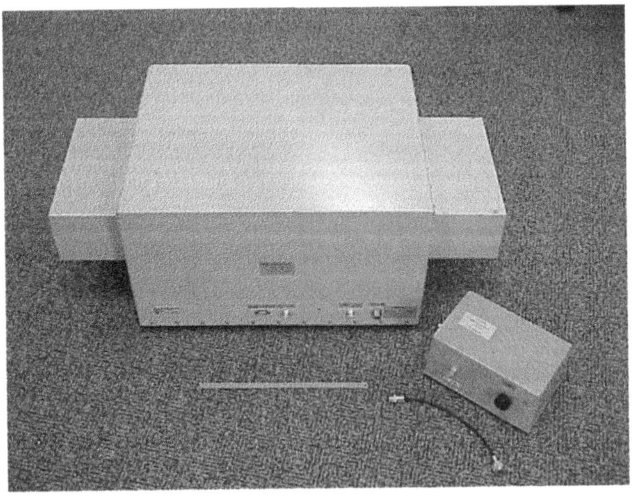

FIGURE 4.28 Probe with integrated Q damping and RF feed unit for drug detection research (with kind permission ©ERA Technology).

4.7.1 Cocaine

Cocaine ($C_{17}H_{21}NO_4$, Mol. Wt. 303.4, $d = 1.25$ g cm^{-3}) forms monoclinic crystals containing one molecule as the repeat unit, so all the NQR lines are singlets. The asymmetry parameter is small (0.04) as in heroin, so ν_+ and ν_- are close to each other in both compounds. The measurements at King's College were performed on a reasonably pure sample according to an elemental analysis, and the results at room temperature are shown in Table 4.4. These results can be compared with measurements undertaken at the Naval Research Laboratory in Washington, D.C. [120] on an "impure" sample of cocaine and one after recrystallization from ethanol and shown in Table 4.5.

Note that the NQR frequencies and spin–lattice relaxation times are not significantly changed, whereas the linewidths (column labeled $\Delta\nu_§$) are very different, being substantially larger in the impure sample. This is a characteristic feature of NQR spectra, which are, in general, inhomogeneously broadened and therefore dependent on the presence of impurities or defects in the crystals, an effect that could be exploited in determining the likely origin of a seizure.

4.7.2 Cocaine Hydrochloride

Cocaine hydrochloride ($C_{17}H_{22}NO_4Cl$, Mol. Wt. 339.9, $d = 1.342$ g cm^{-3}) has a noncentrosymmetric group and may therefore show piezoelectricity. Some published data from King's College, London, and other sources are given in Table 4.6 at various temperatures; the King's College sample was the pure material. Note the rather small temperature dependence of the ^{14}N frequencies, about –0.03 kHz K^{-1} near room temperature, again a characteristic of NQR lines. Both nuclei, ^{14}N and ^{35}Cl, can be used for detecting this material and other drug hydrochlorides. The data were measured using a pulsed Fourier transform spectrometer. All data refer to ^{14}N transitions and the values at 4.2 K were obtained using a SQUID by Yesinowski et al. [224].

TABLE 4.4 ^{14}N NQR Frequencies in Cocaine at Room Temperature

	ν (kHz)	e^2qQ (h kHz)	η	$\Delta\nu_§$ (Hz)	T_2^* (ms)	T_1 (ms)	T_2 (ms)	T_{2e} (ms)
ν_+	3817	5023	0.03	570	0.6			18 ± 2
ν_-	3718		0.09	180	1.7			ND

TABLE 4.5 ^{14}N NQR Data for Pure and Impure Samples of Cocaine

Temp (K)	ν (kHz)	$\Delta\nu_§$ (Hz)	T_1 (ms)	T_2 (ms)	e^2Qq (h kHz)	η
295 (impure)	ν_+: 3816.8	860	210	85	5022.9	0.0395
	ν_-: 3717.6	870	ND	ND		
295 (pure)	ν_+: 3816.8	260	160	81	5022.8	0.0395
	ν_-: 3717.5	270	75	75		

ND, no data.

4.7.3 Heroin (Diamorphine)

Heroin ($C_{21}H_{23}NO_5$, Mol. Wt. 369.4, $d = 1.32$ g cm^{-3}) has been more intensively studied at King's College as part of a research program in collaboration with Defence Science and Technology Laboratory dstl. Two crystal forms, orthorhombic and monoclinic, have been reported in the litreature; it appears that all published NQR data refer to the former, in which only one set of spectral lines should be observed. The experiments used an inspection volume of 13 litres and different samples of heroin containing nominally 50% of the drug. The RF probe and electronics were developed and constructed by ERA Technology. Six different samples of heroin from seizures were studied with varying percentages of active material; two of the samples were diluted with caffeine (itself a nitrogen rich-compound). The excitation frequency was 4.007 MHz with PSL sequences of pulse repetition time of 1.1 ms and a signal acquisition time of 0.512 ms; the Q factor of the coil was 214 and was damped immediately following the pulse so as to reduce the dead time to 0.14 ms. In a typical experiment, a 2.2-kg sample of heroin containing nominally 50% of the drug in a coil volume of 13 litres (a filling factor of 24%) gave a SNR (defined as the ratio of real signal amplitude after Fourier transform divided by rms noise) of 33.4 in a total time of 160 s, constituted of 64 PSL scans each containing 20 echoes, with a 2.5-s delay between each scan. QR sciences also developed a system to detect heroin base, which was trialed at a major airport.

Previously published frequencies and relaxation times are reported in Table 4.7 (row 3) and compared with King's College data in rows 1 and 2 obtained on two samples of high purity on loan from dstl. There is some discrepancy between the two sets, the reasons for which are not understood.

Temperature variation measurements of frequencies and relaxation times have been conducted at Kings College and can be seen in Figure 4.9. The relaxation times are also a function of temperature and are shown in Figures 4.29 and 4.30, which highlight the dependence of T_1on temperature. The

TABLE 4.6 NQR Parameters of Cocaine Hydrochloride

Temp (K)	ν (kHz)	$\Delta\nu_{\S}$ (kHz)	T_1 (ms)	T_2 (ms)	e^2Qq (h kHz)	η
4.2	ν_+ : 965		ND	ND	1186	0.258
	ν_- : 815		ND	ND		
77	ν_+ : 961	5	370	25	1182	0.250
	ν_-: 813	5	360	22		
141	ν_+ : 960	5	820	6	1180	0.252
	ν_- : 811	5	ND	ND		
295	ν_+ : 961	5	700	1.5	1178	0.263
	ν_- : 806	4	2000	> 0.3		

ND, no data.

TABLE 4.7 Room Temperature ^{14}N Data for Heroin base

	ν (kHz)	e^2qQ (h kHz)	η	$\Delta\nu_{\S}$ (Hz)	T_2^* (ms)	T_1 (ms)	T_2 (ms)	T_{2e} (ms)
ν_+	4007	5305	0.021	—	0.26	524	6	12 ± 1
ν_-	3951			740	0.43	400 ± 100	6 ± 4	19 ± 3
ν_+	4006.0	5304	0.021					
ν_-	3950.2							
ν_+	4025	5317	0.028	several		<1000	>0.1	
ν_-	3950			kHz		400	>0.01	

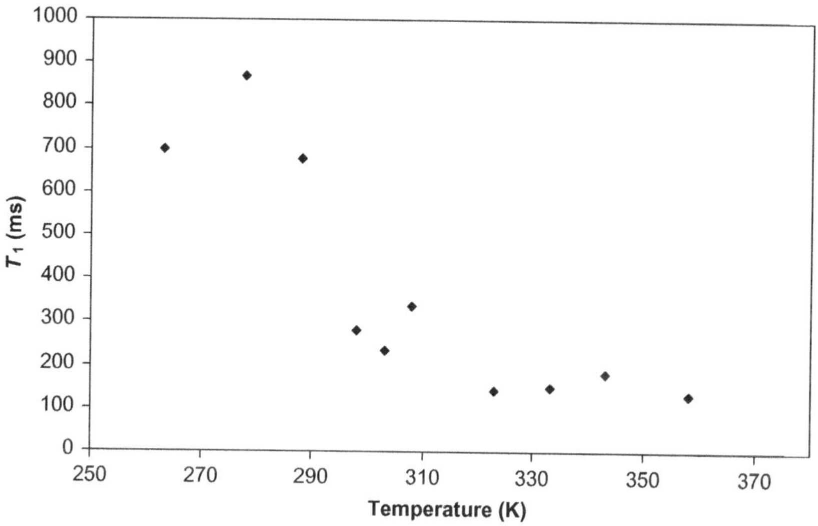

FIGURE 4.29 Temperature dependence of the spin lattice relaxation time, T_1, for the ν_- line of heroin between 260 and 360 K.

multiple-pulse echo decay time T_{2e} varies for different values of the interpulse spacing, 2τ, in a PSL sequence, but only significantly so below values of 0.5 ms as shown in Figures 4.31 and 4.32. A typical spectrum with $2\tau = 0.32$ ms is shown in Figure 4.33 in which both ν_+ and ν_- lines are visible. From Figure 4.9 the average temperature coefficient of the heroin frequencies is approximately -0.14 kHz K^{-1} near room temperature. All the pulses had duration of 10 μs, with a dead time after the train pulses of 126 μs. The data were obtained at 64 points with a dwell of 2.5 μs, and the signal was averaged over 2756 scans with a repeat time of 500 ms. The coil tuning/matching and excitation frequency was 3975 kHz. The spectrum was obtained by echo zero padding the time-domain data to 128 points prior to echo Fourier transform processing. The magnitude signal is shown. The two lines are at 3950.2 and 4006.0 kHz.

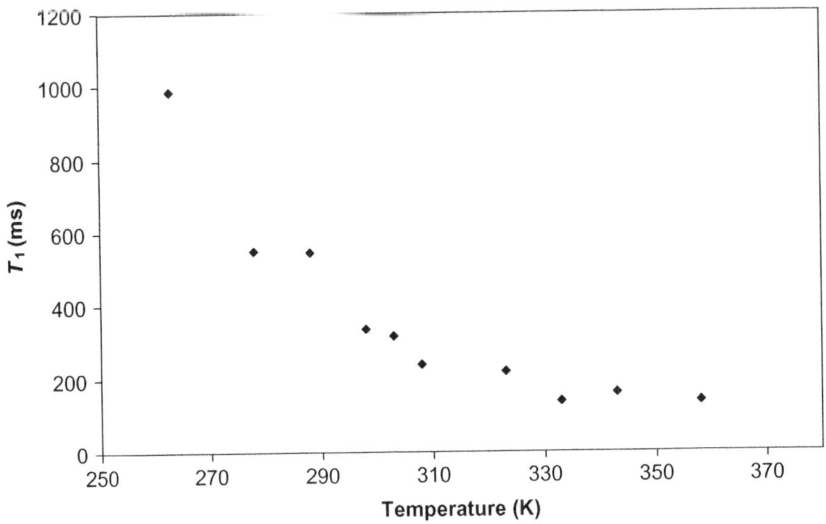

FIGURE 4.30 Temperature dependence of the spin lattice relaxation time, T_1, for the ν_+ line of heroin between 260 and 360 K.

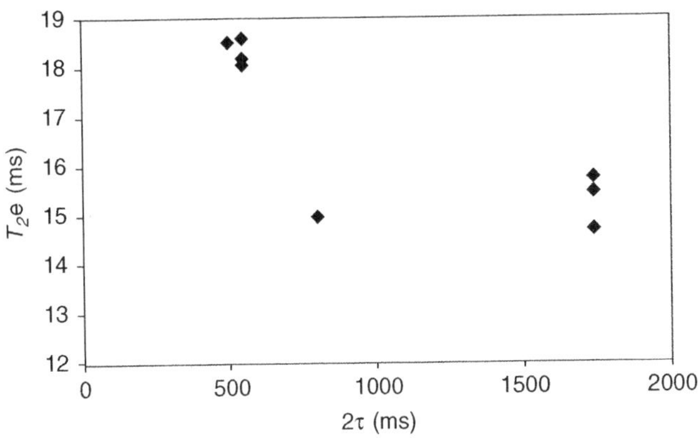

FIGURE 4.31 Dependence of the pulse spin echo relaxation time, T_{2e}, on the interpulse spacing, 2τ, for the 3951-kHz line of heroin at room temperature.

NQR frequencies have been measured at King's College, London, in two other illicit drugs: the ^{35}Cl NQR frequencies in heroin hydrochloride monohydrate and in ecstasy (MDMA hydrochloride) at room temperature have been measured as 1.950 and 2.712 MHz, respectively.

If both signal and noise can be represented by normal distributions about their mean values with equal variances, then detection time can be defined as the total time in minutes estimated to achieve an SNR of 3.3. This

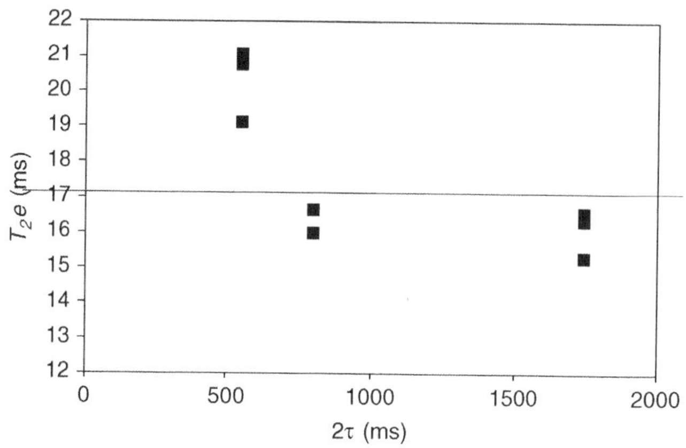

FIGURE 4.32 Dependence of the pulse spin echo relaxation time, T_{2e}, on the interpulse spacing, 2τ, for the 4.007-kHz line of heroin at room temperature.

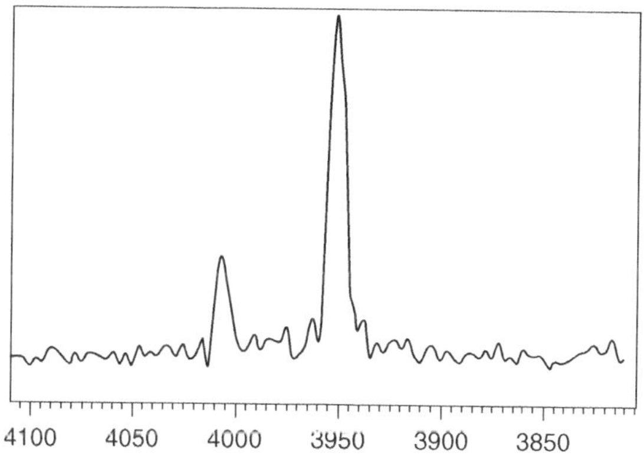

FIGURE 4.33 ^{14}N NQR spectrum of heroin obtained by summing 14 echoes of a PSL sequence with $2\tau = 316\ \mu s$.

corresponds to the so-called detection limit in chemical analysis for a sample containing a net amount of 200 g of illicit material in the given volume. Therefore, at a threshold for a positive response of 1.65, the probability of detection (PD) can be said to be 95% and the false alarm rate (FAR) 5%, that is, in 100 drug-containing items, 95% will be registered as positive. The threshold can be moved to lower or higher values of the SNR if the balance of PD and FAR needs to be changed; for example, an SNR threshold of 2.0 will give a PD of 90% and a FAR of 2%. Smaller quantities of a drug than 200 g can still be detected (an advantage of NQR methods relative to other techniques, such as trace detectors) but will require correspondingly longer

TABLE 4.8 Estimated Detection Times for Cocaine and Heroin

Drug	ν_Q (MHz)	Quantity (g)	V_c (liters)	t_d (s)	Detection Time (min)
Heroin	4.007	200	174	0.9	3
Cocaine	3.817	200	174	0.6	2

detection times. To estimate quantities, it is recommended that the above detection times be lengthened by a factor of between 2 and 3. These detection times could be significantly shortened by the use of more sophisticated data processing methods.

4.8 SUMMARY

The NQR method is noninvasive, highly specific, and signals are generally seen in solids, so suspensions of materials and mixtures with other substances are often unsuitable for detection. Signals have been seen in suspensions of ammonium nitrate (AN) provided the solid is still present and mixtures can be detected as solids, for example, in medicines. Unlike the closely related technique of NMR, no static magnetic field is necessary, so remote materials and large volumes (the largest value known is 8000 litres) can be examined.

The material under investigation may be placed inside a coil (internal field detection) or outside a coil (external field detection). In the case of external fields and small samples of material, the re-radiated signal in the order of 0.1 pT. Internal fields and large samples cause fields in the range greater than 100 pT. This means that the design of the exciting coil and receiver is critical to achieving as high an SNR as possible. Major contributors to noise are the thermal or Johnson noise in the receive coil as well as noise in the first stage of the receive amplifier.

A major challenge in the design of coils for NQR applications is ensuring that the very weak signal from the sample is not masked by the decaying energy from the transmit pulse. High Q factors of the coils mean both slow buildup and a long ring-down time after the transmit pulse ceases, and this energy must be reduced to very low levels, often by means of Q damping circuits, to permit detection of the signal from the sample.

In general, the SNR increases as a function of $\omega^{3/2}$ so that materials whose resonance is below 1 MHz are more difficult to detect than materials whose resonance is above 1 MHz. The physics of detection is complex and RFI as well as signals caused by magnetoacoustic and piezoelectric signals or the ringing signals from the NQR coil complicate system design.

Times for detection depend on the substance and the need to average pulse sequences as well as the requirement to retune the system for different substances. All these factors have limited the widespread acceptance of NQR as a detection technique but some commercialization has taken place.

Considerable efforts have been made to overcome the intrinsic limitations of NQR for land-mine detection, these being listed below:

Low SNR on some materials.
Radio-frequency interference
Long interrogation time for a full range of explosives
Piezoelectric ringing from earth materials
Power budgets

These challenges still remain to be fully solved. NQR techniques have found most success where the sample can be placed inside a coil, and the SNR is much higher than when the sample is external to the exciting coil. NQR offers promise, but much research work remains to be done to bring it to maturity.

CHAPTER 5

Radar Systems

5.1 INTRODUCTION

The most basic radar system is the Doppler radar, is also termed a homodyne radar system, which it transmits a fixed frequency and, by mixing the received signal with the source signal, produces an output whose frequency is proportional to the radial velocity of the target signal. Doppler radar is unable to determine range but can determine velocity toward or away from the radar by suitable design. To determine range requires some type of timing marker on the transmitted waveform, and this can be achieved by either frequency-modulated continuous wave (FMCW), by stepped frequency continuous wave (SFCW), coded modulation (CM), noise modulation (NM), synthesized pulse (SPM), holographic (HM), and amplitude modulation (AM), which is essentially operation in the time domain (TD) using short pulses or impulses. In this chapter, we will consider the most commonly used techniques in turn and discuss the key parameters, which need to be considered in the design and selection process. As long as the bandwidth of the transmitted information is equal, there is no fundamental difference between any of the modulation techniques apart from spatial modulation. The performance of radar systems is then defined by the inherent noise of the receiver, which varies depending on its bandwidth as well as the self-generated clutter of the radar. In this chapter the various modulation techniques are considered and their strengths and weaknesses are reviewed.

5.2 DOPPLER RADAR SYSTEMS

One reason for considering using a Doppler radar is that the many environments are very cluttered, and since Doppler radar detects only movement in a

EM Detection of Concealed Targets, by David J. Daniels

radial direction to the transmitter, the clutter from stationary objects is eliminated. If only velocity is required, Doppler radar is an obvious choice. In addition, Doppler radars are inherently simple to design and build. King [225] discusses the design issues involved with homodyne radar systems, including modulated systems, and is a suitable reference for such work. The choice of frequency of operation is, however, critical because the type of intervening material and its attenuation affects the propagation characteristics of radar in a particular situation. A generic block diagram of a Doppler radar is shown in Figure 5.1. It consists of an oscillator, RF amplifier, coupler, circulator, and antenna, which form the transmit chain. The received signal from the antenna passes into the mixer where it is down-converted to a baseband frequency, amplified by a baseband amplifier, and then bandpass filtered to remove out of band low-frequency and high-frequency signals. It is then fed into a signal-processing stage, which could comprise an A/D converter and then signal processor.

The basic equation for the Doppler frequency shift is given by

$$f_d = \frac{2 v_r f_0 \cos \psi}{c} \tag{5.1}$$

where v_r = radial velocity
f_0 = transmitted frequency
ψ = angle between the target vector and the radial direction

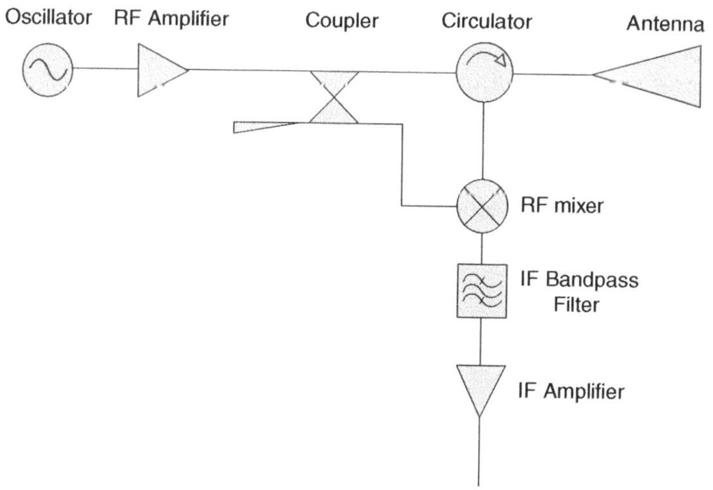

FIGURE 5.1 Doppler radar system.

Doppler radar systems are very useful for measuring human movement, respiration, and heartbeat and with a suitable choice of center frequency can be used to detect humans concealed in buildings or containers. The frequency of operation of these systems ranges from X-band 8 to 12.5 GHz (λ = 3.75 to 2.4 cm), down to L-band 1 to 2 GHz (30 to 15 cm), and P-band 0.3 to 1.0 GHz (100 to 30 cm). The lower the frequency, the greater the penetration but the lower the baseband frequency, which makes processing more challenging when subhertz frequencies are involved.

It is useful to understand the characteristics of human movement in order to determine the operating parameters of a radar design. Typical velocities are given in Table 5.1, and it can be seen that the velocity of breathing and heart movement is the lowest encountered. The typical Doppler frequencies generated as a function of transmitter frequency and radial velocity of 0.01, 0.1, 1, and 10 m s^{-1} are shown in Figure 5.2.

At 1 GHz the Doppler frequency for human breathing is less than 0.1 Hz, and this low frequency poses challenges for the design and operation of the radar in that the radar must remain very stable and noise free while measurements are being taken. In addition, the signal generated by observers and detected in the back lobes of the antenna may be larger than that of the subject. Much work has been carried out on such Doppler radars, and Nguyen et al. [226] describe and quantify the three main sources of baseband noise (thermal, residual phase, and Flicker) affecting receiver sensitivity in a direct conversion microwave Doppler radar. Flicker noise canceling techniques may drastically reduce the power requirement for heart rate signal detection by as much as a factor of 100. In another study Nguyen et al. [227] consider the problem detecting heart motion of both the radar operator and a subject behind a wall and show preliminary simulations using a blind source separation (BSS) algorithm to confirm the possibility of distinguishing these two heart–motion signals of widely different strength. Bilik and Tablikian [228] addressed the challenge of target classification for ground surveillance Doppler radars by means of signal classification, and Massagram et al. [229] investigated a quadrature Doppler radar system utilizing arctangent demodulation with dc offset compensation techniques. Other investigators have developed experimental systems for the detection of human heartbeat and respiration for people

TABLE 5.1 Velocities for Human Movement

Parameter	Velocity (m s^{-1})
Hand movement	5
Rapid body movement	3
Walking	0.5
Heart movement	0.45
Breathing	0.01

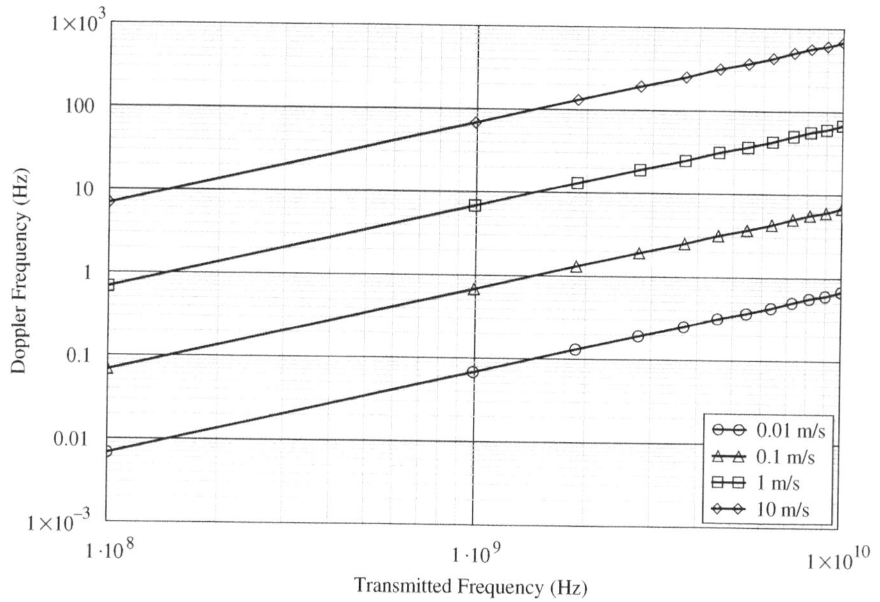

FIGURE 5.2 Doppler frequency (in Hz) versus transmitted frequency (in Hz) for radial velocities of 0.01 m s^{-1} to 10 m s^{-1}.

buried in earthquake debris or behind walls and various studies [230–242] all describe various systems and experimental results.

Movement of the body and limbs is much easier to detect, and Smith [301] Khlopov et al. [243] report that backscattering from the body provides the dominant return, while backscattering from hands and legs give a periodic frequency modulation in accordance with the movement. Khlopov and co-workers modeled the RCS of signals scattered by moving humans as follows:

$$\sigma(t) = \exp[j(\omega_0 + \omega_d)t]\sigma_c \left[1 + \alpha \frac{\sin \omega_d \cos\left(\frac{2\pi t}{T}\right)}{\omega_d \cos\left(\frac{2\pi t}{T}\right)} + \beta \frac{\sin\left(\frac{\omega_d I_h}{I_n}\right) \cos\left(\frac{2\pi t}{T}\right)}{\left(\frac{\omega_d I_h}{I_n}\right) \cos\left(\frac{2\pi t}{T}\right)}\right] \quad (5.2)$$

where $\omega_d = 4\pi v/\lambda$

v = velocity of person

σ_n = RCS of a leg

σ_h = RCS of a hand

σ_c = RCS of a body and head

$\alpha = \mathrm{RCS_{hand}/RCS_{total}}$

$\beta = \mathrm{RCS_{total}/RCS_{total}}$

l_h = length of hand
l_n = length of leg

Such a model also can be used for description of signals scattered not only by persons but also by animals. Khlopov and co-workers note that there is a correlation between height h and length of pace, hence the measurement of the Doppler frequency and rate of walking enables an estimation of the length of pace and height of the target, which enables humans and animals to be distinguished.

The power received by a radar system in free space is given by

$$\frac{P_r}{P_t} = \frac{G^2 \lambda^2 \sigma}{(4\pi)^3 R^4} \tag{5.3}$$

where P_r = received power, in watts
P_t = transmitted power, in watts
G = antenna gain
λ = wavelength, in meters
σ = radar cross section of target, in meters squared
R = range in meters

As an illustration the received signal from a subject behind a wall can be estimated using the following assumptions:

$P_t = 50$ mW
$G = 2.5$
$\lambda = 0.3$ m
$\sigma = 0.001$ m^2

In Figure 5.3 the upper line (crosses) represents the received signal in the case of no additional attenuation due to intervening wall material. The lower line (circles) represents the received signal in the case of additional attenuation of 10 dB per meter due to intervening material. The horizontal line represents the minimum detectable signal due to thermal noise in the receiver and sets the range achievable by the radar. Thus detection of a target (moving in the case of a Doppler radar) can be achieved behind a wall.

Doppler radar systems can be designed to detect the direction of approach of the target by means of an additional mixer as shown in Figure 5.4. which is fed with the source signal with a 90° phase shift compared with the original mixer. The phase of the output from the phase-shifted channel B will lead that from channel A when the target is approaching and lag when the target is receding.

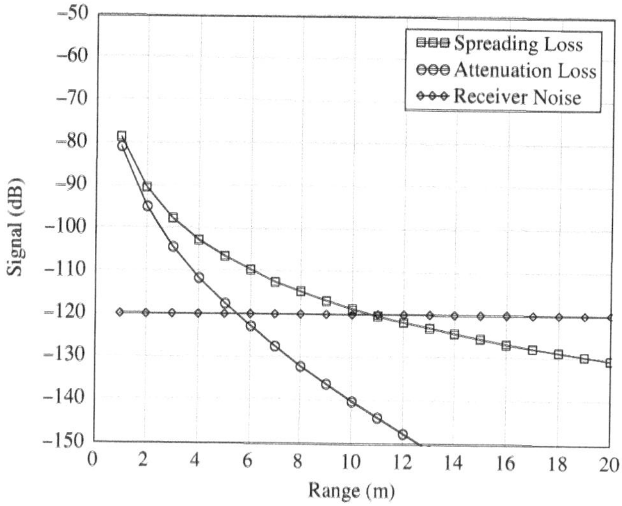

FIGURE 5.3 Calculated received signal levels from 1-GHz Doppler radar.

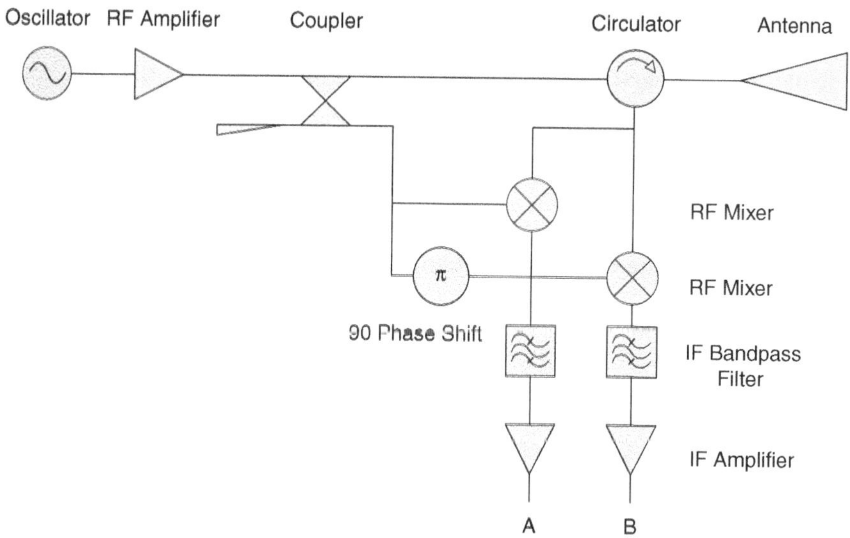

FIGURE 5.4 Doppler radar capable of detecting direction of approach.

5.3 FREQUENCY-DOMAIN RADARS

5.3.1 Introduction

Frequency-domain radars use two or more frequencies to provide range information. The most often encountered radars are the FMCW radar or the SFCW radar, and the principles and design of these are discussed in this chapter. The most basic frequency-domain radar is the two-frequency Doppler radar, which can provide range information to the target as well as information on the radial velocity of the target.

5.3.2 Two-Frequency Doppler Radar

The two-frequency Doppler radar is the most basic ranging radar and uses the two frequencies to measure the range to the target by measuring the down-converted phase difference between the two Doppler outputs. A very simple system design is to employ to separate Doppler radar modules and then measure the phase difference between the Doppler outputs as described by Araki et al. [244]. The distance to the target is given by

$$D = \frac{c\Phi}{4\pi(f_1 - f_2)} \tag{5.4}$$

where Φ = phase difference between the two Doppler output frequencies

f_1 = frequency 1

f_2 = frequency 2

Note that because the phase is modulo values of 2π, the maximum unambiguous range is given by

$$R_u = \frac{c}{2\Delta f} \tag{5.5}$$

This implies that the difference in frequencies should be small if a reasonable range is desired. For example, a range of 20 m would require Δf to equal 7.5 MHz. This requires the filters used in the system shown in Figure 5.5 to have Q values in excess of 140. For example, a system able to penetrate through buildings would have frequencies of 1 and 0.9925 GHz, respectively. Alternatively, the same system could be designed using two separate Doppler radar modules and antennas and a phase detector to measure the range.

The accuracy of the ranging information can be improved by increasing the frequency separation between the two oscillators but note that the range ambiguity is then directly related to the number of modulo values of phase, and a sequential ranging measurement is required. This leads to the concept of the stepped frequency radar in which the carrier frequency is stepped in known increments and is described in the following section.

5.3.3 Stepped Frequency Radar Systems

A stepped frequency continuous-wave radar radiates a sequence of frequencies ranging over a defined bandwidth in regular increments. Note that as any

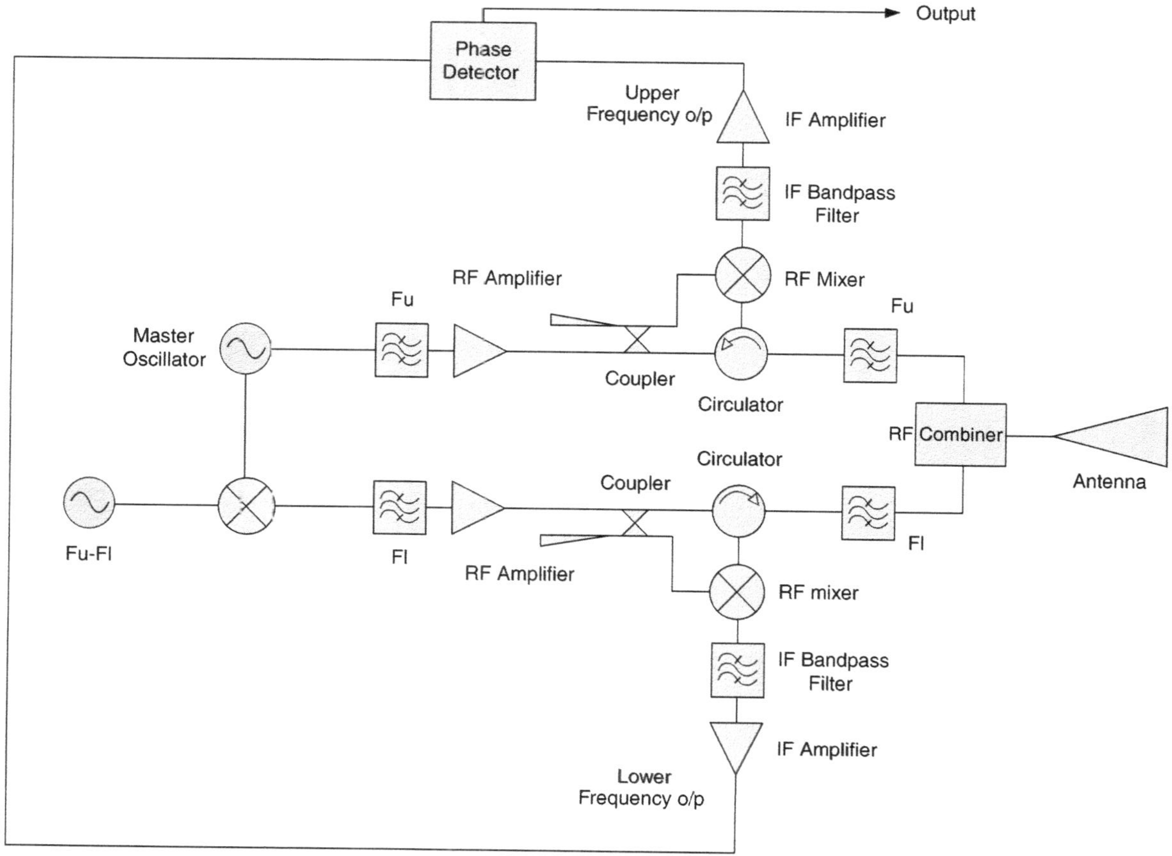

FIGURE 5.5 Generic two-frequency Doppler radar using a single antenna.

repetitive pulsed signal can be transformed to a frequency-domain representation, which consists of line spectra whose frequency spacing is related to the pulse repetition rate and envelope is related to the pulse shape. Hence, a repetitive impulsive waveform can be synthesized by transmitting a sequential series of individual frequencies whose amplitude and phase is accurately known. An analysis of the design of hardware and signal processing for a SFCW GPR system is given in the literature [245–249]. The main advantages of the stepped frequency continuous-wave radar are its high dynamic range (>150 dB) and low noise floor as well as the ability to avoid certain frequencies when transmitting, thus making compliance with licensing requirements much easier than most other types of modulation.

Two forms of the synthesized radar can be considered. The first and simplest system is a stepped frequency continuous wave as shown in Figure 5.6. The second form is more complex in that each individual frequency is appropriately weighted in amplitude and phase prior to transmission.

In both cases the difference frequency is composed of contributions from all targets up to and beyond the maximum ambiguous range given by

$$R_{\text{amb}} = \Delta R\, N \tag{5.6}$$

where ΔR is the range resolution and N is the number of frequency steps. The radar radiates a sequence of N frequencies, and the amplitude and phase of

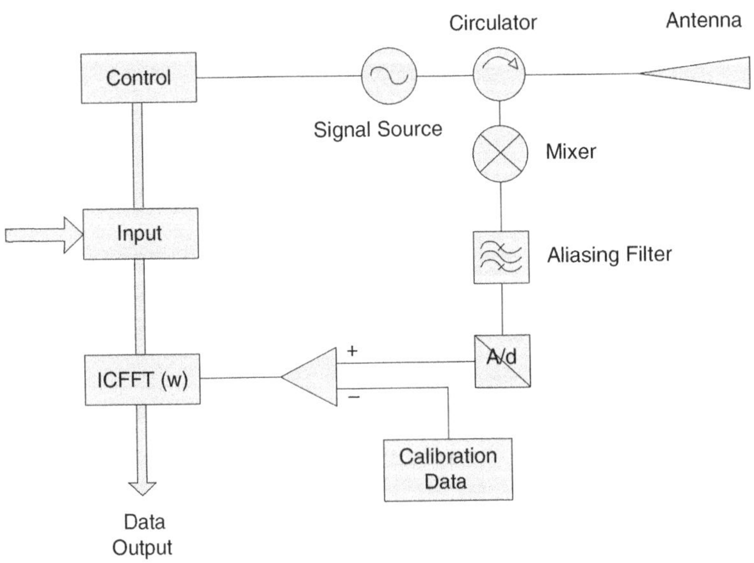

FIGURE 5.6 Block diagram of a stepped frequency radar system.

the received and down-converted signal is stored. A complex inverse fast Fourier transform or equivalent algorithm is then used to produce a time-domain version of the reflected signal.

The range resolution is given by

$$\Delta R = \frac{c}{2N\Delta f} \tag{5.7}$$

For example, a radar using 256 steps with frequency increment of 5 MHz would achieve a range resolution in air of

$$\Delta R = 0.117\,\text{m} \tag{5.8}$$

and an maximum ambiguous range

$$R_{\text{amb}} = 29.95\,\text{m} \tag{5.9}$$

The signal transmitted by the radar is given by

$$E'_T = E_T \exp(-j\omega t) \tag{5.10}$$

Then in the case of a lossy medium and a single target, the received signal can be considered as

$$E_R = \frac{E_T}{R^2}\sigma \exp\{j(2kR - \omega t)\} \tag{5.11}$$

where R = range to the target
s = target scattering cross section
k = propagation constant

The received signal can be represented by a phasor as shown in Figure 5.7. As the operating frequency is changed such that k changes by Δk, the phasor rotates by an angle equal to $2\Delta kR$.

Evidently, the amount of rotation is related to the target range. If the transmitted signal is incrementally increased, the frequency of the received signal will be related to the range, taking into account the propagation constant.

$$E'_R = \sum_{n=0}^{N-1} E_{rn} \tag{5.12}$$

For multiple targets the received signal becomes the phasor sum of all contributions as shown in Figure 5.8

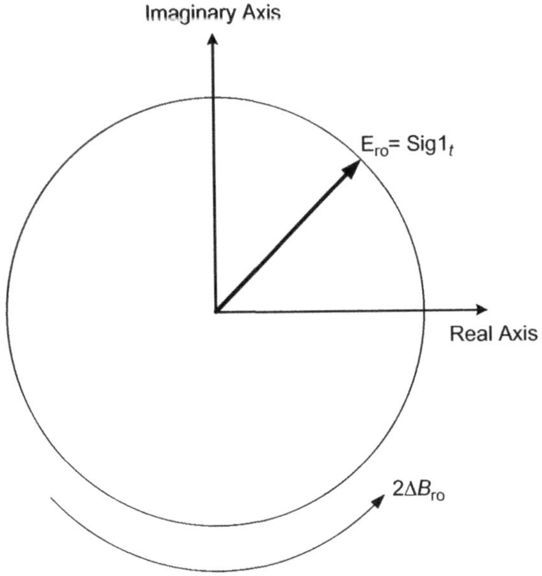

FIGURE 5.7 Phasor representation of measured signal from a single target.

The range to each target can be determined by performing a suitable transform with respect to the steps of frequency. The received signal can be expressed as

$$E_n = \sum_{k=0}^{N-1} \frac{\sigma_k E_T}{r_k^2} \exp\{c/j4\pi[(f_0 + n\ \Delta f) r_k]\} \tag{5.13}$$

where r_k = distance to the kth target
σ_k = target scattering coefficient
f_0 = start frequency
Δf = frequency increment
n = number of frequency steps

This equation can be seen to take the form of a discrete Fourier transform and, therefore, the scattering magnitude of E_n can be obtained using a fast Fourier transform to indicate depth, in quantized steps, of a target.

A synthesized system may be operated such that the received signal may be integrated as long as convenient in relation to the radar survey speed. Typically, each frequency might be transmitted for a time of 100s and the resulting intermediate frequency integrated for this interval thus providing a receiver bandwidth of

$$B = 1/t \tag{5.14}$$

which for this time would equal 10 kHz.

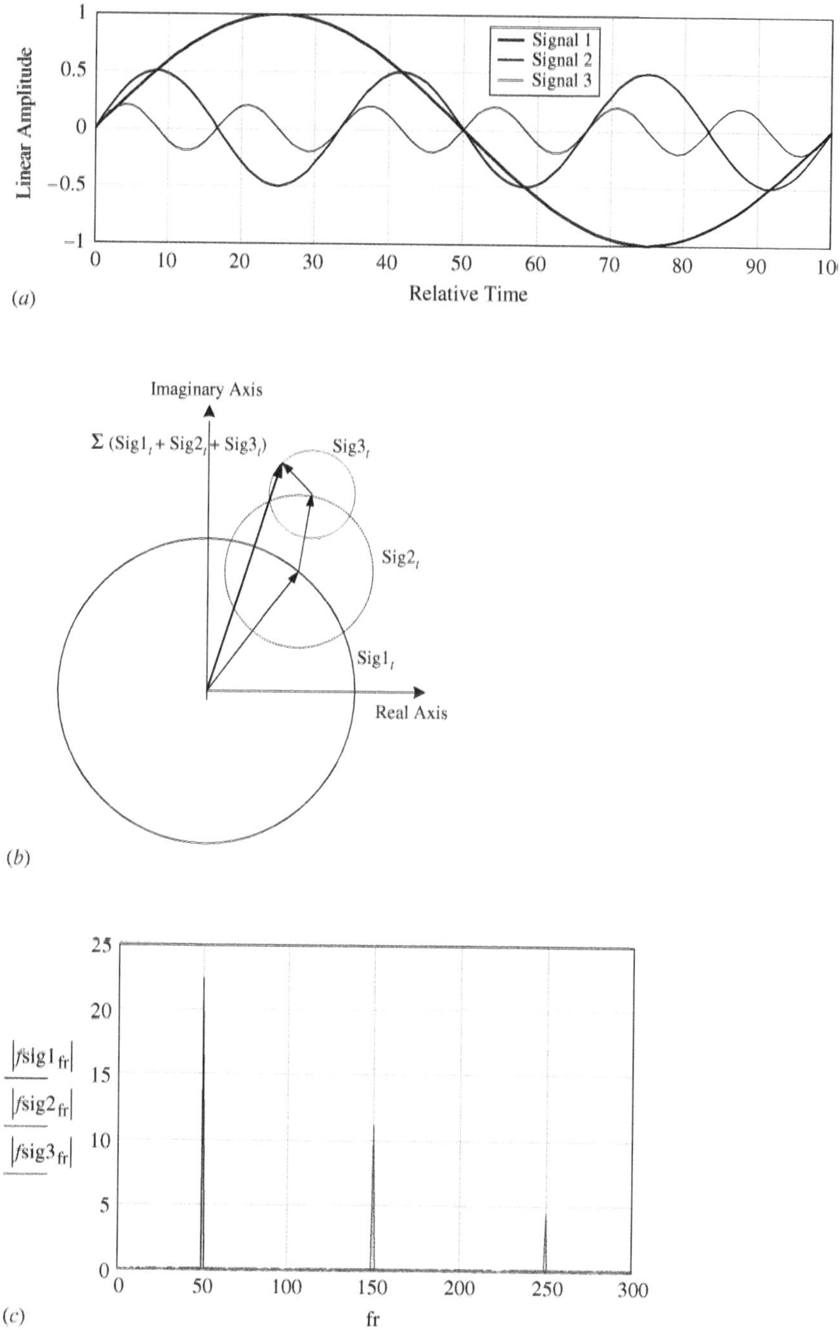

FIGURE 5.8 Phasor representation of the received signal from multiple targets in different domains: (*a*) amplitude–time, (*b*) phasor, and (*c*) amplitude–frequency representations.

Evidently, the receiver thermal noise over this bandwidth is very much lower than that of the receiver of the time-domain radar, and an improvement of typically 50 dB in noise performance compared with the latter can be achieved. In the case of synthesized radar simple antialiasing filters cannot remove contributions to the received signal outside the maximum ambiguous range. It is therefore important to choose operating parameters that minimize the aliasing effect. One method is to iteratively determine the range gate of the target. The initial measurement is taken with low resolution and large maximum ambiguous range. The range to the target is determined and the resolution is then increased.

Normally, the radar is calibrated, both to establish a reference plane for measurement as well as to reduce the effect of variation on the frequency characteristics in components and antennas. The radar system will introduce additional phase shifts on the transmitted and received system. The electrical lengths of the signal paths to the antennas and the effective radiation phase center of the antennas will cause these. This means that the phases of the transmitted and received signals will be different at each integer frequency and will require compensation.

One method of calibrating the radar [250] is to place a known reflector, such as a metallic plane, which is suitable at a defined distance from the antennas. The compensation factors can be adjusted such that the indicated range then equates to the actual range. Essentially,

$$E_n = \frac{E_T \sigma}{r_{\mathrm{ref}}^2} \exp\{j(2k_{\mathrm{ref}} - \omega t) + j\varphi_{\mathrm{ref}}\} \tag{5.15}$$

Evidently, it is important that the recorded values of amplitude and phase at each frequency are accurately related. Any temporal variation in the system characteristics, which degrades the system calibration, will, of course, reduce the resolution and accuracy of measurement. In addition, the repeatability of the frequency is important and must be such that the calibration remains valid.

The main advantages of a stepped frequency continuous-wave radar are its ability to adjust the range of frequencies of operation to suit the material and targets under investigation, a higher mean-radiated power level per spectral line and the ability to integrate the received signal level, hence improving the system sensitivity.

The calibration of the radar does, of course, depend on stable system characteristics and antenna parameters that are invariant with front surface–antenna spacing. The synthesized pulse radar is a variant of the stepped frequency continuous-wave radar in that the relative amplitudes and phases of the transmitted frequencies are adjusted on transmission in order to synthesize the desired pulse waveform.

A significant advantage for this approach is that it is possible, within limits, to take account of the frequency characteristics of the antennas. For example, weighting the amplitude of the low-frequency spectral lines can compensate the low-frequency cutoff point of an antenna. However, the requirement to maintain an accurate phase relation between each of the spectral lines is difficult to achieve in real time. It is generally easier to carry out computation on down-converted and recorded data; hence, the stepped frequency continuous-wave radar is a more economic design option.

5.3.4 Frequency-Modulated Continuous-Wave Radar

Frequency-modulated continuous-wave (FMCW) radar systems have been used in preference to AM systems because their dynamic range is potentially much greater, because it is possible with stepped frequency systems to shape the transmitted spectrum to meet license requirements, and at high RF frequencies it is easier to design and build radar systems.

It is useful to state the relationship between a time-domain function $f(t)$ and its representation in the frequency domain $f(\omega)$. Fourier transform methods can be used to show that

$$\mathrm{F}(f(t)) = f(\omega) \tag{5.16}$$

that is,

$$f(\omega) = \frac{1}{2\pi} \int_{-\infty}^{\infty} f(t) e^{j\omega t}\, dt \tag{5.17}$$

The fast Fourier transform (FFT) is often used to compute the transform, but certain well-established restrictions must be borne in mind due to its finite limitations.

A typical block diagram of generic dual-polarized FMCW radar is shown in Figure 5.9. The main advantages of operation in the frequency domain are the wider dynamic range, lower noise figure, and higher mean powers that can be radiated. In addition a much wider class of antennas, that is, horns, logarithmic, and the like, is available for use by the designer.

An FMCW radar system transmits a swept carrier frequency usually by means of a oscillator over a chosen frequency range on a repetitive basis. The received signal is mixed with a sample of the transmitted waveform and results in a difference frequency that is related to the phase of the received signal, hence its time delay and range of the target. The difference frequency or intermediate frequency (IF) must be derived from an complex or I/Q mixer pair if the information equivalent to a time-domain representation is required as a single-ended mixer only provides the modulus of the time-domain waveform.

In the radar shown in Figure 5.9, the master oscillator is used to drive a sweep generator, which can be realized using either a VCO or direct digital synthesis (DDS). This applies a linear sweep voltage to a 7-GHz oscillator, and it in turn it amplified and multiplied by 13 to provide a 1.3-GHz bandwidth sweep signal from 92.95 to 94.25 GHz. The output from the multiplier is passed through an isolator to avoid loading the multiplier and then bandpass filtered.

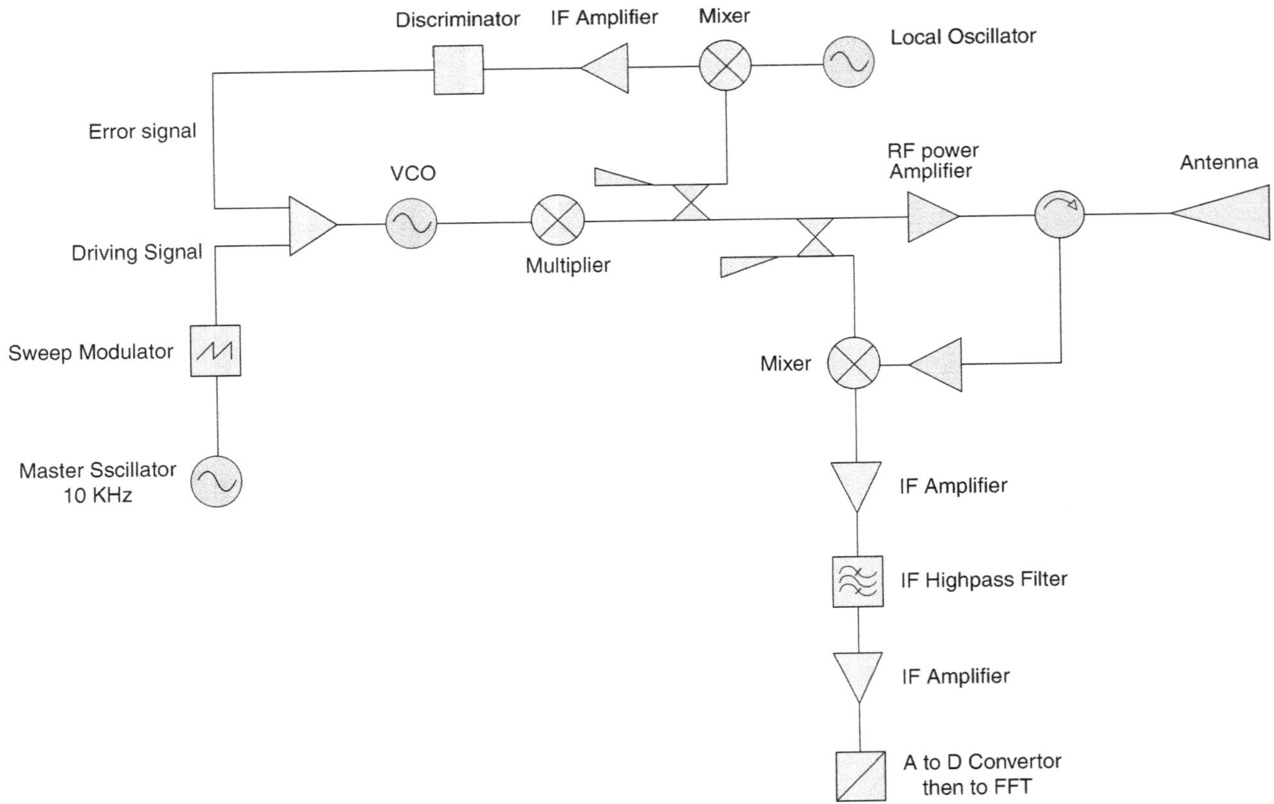

FIGURE 5.9 Block diagram of a generic FMCW radar system with discriminator linearization.

An RF power divider is used to provide the local oscillator signals to the two balanced mixers. An output from the power divider is amplified and then passed through a circulator and orthomode transducer to the antenna. The circulator enables the received signal from the antenna to be applied to the mixer to generate a co-polar signal, and the orthomode transducer is used to generate a cross-polar signal to the second mixer. The complex IF signals from the mixers are filtered and analog-to-digital converted to provide an amplitude range for both co- and cross-polar target signatures.

In an FMCW radar the transmitter frequency is changed as a function of time in a known manner as shown in Figure 5.10. A target return will exist at a time T_d given by

$$T_d = 2R/c \tag{5.18}$$

where R is the range in meters and c is the velocity of light in meters per second.

The choice of modulating waveform defines the resultant IF spectrum, and it is desirable to minimize the bandwidth of the IF spectrum due to a single target in order to optimize the target resolution. If this target return signal is mixed with the transmitted signal, a beat frequency, termed the IF, will be produced. This will be a measure of the target range. If the transmitted signal is modulated with a triangular modulating function at a rate over a range, then the intermediate frequency is given by

$$f = \frac{4Rf_m\Delta f}{c} \tag{5.19}$$

where f_m is the modulation frequency in hertz and Δf is the frequency deviation in hertz.

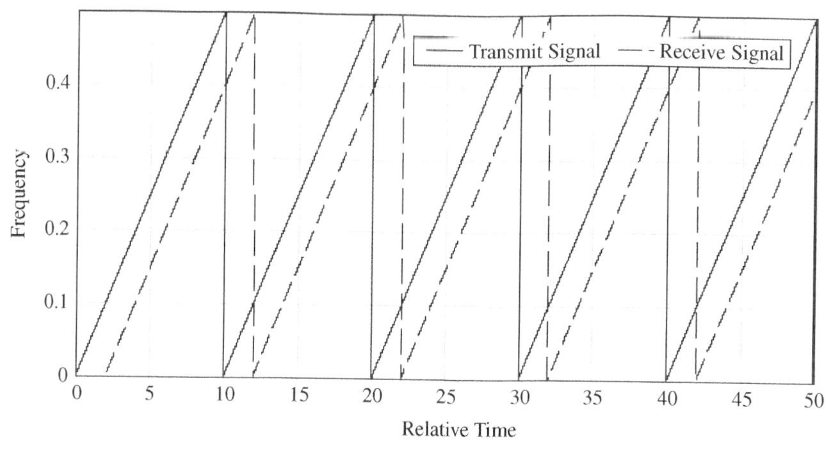

FIGURE 5.10 Typical FMCW sweep waveforms showing transmitted and delayed sweeps with relative frequency y axis and relative time on the x axis.

The RF signal is thus swept from f_{lower} to f_{upper} as shown in Figure 5.11, which shows a short-time Fourier transform RF. Relative frequency is shown on the y axis and relative time on the x axis. The down-converted IF output is shown in Figure 5.12, and it is important to excise the signal at the sweep turn-around to avoid the generation of spurious frequencies. This can be analyzed as follows.

If the case of a single target is considered, then the IF waveform as a function of time would be given by

$$f_{IF}(t) = A \sin \omega_{if} t \tag{5.20}$$

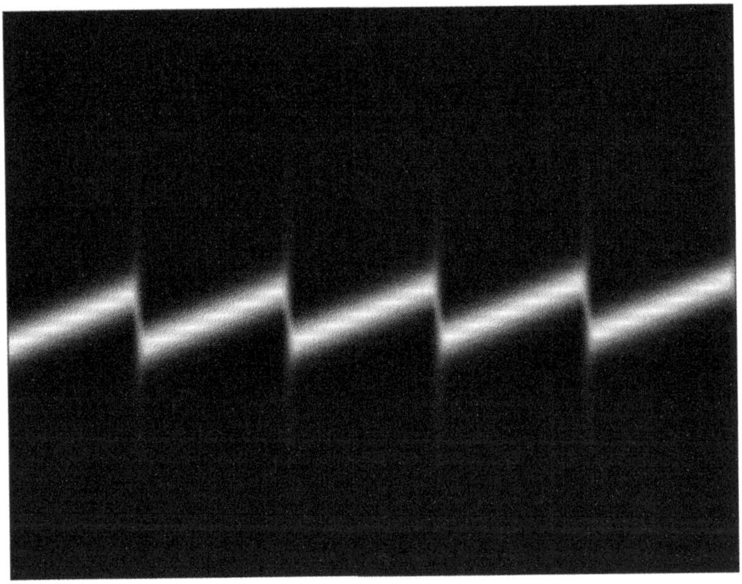

FIGURE 5.11 STFT of the transmitted RF from an FMCW radar with relative frequency y axis and relative time on the x axis.

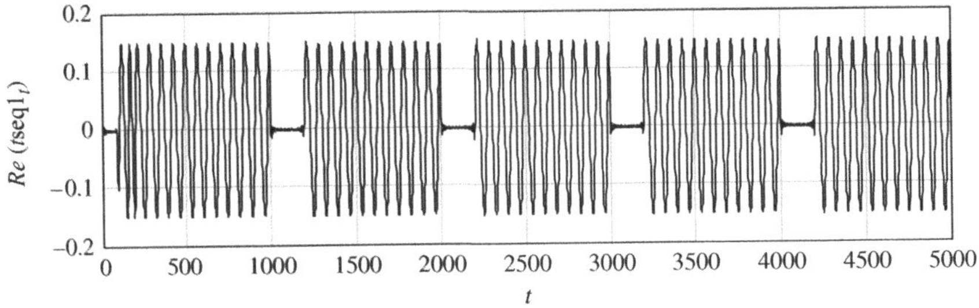

FIGURE 5.12 Generic IF waveform with excising of the turn-round signal.

The repetitive nature of the RF sweep effectively convolves the basic IF waveform with line spectra

$$f'_m(t) = \mathrm{F}(A \sin \omega_{\mathrm{if}} t) \times \mathrm{F}(f_m(t)) \tag{5.21}$$

The Fourier transform of $f_{\mathrm{IF}}(t)$ *is*

$$\mathrm{F}[A \sin(\omega_{\mathrm{if}} t)] = \mathrm{F}\left[A\left(\frac{e^{j\omega_{\mathrm{if}} t} - e^{-j\omega_{\mathrm{if}} t}}{-j2}\right)\right] \tag{5.22}$$

$$= -A[j\pi\delta(\omega - \omega_{\mathrm{if}}) + j\pi\delta(\omega + \omega_{\mathrm{if}})] \tag{5.23}$$

The Fourier transform of $f'_m(t)$ can be derived after rearranging the integration period:

$$\begin{aligned} f'_m(t) &= 1 \quad \text{for } 0 < t < (T - \tau) \\ &= 0 \quad \text{for } (T - \tau) < t < (T') \end{aligned} \tag{5.24}$$

Rearranging the time axes to make

$$f'_m(t) = 1 \quad \text{for} \quad (-T/2 + \tau/2) < t < (T/2 - \tau/2) \tag{5.25}$$

$$= 0 \quad \text{for} \quad (-T/2) < t < (T/2 + \tau/2), (-T/2 - \tau/2) < t < (T/2) \tag{5.26}$$

$$\mathrm{F}\{f'_m(t)\} = \frac{A}{T} \int_{T/2+\tau/2}^{T/2-\tau/2} \exp(jn\omega_m t)\, dt = \frac{A}{T} \int_{T'/2}^{T'/2} \exp(jn\omega_m t)\, dt \tag{5.27}$$

where n is the harmonic number.

$$\mathrm{F}\{f'_m(t)\} = \frac{AT'}{T} \frac{\sin\{(n\omega_m T'/2)\}}{n\omega_m T'/2} \tag{5.28}$$

where T' is $T - \tau$.

$$\mathrm{F}\{f'_m(t)\} = \frac{A(T - \tau)}{T} \frac{\sin\{[n\omega_m(T - \tau)/2]\}}{n\omega_m(T - \tau)/2} \tag{5.29}$$

Hence, the IF spectrum is given by

$$f_{if} = A \times \left\{ [-j\pi\delta(\omega-\omega_{if}) + j\pi\delta(\omega+\omega_{if})] \otimes \left[\frac{(T-\tau)}{T} \frac{\sin\{[n\omega_m(T-\tau)/2]\}}{[n\omega_m(T-\tau)/2]} \right] \right\} \tag{5.30}$$

It can be seen that the periodic IF signal consists of an envelope function (sin x/x) enclosing a line function. In essence, the FMCW radar measures the phase of the IF signal, which is directly related to target range. The frequency of the IF signal can be regarded as a measure of range. Normally, the IF waveform is windowed before it is inverse transformed and a typical filtered mixer output from a single target would appear as shown in Figure 5.13.

An inverse complex frequency–time transform can reproduce a time-domain equivalent to the impulse radar, but most frequency-domain radars produce a signature as shown in Figure 5.14. The effect of windowing the IF waveform is significant, and the unwindowed case gives rise to the well-known sinc [sin $(x)/x$] function. This limits the dynamic range of the receiver, whereas a windowed case can potentially achieve a better dynamic range albeit at the disadvantage of reduced resolution as shown in the lower part of the graph. It can be seen that the close-in side lobes are some −15 dB for the rectangular window and decrease to less than −40 dB for the Hamming window. This enables smaller targets close in range to a larger target to be detected.

The range resolution of an FMCW radar is inherently given by

$$\Delta R = \frac{c}{2B} \tag{5.31}$$

where B is the bandwidth of the FM sweep.

However, this is not always achieved because the FMCW radar system is particularly sensitive to certain parameters as noted by Ji-Chang [251] as well as Hua et al. [252]. It requires a high degree of linearity of frequency sweep with time to avoid spectral widening of the IF and hence degradation of system

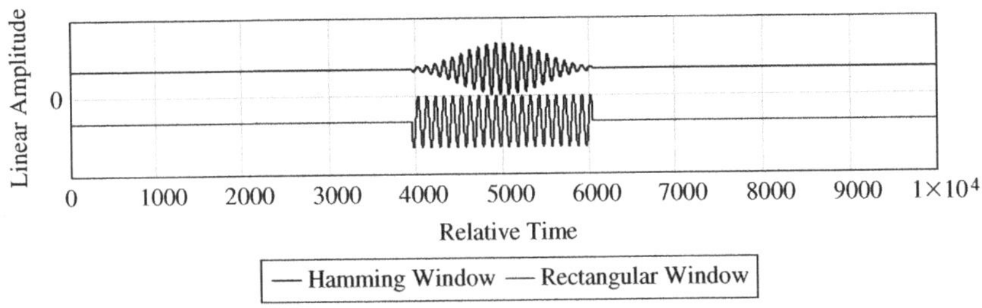

FIGURE 5.13 IF waveforms showing Hamming windowed time sequence.

resolution [253–255]. However Griffihs [302, 303] notes that large sweep linearity errors have can be tolerated more if they are slowely varying. This effect can be illustrated by considering the case of FMCW radar with a sinusoidal nonlinearity as shown in Figure 5.15, and the comparison of the IF spectrum is shown for perfect (zero nonlinearity) and 2% nonlinearity (Figure 5.16). The main effect is to broaden the width of the IF spectrum as the

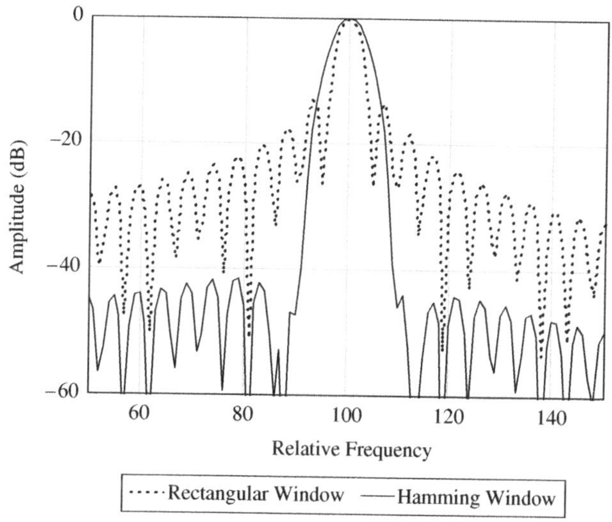

FIGURE 5.14 Spectrum of IF waveforms showing unwindowed (upper) and Hamming windowed (lower).

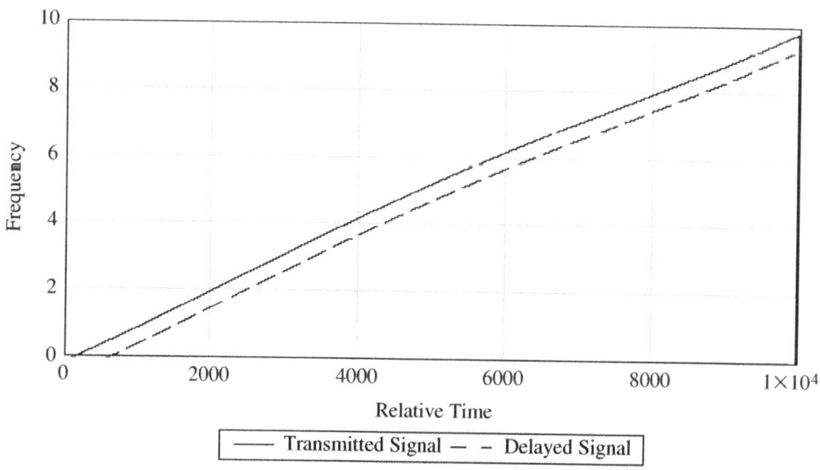

FIGURE 5.15 FMCW sweep waveforms showing nonlinear transmitted and delayed sweeps with relative frequency *y* axis and relative time on the *x* axis.

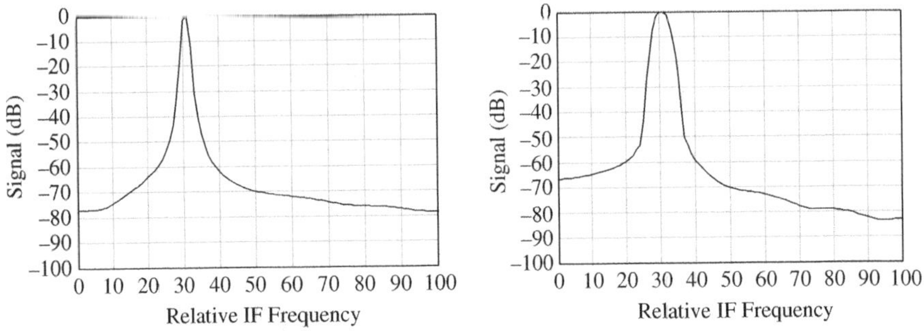

FIGURE 5.16 IF power spectrum for zero nonlinearity (lhs) and 2% sweep nonlinearity (rhs).

extent of nonlinearity increases and increase the side-lobe level. Practically, a useful system should aim to keep all nonlinearities at less than 0.1%.

Dennis and Gibbs [256] show graphs of the ratio of peak/side-lobe level as a function of linearity for various target ranges. The type of nonlinearity also affects the IF spectrum with small-amplitude periodic variations having a significant effect. Brooker [257] considers this design issue in a study that provides a useful summary of various well-established means of controlling nonlinearity. The two main methods are based on lookup table linearization of the sweep control voltage and or a feedback loop to control the error between the demanded rate of change of frequency and that produced by the oscillator.

Linearization of the sweep is achieved by three well-established methods: correction by means of a analog correction for a known nonlinearity, a digital lookup table to correct the measured nonlinearity, and a dynamic correction technique using a delay line discriminator. The main difficulty with the first two methods is that the nonlinearities are often temperature dependent, which then requires either temperature stabilization of a measurement of device temperature and a set of characteristics to cover the required temperature range. This is easier to achieve with a digital lookup table.

A basic analog correction circuit is shown in Figure 5.17 and comprises a source oscillator, which feeds three circuits: one to provide a basic sweep, one to provide a nonlinear voltage to correct the main VCO oscillator nonlinearity, and the third to compensate any sinusoidal variation across the sweep. The fourth element sets the dc central operating point on the VCO.

An alternative approach is a digital lookup table as shown in Figure 5.18 in which the corrected values are stored in a matrix in a read-only memory. This provides an output to digital-to-analog converter (DAC), which controls the VCO. The DAC may have a filtered output to remove the voltage steps if the DAC resolution is low. A temperature sensor is used to control a memory address pointer to access the correct values for a given temperature.

The delay line discriminator is a dynamic method of achieving linearization and is often used in FMCW radar systems. It is based on the use of an additional

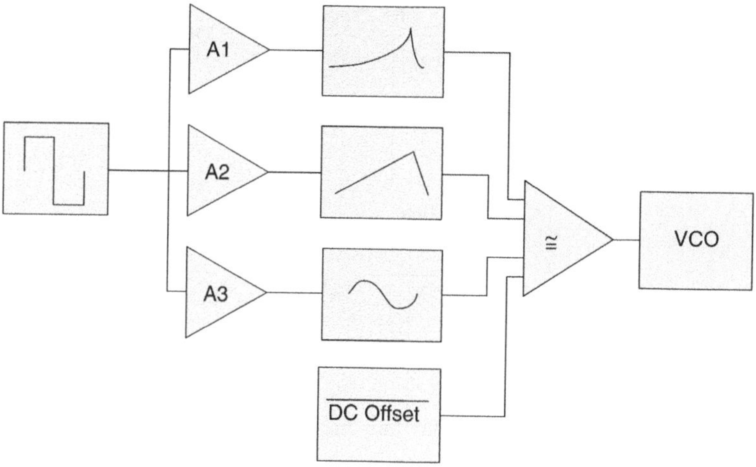

FIGURE 5.17 Analog sweep correction circuit.

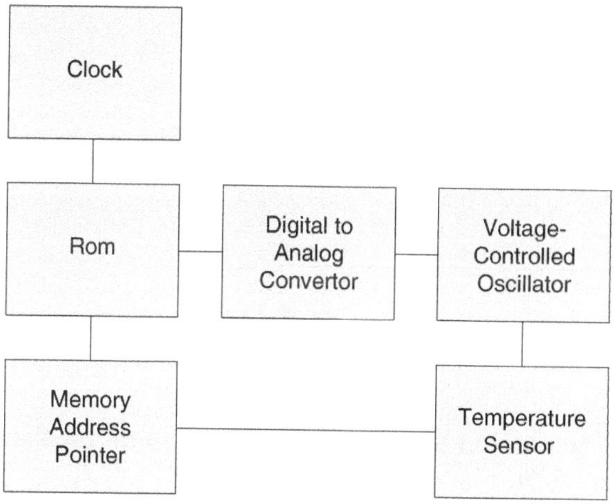

FIGURE 5.18 Digital lookup table correction.

mixer and delay line to create a radar return at a known range, the latter being set by the delay. The delay line discriminator provides a differentiated output in the frequency domain, that is, a constant output frequency if the frequency–time characteristic is linear. Clearly the loop bandwidth of the system will define the ability to dynamically correct for nonlinearities. Johnson [258] and Brooker used a surface acoustic wave, delay line discriminator to provide linearization of a millimeter-wave FMCW radar (Figure 5.19).

A further parameter, which needs to be carefully controlled, is the purity of the spectral output. Both the phase noise spectrum and in-band harmonics

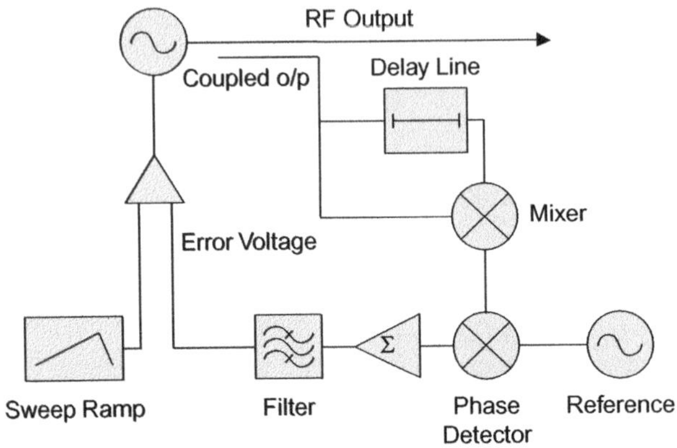

FIGURE 5.19 Delay line discriminator linearizer.

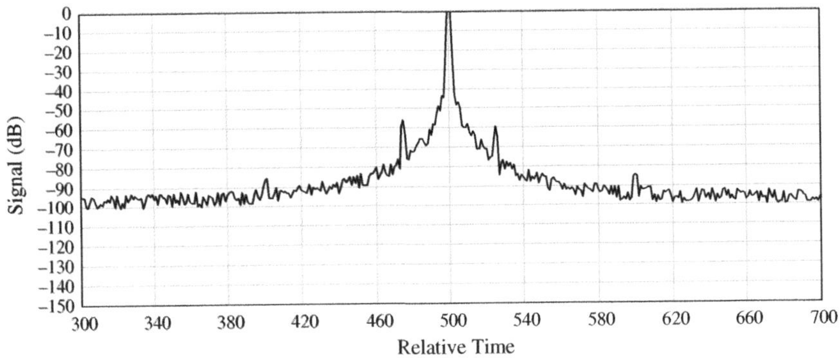

FIGURE 5.20 Typical oscillator spectrum showing noise and spurs.

(spurs) should be reduced to very low levels. A typical oscillator spectrum is shown in Figure 5.20.

In the case of in-band harmonics, the effect is to generate clutter, while in the case of phase noise the effect is to reduce the sensitivity of the radar for adjacent targets. The amplitude of the transmitted signal should also be leveled over the band of frequencies that are radiated, otherwise the variation will modulate the IF output.

The oscillators that provide the sweep for an FMCW radar system may be generated by a VCO, which is driven by a voltage generated from a lookup table that is precompensated for nonlinearities, or alternatively a digital sweep generator produces a linear ramp signal using DDS techniques that is applied to a coarse tuning port of a high-frequency tunable oscillator, such as a YIG tunable oscillator (YTO). A PLL has as its input an accurate linear swept-frequency sinusoid from a DDS and a linear swept-frequency output signal from the YTO to produce an error correction signal that is applied to a fine-tuning port of the high-frequency tunable oscillator. The error correction signal compensates for any nonlinearities introduced into the linear swept-frequency output signal by the high-frequency tunable oscillator.

Morgan et al. [259] compares the performance of a VCO and DDS and the effect on the range performance and concludes that a DDS solution offers better phase noise and better linearity, although spurious returns at multiple range are some 20 dB higher than the VCO solution.

The leakage that occurs between the transmitter and receiver antennas or via the circulator can have the unwanted effect of biasing the mixer into saturation and hence cause a reduction in system sensitivity and dynamic range. This problem can be partially solved by using an additional path to introduce a phase cancellation to reduce the leakage, but this tends to be limited to narrower bandwidth systems. Another unwanted effect that for large targets the phase noise of the source then appears in adjacent range bins, completely saturating these and dominating any return echoes from smaller targets. In the worst case for close-in targets this appears as radial lines on a PPI display and can increase the noise level over all the range bins at that angle on the display.

Other features of an FMCW radar system must also be considered. Changes in VSWR of the microwave components over the range of swept frequencies may cause "frequency pulling" of the transmit oscillator unless high levels of isolation are built in. Changes in VSWR can be caused by variations in antenna-to-subreflector spacing or by changes in the characteristics of components such as the circulator or mixer.

The amplitude–frequency transfer characteristics of all of the components in an FMCW radar system should be substantially flat. Ideally, amplitude ripple levels should be less than ± 0.25dB otherwise the radiated waveform will exhibit an amplitude modulation that will cause spectral spreading of the IF waveform with a resultant loss of solution and system performance.

The greater dynamic range of the FMCW radar is a significant advantage provided that the sweep linearity can be maintained and the spectral broadening and side lobes of the IF envelope are minimized. The output from the IF will, in the case of the antenna transmitting into an infinite lossy dielectric half space, contain signals due to imperfections in the radar system itself. These will be chiefly caused by, in order of significance, reflected energy from the antenna mismatch, leakage from port 1 to port 3 of the circulator, and leakage via the directivity of the coupler.

As most of these signals will be at very low frequencies, a high-pass filter in the IF signal path can be used to reduce their effect. The same filter can be used to compensate for the spreading loss in free space encountered by the radiated and reflected signal. In voltage terms an R^{-2} variation translates to a high-pass filter with a 12-dB per octave attenuation characteristic. In the case of a complex mixer with I/Q outputs, similarity of the mixer characteristics is important and a well-matched pair should be selected.

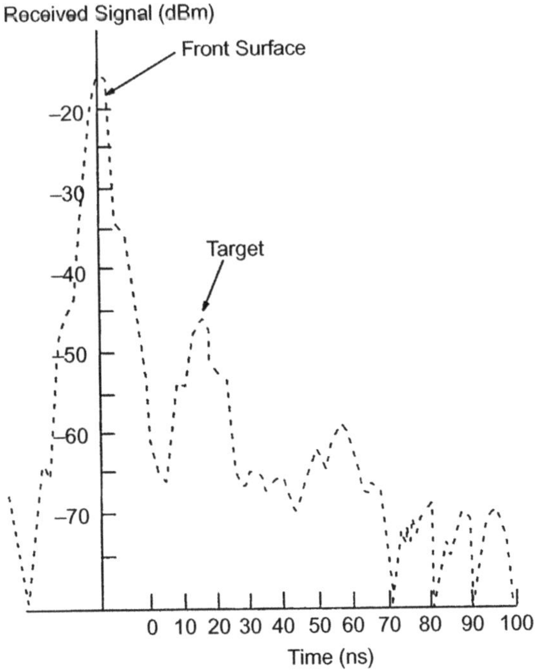

FIGURE 5.21 Typical FMCW radar received signal versus range in meters (with kind permission IET).

The complex output from the mixer consists of a set of frequencies representing reflections from individual targets. If an inverse Fourier transform is carried out, then a time-domain representation of the target reflections can be generated. However, it is well known that the resolution of the fast Fourier transform is suboptimum for most spectral analysis, and a range of alternative transforms can be used to obtain improved resolution. In a classic study Kay and Maple [260] discussed alternative transform techniques and showed the following methods could provide improved performance compared with the FFT. Other issues that affect FMCW radars are close-in clutter, and this is considered by Al-Attar et al. [261] as shown in Figure 5.21.

5.4 HARMONIC RADAR

Harmonic radar is used for the detection of concealed electronic devices and exploits the rectifying properties of different types of junctions to detect the re-radiated harmonics from an illuminated target. Its clutter rejection is very good, but other effects limit its range particularly when radiated power is limited. A typical continuous-wave harmonic radar system is shown in

Figure 5.22. It is also possible to design harmonic radars that are pulse modulated, but for ease of explanation a continuous-wave design is appropriate. The radar consists of a master oscillator, which is modulated with a 1-kHz audio tone in this example, followed by an RF amplifier, isolator, high Q bandpass filter, and RF power amplifier. In the system shown in Figure 5.22 the same antenna serves for transmit and receive, but some variants use separate antennas for the second and third harmonic channels. The return signal from the circulator is fed to the receive channels via a high-pass filter to remove any residual fundamental. Each receiver channel follows a similar design using a multiplied version of the fundamental frequency to mix with the received harmonic and hence down-convert to baseband to extract the 1-kHz audio modulation.

The principle of the harmonic radar is related to the properties of junctions that have a nonlinear I/V characteristic. The two most common are the metal-to-metal junction and the semiconductor junction. The latter is found in electronic devices and hence the harmonic radar can be used to detect concealed electronic devices. However, unless well designed and constructed, it suffers from false alarms due to responses from metal-to-metal junctions.

The I/V characteristics of a semiconductor rectifying junction is given by Eq. (5.32) and is valid for $V>0$:

$$I_{\text{sd}} = I_{\text{sat}} e^{qV/kT} \quad \text{for} \quad V \geq 0 \tag{5.32}$$

The I/V characteristics of a metal-to-metal rectifying junction is given by Eq. (5.33) and is valid for all values of V:

$$\begin{aligned} I_{\text{mmd}} &= I_{\text{sat}} e^{qV/kT} \quad \text{for} \quad V \geq 0 \\ I_{\text{mmd}} &= I_{\text{sat}} e^{-qV/kT} \quad \text{for} \quad V \leq 0 \end{aligned} \tag{5.33}$$

The characteristics are shown in Figure 5.23.

When a voltage is applied to the junctions, the current that flows is shown in Figure 5.24, and it can be seen that the characteristics are nonlinear. This in turn creates a series of harmonics when the junction is subjected to radiation at a fundamental frequency.

The simulation shown in Figure 5.24 models the current flow as a power series expansion up to the fifth power. The Fourier transform of the rectifying junction currents is shown in Figure 5.25, and it can be seen that the semiconductor response contains both odd and even harmonics, whereas the metal-to-metal junctions generate predominantly odd harmonics.

Semiconductor junctions can be identified using even harmonics, most usually the second, whereas metal-to-metal junctions can be identified using

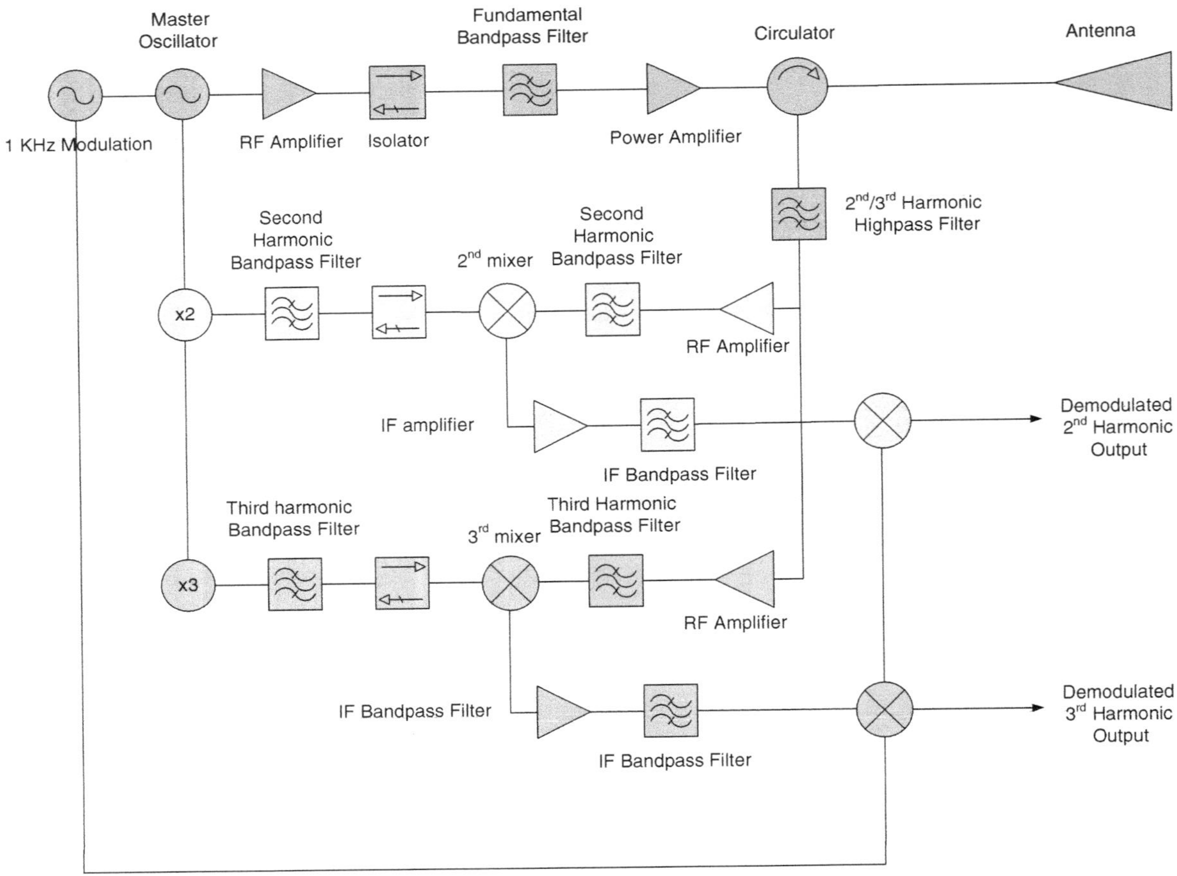

FIGURE 5.22 Typical harmonic radar system.

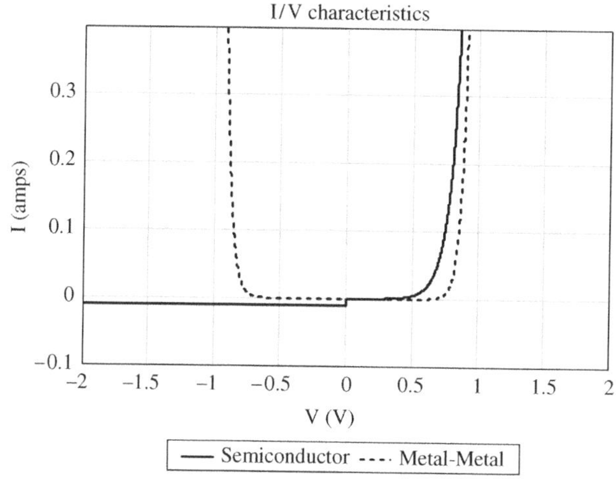

FIGURE 5.23 I/V characteristics of rectifying junctions.

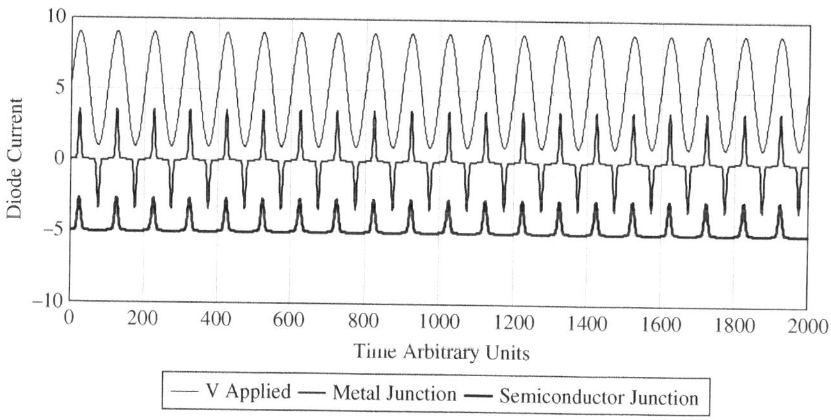

FIGURE 5.24 Current flow in rectifying junctions.

odd harmonics, most usually the third. Flemming et al. [262] considered the basic theory of radar target scattering cross section and noted that under small signal conditions the harmonic current induced in the target is proportional to the nth power of the applied voltage. This leads to a definition of the nth harmonic radar cross section as follows:

$$\sigma_n = \frac{P_{\text{rerad}}}{F_i^n} \tag{5.34}$$

where F_i^n is the incident power flux in Watts m^{-2}.

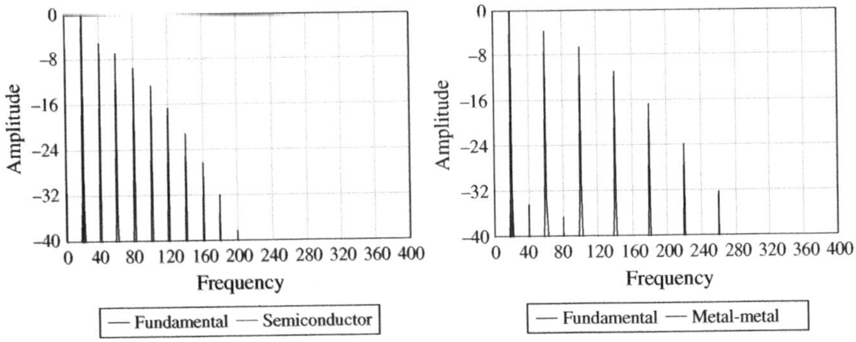

FIGURE 5.25 Spectra generated by semiconductor and metal-to-metal junctions.

The harmonic cross sections of the target depend on whether it is intentional or a target of opportunity. Targets used for tagging often consist of a diode coupled to a small antenna, and Shefer et al. [263] report a second harmonic radar cross section of 1000 $W^{-1}\ m^4$ for a diode coupled to a high-gain antenna for use as a transponder. Much use has been made of such transponders to track the dispersal of tagged insects, and O'Neal et al. [264] report detection ranges up to 900 m with a transmitted wavelength of 3.2 cm at 25 kW using tags 16 mm long and weighing 0.8 to 12 mg using diodes arranged at the center of the length of wire. Colpitts and Boiteau [265] reported on the harmonic radar transceiver design: miniature tags for insect tracking.

Fleming noted that the second harmonic radar cross section of printed circuit boards lay in the range of 10^{-5} to $10^{-5}W^{-1}m^4$, while small metal assemblies such as cable shackles, key rings, and vehicle components were in the range of 10^{-7} to $10^{-14}\ W^{-2}\ m^4$. It is also possible that in the case of the well-known rusty bolt effect (a metal-to-metal rectifying junction) that mechanical movement may also change the target scattering cross section as a function of mechanical movement and time. The calculated spectral response of metal-to-metal junctions shows a difference between the second and third harmonic of −30 dB and Fleming reports a difference of −40 dB. The key driver of performance of harmonic radars is the range law, which adversely affects the range at which the higher order harmonics can be detected.

The U.S. Army developed radars in the mid-1970s based on the harmonic radar principle, and the METRRA system described by Elsner [266] radiated a set of frequencies of 230, 308, 436, 550, and 737 MHz at 100 W continuous wave (CW) and 5 kW peak and received at 690, 923, 1308, and 1652 MHz. The power density at the target (small) could be set from 0.02 W m^{-2} up to 10 W m^{-2} for target ranges of around 6 m and 0.87 W m^{-2} for large targets at ranges of up to 18 m.

Bucci et al. [267] measured the harmonic radar cross section of a diode using a fundamental of 4 GHz and reported a second-harmonic cross section of $2 \times 10^{-5} m^2$ at an incident power density of 1×10^{-2} W m^{-3} and a third-harmonic cross section of $1.5 \times 10^{-7}\ m^2$ at an incident power density of 1×10^{-3} W m^{-3}. Harger [268] developed a theory for the radar transmission equation for harmonic radar operating over a planar, finite dielectric earth through foliage derived for various classes of nonlinear scatterers. He noted that the maximum detection range of a ground-based system is most sensitive to polarization, transmit antenna height, and transmit wavelength; moderately sensitive to transmit power and transmit antenna area; and least sensitive to receive antenna area, harmonic scattering cross section, and mode of data processing. Also, there seems to be a critical path distance through foliage; at distances less than this, small wavelengths are desirable, and, conversely, the upper transmit frequency limit may be set by nonlinear scatterer response.

The conventional radar equation for a linear scattering target is given by

$$\frac{P_r}{P_t} = \frac{G^2 \lambda^2 \sigma}{(4\pi)^3 R^4} \tag{5.35}$$

In the case of the harmonic radar the target scattering cross section now depends on a number of additional factors such as incident power level and frequency.

The received power can be expressed as

$$P_r = \frac{P_t^n G^{n+1} \lambda^2 \sigma_n}{(4\pi)^{n+2} R^{(2n+2)}} \tag{5.36}$$

The normalized relative received signals in decibels as a function of range are shown in Figure 5.26, and it can be seen that the second- and third-harmonic re-radiated signals follow inverse sixth and eighth power laws, respectively. Practical limitations of hardware usually limit these to the second and third harmonics. It can be seen that the fall off in signal of the third harmonic is significant and severely limits range. This leads to a design and build challenge because the radar must have high sensitivity for the harmonics and ensure that the antenna is virtually completely free of contacts that could form junctions. The harmonic radar uses synchronous detection techniques and isolation between the fundamental, and the harmonic channels are achieved by means of filtering. The receive channels must achieve the highest sensitivity, and the noise figure of these must be as low as possible. If the bandwidth of the latter can be restricted to 1 kHz, the sensitivity can be on the order of −30 dBm, which is less than the theoretical value because of limitations in components and constructional techniques.

Harmonic radars require very careful specification of component characteristics. Isolation and filtering are key issues and the use of double-screened cables as well as individual screening for the transmitter and receiver modules are essential. Fleming et al. [262] note many of the requirements for cables and components in order to achieve a harmonic rejection of over 200 dB. Many harmonic radars used for search operations operate at a fundamental frequency

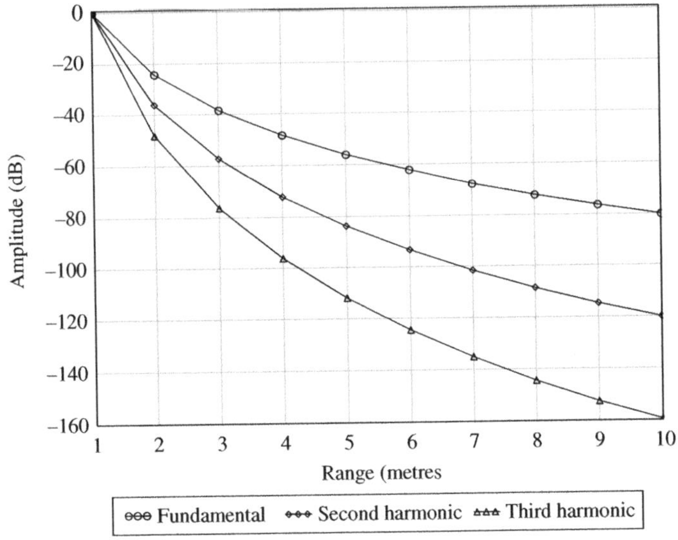

FIGURE 5.26 Harmonic radar range law for second- and third-harmonic received signals.

of below 1 GHz, using third harmonics up to 3 GHz, although Fleming et al. [262] reported decreasing signals from printed circuits above 1 GHz and Harger [268] predicted a cutoff in metallic junctions above 1.5 GHz.

5.5 NOISE RADAR

Noise-modulated radar offers some very attractive possibilities to the designer of GPR systems. The radiated power is evenly spread throughout the spectrum and the receiver is less susceptible to interference. However, until recently such systems were relatively rare. Developments over the last few years are changing that situation and more efforts are being put into the development of noise radar systems.

The basic principle of operation of noise radar is that of a correlator. The radar transmits a noise signal and the received signal is a time-delayed version τ_s of the transmitted signal. In the receiver the transmitted signal is used via a variable delay τ_v to cross correlate the received signal.

The received signal can be written as

$$v(r) = \sigma_t v(t - T_s) + vc(t) \tag{5.37}$$

where $v(t - T_s)$ = transmitted signal reflected by the target at time T_s

σ_t = target reflectivity

$vc(t)$ = reflections from clutter

The cross correlation can be written as

$$R(\tau) = \int_{0}^{T_i} v(r)v(t - T_s)\,dt \tag{5.38}$$

where T_i is the integration time.

The noise source is usually best described as band-limited noise, as it is impractical to generate white noise. Graphs of white noise and band-limited noise are shown in Figures 5.27 and 5.28.

An example of the cross-correlation output of noise radar using the noise shown in Figure 5.29 is shown in Figure 5.30. As the bandwidth is reduced, the resolution of the radar is commensurately reduced; and as the integration time is reduced, the dynamic range of the radar is reduced. The generic system design of a noise radar is shown in Figure 5.31.

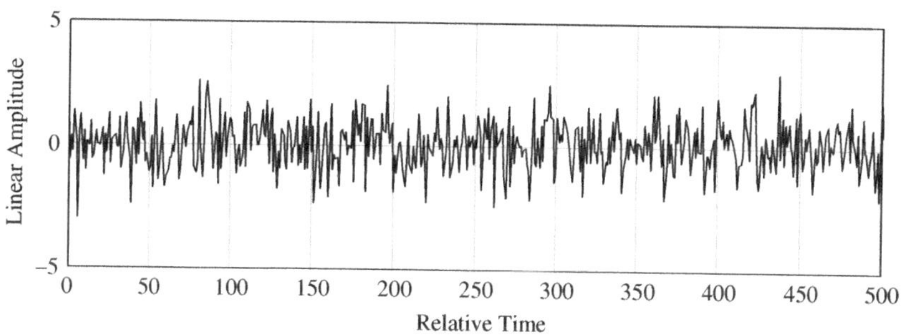

FIGURE 5.27 Gaussian white noise with time in arbitrary units.

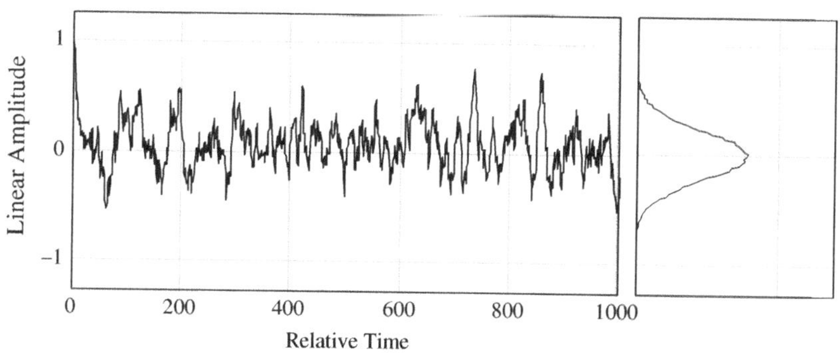

FIGURE 5.28 Band-limited Gaussian noise and associated probability density with time in arbitrary units.

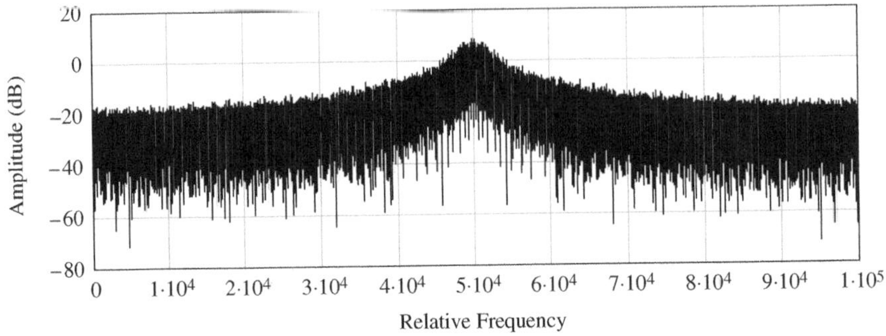

FIGURE 5.29 Spectrum of noise (in dB) versus relative frequency Hz.

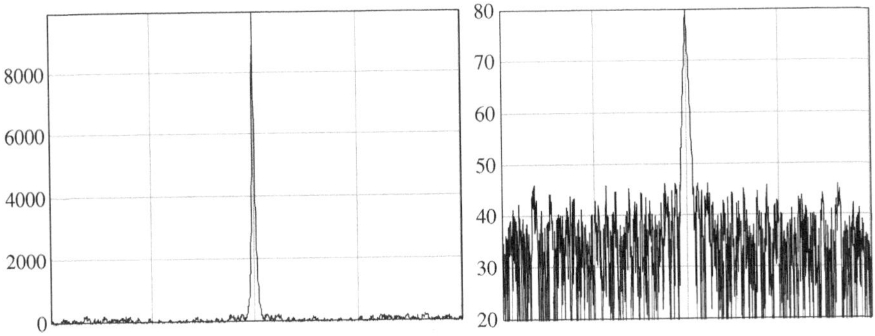

FIGURE 5.30 Cross-correlation outputs of a noise radar plotted on a linear amplitude scale (lhs) and in decibels (rhs) for arbitrary time/range.

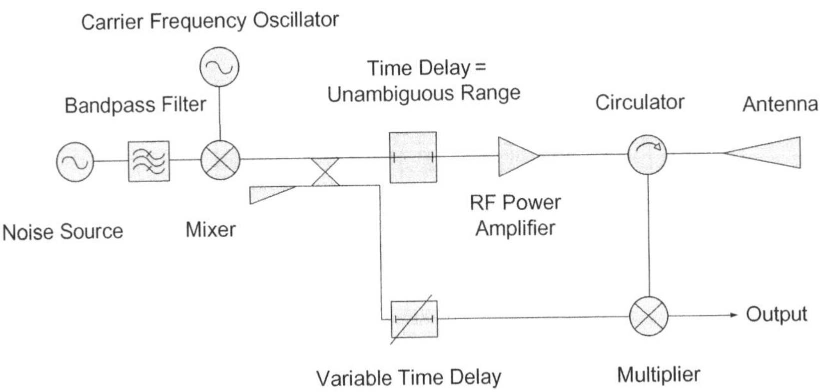

FIGURE 5.31 System diagram of generic noise radar

Guosui et al. [269] describe the development of random signal radar (RSR) over the past 30 years. They review conventional methods of implementing such as correlation, spectrum analysis, and autocorrelation and consider the advantages and disadvantages of noise frequency modulation cw radar and random binary phase-coded cw radar.

Some of the key design issues in terms of noise radar are the ambiguity function and the range side-lobe suppression. Dawood et al. [270] consider the underlying issues in terms of optimizing the ambiguity function. They developed an ultra-wideband (UWB) random-noise radar system that transmits an UWB random-noise (Gaussian) waveform with uniform power spectral density (PSD) in the 1- to 2-GHz frequency range. They showed that for a random-noise radar a correlator matched to the transmit process is required and that for a UWB transmit random process, the compression or stretch due to the range rate on the envelope of the return process cannot be ignored.

When applied in airborne imaging surveillance, UWB random-noise radars have their special merits. Xu and Narayanan [271] showed that because of the randomness and the ultra-wide bandwidth of the transmit-and-receive signals, such radars can be used for covert detection and identification and are less susceptible from hostile detection and jamming while preserving very high range resolution. However, the images are plagued with artefacts caused by high range side-lobes. Xu and Narayanan [271] proposed a new technique for the range side-lobe suppression of UWB random-noise radar, which combines median and apodization filtering.

Axelsson [272] considered the ambiguity functions of noise radar systems and noted that the range ambiguity is suppressed. His study discussed a number of issues related to noise radar and points out that the output correlation integral is accompanied by a noise floor, which limits the possible side-lobe suppression as determined by the time–bandwidth product. In a later study Axelsson [273] considered the issues of the noise floor limiting the possible side-lobe suppression that accompanies the correlation integral. In strong clutter with dominant reflectors the induced noise floor can be too high, and further suppression was achieved using mismatched filtering applied to suppress the side lobes of random noise radar. He describes an iterative subtraction algorithm for cancellation of noise floor due to dominating reflectors.

Gray [274] describes a noise radar system consisting of an array of K transmit antennas and M receive antennas. Two implementations are considered: In the first, a single noise source is, after appropriate phase shifting, transmitted; in the second approach K-independent noise sources are transmitted. The two approaches are compared and an architecture for the latter that no longer requires steering phases on transmission is presented.

A key component development has been by Juergen Sachs of the Technical University of Ilmenau, Germany, and the Meodat Company, which have jointly developed an integrated circuit SiGe-technology radar "chip." This offers considerable potential for noise radar systems, and further developments from that described by Daniels [275] are provided by Sachs et al. [276].

5.6 SPATIAL MODULATION

Single-frequency methods of imaging are based on the technique of viewing the target from a number of physically different positions in an aperture over the target and recording the amplitude and phase of the received signal and then mathematically reconstructing an image of the target. The process may be either holographic, whereby the recorded field represents the scattered field at a plane, or tomographic, whereby the recorded field is that of the transmitted field through the target at all angles of view.

The drawbacks to the process are that accurate positioning of the antenna elements is required, and this usually involves mechanical positioning of either the transmitter and receiver elements or both. The accuracy of measurement in terms of amplitude and phase are key factors in the quality of the image reconstruction. This imposes a practical limitation on operational situations but may be useful either where the position of the target can be controlled or the target can be viewed at all sides as in a tomographic image. When the target has multiple layers of different velocities of propagation, as might be the case for a human, the solution of the forward propagating wave paths as a function of view angle is nontrivial

The differences between optical holography and radar holography appear to lie in the way in which the wave field is recorded. In the case of the optical hologram, the image plane comprises the forward scattered field as shown in Figure 5.32. In the case of a radar system, the recorded field comprises the backscattered field, and this may, of course, not be the same as the forward scattered field.

Pasmurov and Zimoviev [277] describe the fundamentals of radar imaging and holography and consider the relationship of SAR processing to both holographic and tomographic imaging. They point out that the SAR process can be regarded as equivalent to that of a reduced microwave hologram of the wave field along the azimuth direction, that is, the flight path. In optical holography the recorded image is comprised of reference wave field and a scattered wave field from the target. If the original source wave field illuminates the recorded image, the diffracted wave field is identical to the originally scattered wave field from the target.

Spatial modulation is generally used at close range to the target and hence many of the antennas have relatively low gain and hence a wide beamwidth, which results in poor resolution in either the x or y dimensions. Synthetic aperture methods aim to increase x or y resolution by synthetically generating an antenna with a large aperture and consequent reduced beamwidth.

The synthetic aperture array must consist of a minimum number of samples in order to avoid aliasing. The minimum number of samples is given by Osumi and Ueno [278].

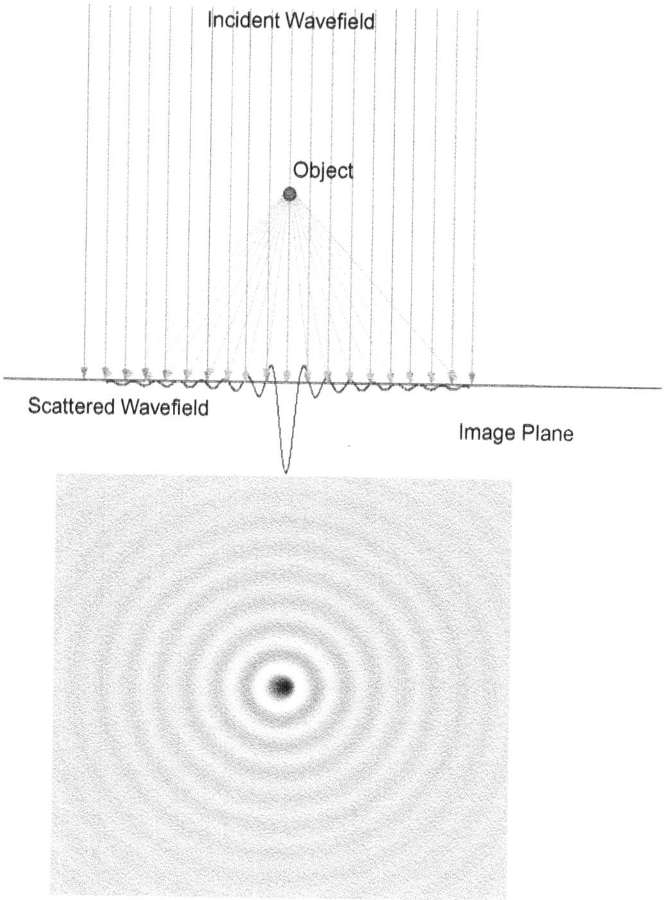

C1

FIGURE 5.32 Holographic recording.

$$N = \frac{2R\beta^2}{\lambda} \tag{5.39}$$

where β = antenna beam width in radians
R = range to the target
λ = radiated wavelength

There are several variations on the synthetic aperture method that can be considered. When a single frequency is used, holographic methods can be employed to generate an image. The holographic method records the amplitude and phase of the received signal in a plane over the target. This function is then correlated with a test function that is set to provide a maximum value of the cross correlation when reflection occurs and to be zero otherwise.

The holographic image reconstruction method is defined in the two-dimensional case by the correlation between the test function $h(x,t)$ and the received signal $v_r(x_r,t)$, where x is the coordinate vector of the imaging point and x_r is the coordinate vector of the receiving point, as shown in Figure 5.33, and $v_r(x_r,t)$ is the signal received at the receiving point.

The linear operation to reconstruct the image is defined by Osumi and Ueno [278]:

$$b(x) = \iiint_{-\infty}^{\infty} v_r(x_r, t)\ h(x - x_r t)\ dt\ dx_r\ dy_r \tag{5.40}$$

As the received signal can be considered to be

$$v_l(x_r, t) = \frac{1}{2\pi} \int_{-\infty}^{\infty} V(x_r, \omega) \exp\ j(\omega t)\ d\omega \tag{5.41}$$

$$v_r(x_r, t) = \frac{1}{2\pi c} \iint \frac{\cos\theta}{r^2} \sigma(x) \frac{d}{dt} \frac{(t - 2r)}{c} dx\ dy \tag{5.42}$$

where $\sigma(x)$ is the reflectivity of the target.

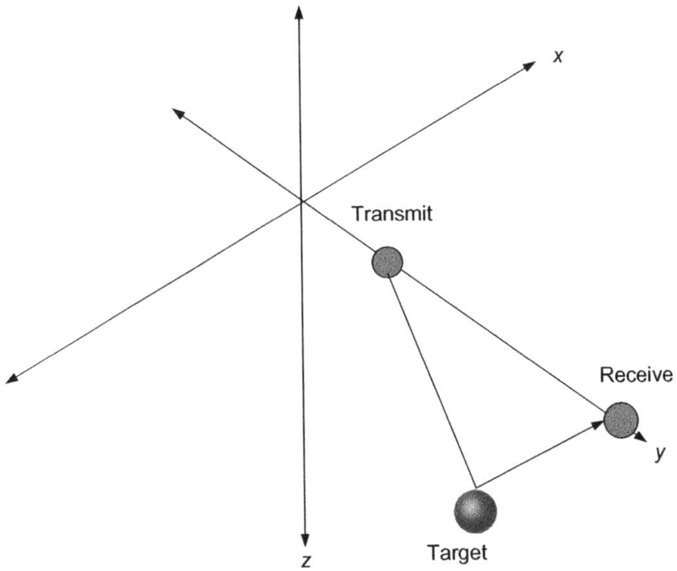

FIGURE 5.33 Coordinates of imaging system.

Hence the image function of the object T

$$b(x) = \frac{1}{2\pi c} \iint_s \sigma(x_r) b_0(x - x_r)\, dx\, dy \tag{5.43}$$

where

$$b_0(x) = \iiint \frac{\cos\theta}{|x - x_r|^2} \frac{d}{dt} \frac{(t - 2|x - x_r|)}{c} h(x_r, t)\, dx\, dy \tag{5.44}$$

The effect of material attenuation is significant as the general effect is to apply a windowing function across the recording aperture, thus limiting its useful size in relation to sharply focused images as shown in Figure 5.34.

In addition, the effect of both material attenuation and relative permittivity on the antenna beam width should be considered. As the values of loss and relative permittivity increase, the beam width of the antenna reduces, and this

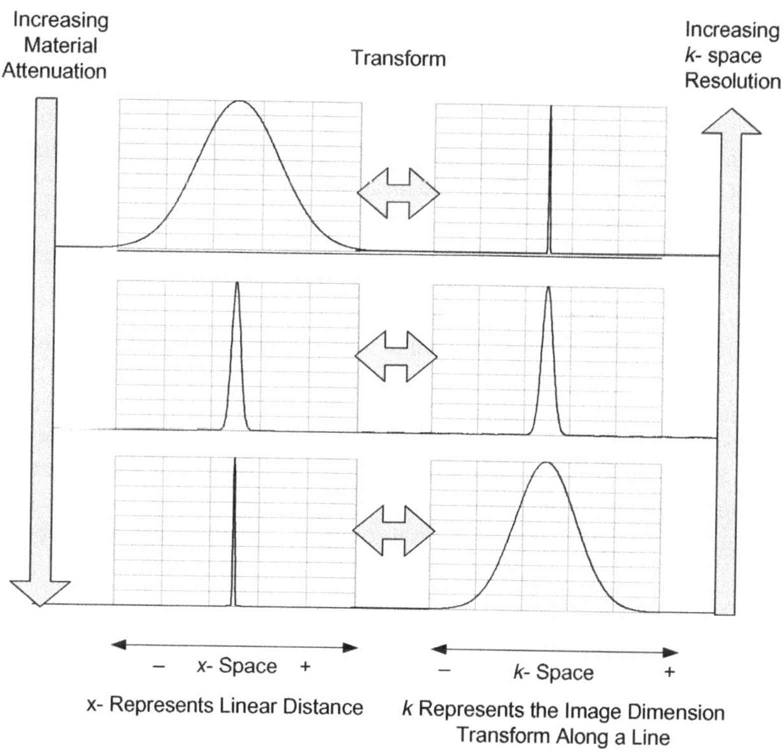

FIGURE 5.34 Effect of material attenuation on radar synthetic aperture image.

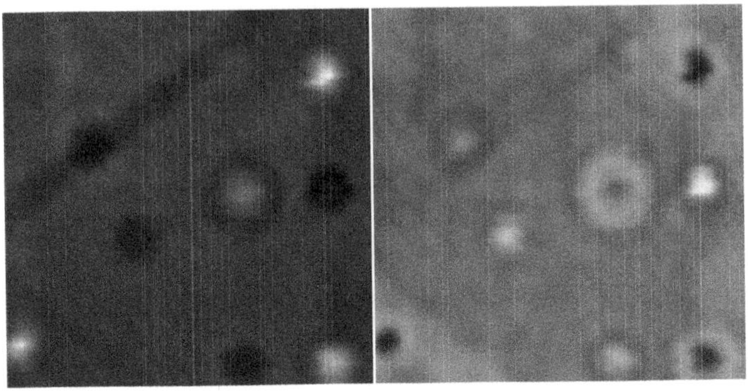

FIGURE 5.35 Holographic images of coins and wires at 4 and 3.6 GHz (Ivashov).

degrades the gain of the synthetic aperture. In general, synthetic aperture methods are most useful in lower loss materials.

Ivashov [279] notes that holographic radar receives a signal amplitude that decreases with the range law and material attenuation, hence penetration depth, depends on the attenuation in the surveying medium, and the target reflectivity with the added complication that targets at short range will obscure targets at deeper range in the recorded images. This phenomenon may be exploited as the ability to detect targets at very short ranges is far better than conventional radar systems.

The results shown in Figure 5.35 show images at two different frequencies. The phase Φ_t of the reflected signal depends on the distance to the target Φ_l and the phase shift at the target Φ_r. In the mixer of the receiver and at certain wavelengths, the amplitude of the received signal will be zero because the phase of the reference and received signal may be equal, hence it is necessary to use several wavelengths to reconstruct the image.

Popov et al. [280] notes the difficulties of solving the inverse scattering problems for both one-dimensional and three-dimensional situations. Huyghen's principle defines the solution of the inverse problem of a monochromatic sounding to be ambiguous, as the field created by the secondary sources on any wave front cannot be distinguished from that of the initial sources. However, in some particular cases additional a priori information allows the ambiguity of the problem to be reduced and the characteristics of the scattering object to be determined accurately enough for practical purposes.

5.7 AMPLITUDE MODULATION

A generic radar system is shown in Figure 5.36 and comprises a synchronizing pulse generator, which controls a transmit–receive (TR) switch as well as providing modulation of the carrier frequency of the main oscillator. The radar is monostatic as it uses only one antenna and the TR switch, which is generally

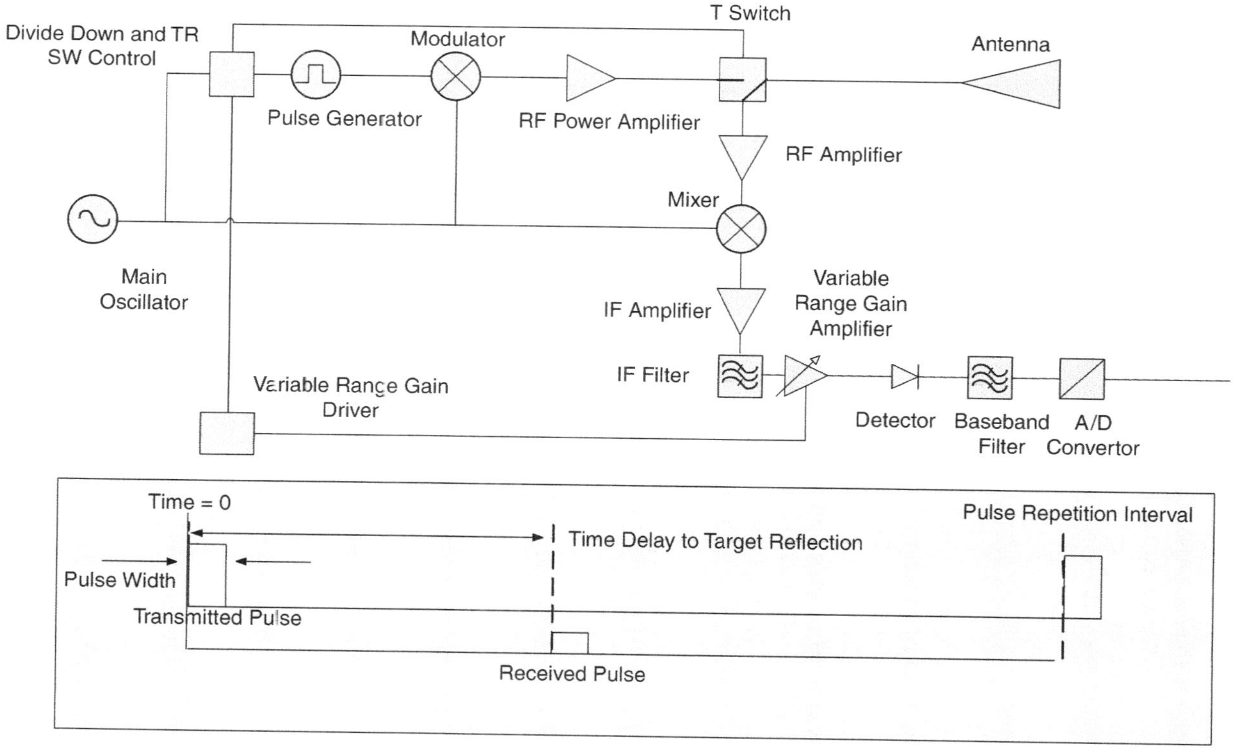

FIGURE 5.36 Generic radar system.

solid state because the transmitted power is in the range of 0.1 to 100 W, is used to enable the transmit pulses to pass to the antenna while isolating the receiver.

The pulse duration defines the range resolution and the pulse repetition interval defines the unambiguous range. Assuming all radar design factors have been optimized, then a pulse radar system should be able to resolve targets separated at a distance equal to half the pulse duration. Hence the range resolution is given by

$$\Delta R \geq \frac{c\pi\tau_p}{2} \tag{5.45}$$

where ΔR is the range resolution in meters and τ_p is the pulse duration.

The range to most concealed targets rarely exceeds 1 km, can be on the order of 100 m, and very often lies in the range 1 to 10 m. Note that if the propagation path is in a dielectric, the velocity of propagation is slowed and the pulse repetition interval needs to be increased in proportion to the reduction in velocity.

The radar range resolution in free space is shown in Figure 5.37 and it can be seen that for a pulse duration of 1 ns the best range resolution is 0.15 m or

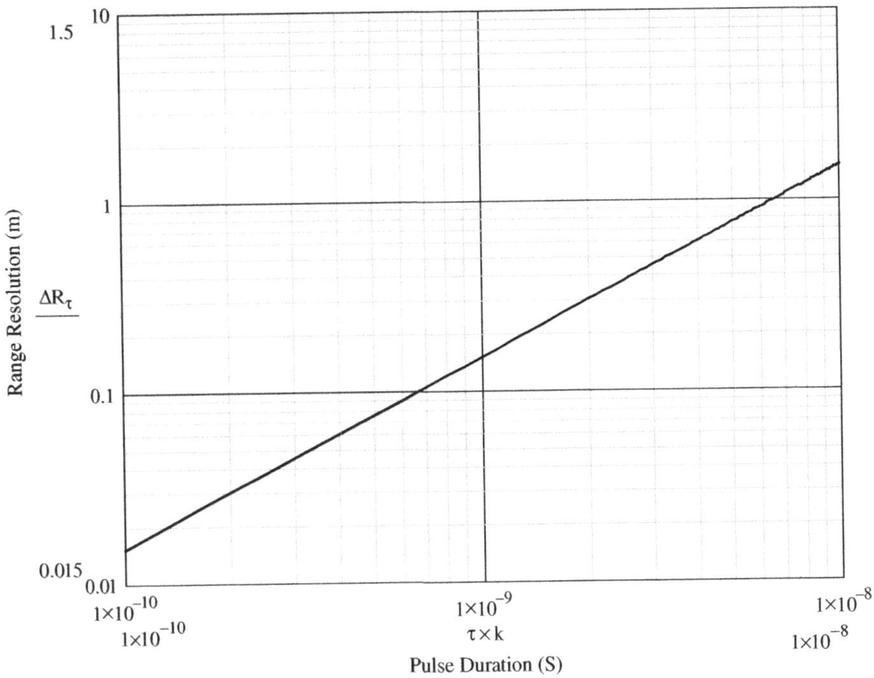

FIGURE 5.37 Radar range resolution in free space in meters versus pulse duration in seconds.

greater. This leads to a different architecture for radar systems with high range resolution, and the majority of time-domain radar systems have to use high-speed sampling techniques in order to achieve the required performance. The majority of time-domain radar systems have used impulses of RF energy variously described as baseband, video, carrierless, impulse, monocycle, or polycycle. The simplified general block diagram of an amplitude-modulated system is shown in Figure 5.38.

The radar is controlled by a master clock, which after a suitable time delay triggers a pulse generator, which generates pulses, typically of amplitude within the range between 10 and 200 V with a pulse width within the range between 200 ps and 50 ns at a pulse repetition interval of between several hundred microseconds to one microsecond or less, depending on the system design. These are applied to the transmit antenna. It is quite feasible to generate pulses of several hundred kilovolts albeit at long repetition intervals. A portion of the transmit signal is usually tapped off to provide an input to the sampling control to ensure timing stability of the signals fed to the sampling head. The radiated pulses propagate and are reflected from the target and are received by the receive antenna.

The repetitive nature of the impulses causes a line spectrum in the frequency domain, and the separation in frequency is related to the repetition interval in that $f = 1/t$, where t is the interval in seconds. The envelope of the spectrum is defined by the characteristics of the impulse and in the case of the example shown is that of a Ricker wavelet. The latter is defined as the second differential of a Gaussian function. Examples of the time domain and frequency domain representations are shown in Figures 5.39 and 5.40.

The output from the receive antenna is applied either to a flash A/D converter or a sequential sampling receiver. If the pulses are of sufficiently long duration (>10 ns) and the radar repetition rate is slow (>100 ms) a flash A/D converter can be used. When the pulses are shorter <10 ns and the radar repetition rate is faster (<100 ms), then a sequential sampling receiver is used. This normally consists of an ultra-high-speed sample-and-hold circuit. The control signal to the sample-and-hold circuit, which determines the instant of sample time, is sequentially incremented each pulse repetition interval. For example, a sampling increment of $t = 100$ ps is added to the previous pulse repetition sampling interval to enable sampling of the received signal at regular intervals as indicated below.

$$T_s = T + nt' \quad \text{for} \quad n = 1 \text{ to } N \tag{5.46}$$

where T = pulse repetition time
t' = sampling interval
N = total number of samples

Certain important limitations in terms of sampling interval should be noted. From the sampling theorem, the sampling interval must be

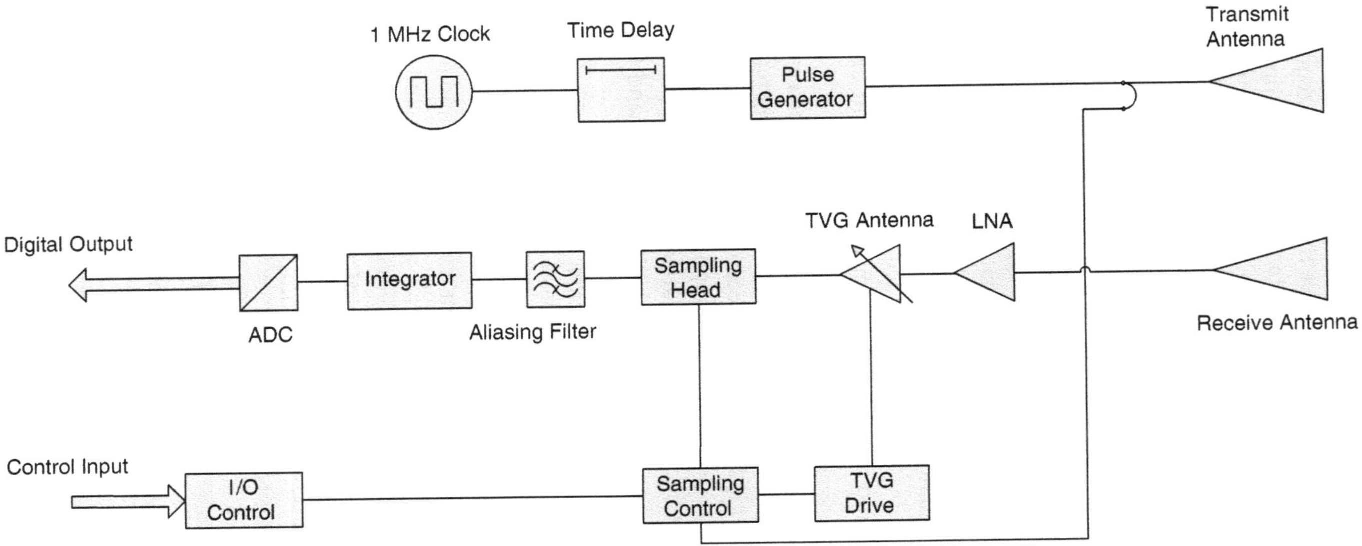

FIGURE 5.38 Generic system diagram for time-domain radar.

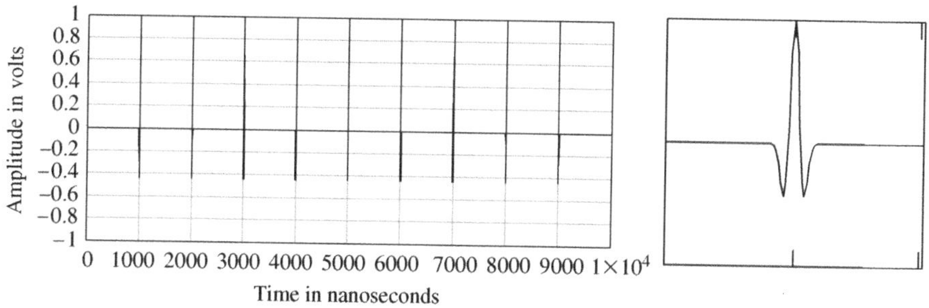

FIGURE 5.39 Simulated sequence of impulses with expanded single impulse.

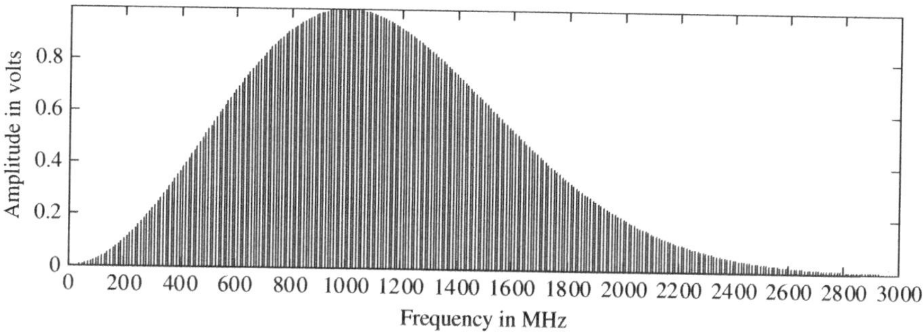

FIGURE 5.40 Amplitude spectrum of radiated impulse sequence.

such as to comply with the Nyquist criterion, where B is the bandwidth of interest;

$$t' \leq \frac{1}{2B} \tag{5.47}$$

In practice, a greater number of samples is normally required for accurate reconstruction and the sampling interval is generally taken as

$$t' \leq \frac{1}{5B} \tag{5.48}$$

Sampling as referred to earlier is a time-extending process with which a high-frequency repetitive signal is duplicated at a lower repetition rate. This type of sampling, where each sample is taken at a fixed frequency with the period of time between samples remaining constant, is known as coherent sampling. The principle of the sampling receiver is, therefore, a down conversion of the RF signal in the nanosecond time region to an equivalent version in the micro or millisecond time region. The incrementation of the sampling interval is

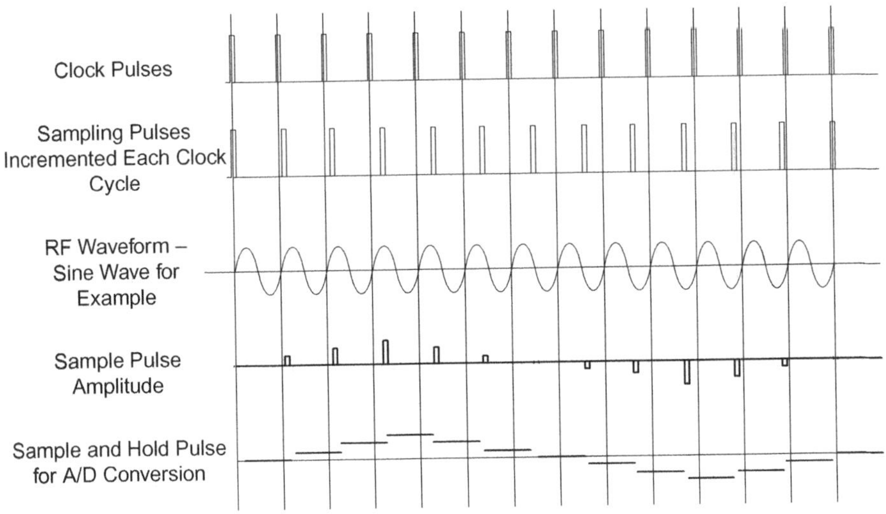

FIGURE 5.41 Sequential sampling receiver timing.

terminated at a stage when, for example, 256, 512, or 1024 sequential samples have been gathered. The process is then repeated. There are several methods of averaging or "stacking" the data; either a complete set of samples can be gathered and stored and further sets added to the stored data set or alternatively the sampling interval is held constant for a predetermined time to accumulate and average a given number of individual samples.

The first method needs a digital store but has the advantage that each waveform set suffers little distortion if the radar is moving. The second method does not need a digital store and a simple low-pass analog filter can be used. However, depending on the number of samples that have been averaged, the overall waveform set can result in being "smeared" spatially if the radar is moving at any speed.

Typical flash A/D converters feature large signal bandwidths of many hundreds of megahertz, sampling jitter less than 5 ps and 8-bit resolution. At bandwidths over 500 MHz, typical sampling resolutions are 4 bits and a more complex system architecture is found. In general most current generation impulse radars use high-speed A/D conversion receivers for bandwidths below 200 MHz where greater resolution can be achieved. An alternative receiver architecture is based on sub dividing the RF band and mixing the individual band to separate intermediate frequency bandwidths of 200 MHz that can be then separately A/D sampled. Alternative methods of data acquisition are based on high-speed analog-to-digital converters (ADC) or the cross-correlator receiver. There are several methods of acquiring the high bandwidth RF signal output from the receiver; direct ADC using high-speed (flash) A/D converters, frequency selection followed by high-speed ADC, or sequential sampling.

The stability of the timing increment is very important, and, generally, this should be 10% of the sampling increment; however, practically stability on the order of 10 to 50 ps is achieved. The effect of timing instability is to cause a distortion, which is related to the rate of change of the RF waveform. Evidently, where the RF waveform is changing rapidly, jitter in the sampling circuits results in a very noisy reconstructed waveform. When the rate of change of signal is slow, jitter is less noticeable. Normally, control of the sampling converter is derived from a sample of the output from the pulse generator to ensure that variations in the timing of the latter are compensated automatically.

The key elements of this type of radar system are the impulse generator, the timing control circuits, the sampling detector, and the peak hold and ADC. The impulse generator is generally based on the technique of rapid discharge of the stored energy in a short transmission line. One method of achieving this is by means of a transistor operated in avalanche breakdown mode used as the fast switch and a very short length of transmission line. However, the avalanche process is statistical by nature and is accompanied by jitter, and step recovery diodes provide a much better jitter performance.

The high-speed sampling approach conventionally used to display fast waveforms produces a low SNR because the spectrum of the sampling pulse is a poor match for that of the received pulse. Being an essentially nonselective filter, it allows large amounts of noise energy to enter the receiver. Also, the sampling circuit tends to add milliamp-level unbalanced currents as well as sampling pulse noise to its output. Although a quite acceptable trade-off for usual laboratory purposes, this may be unacceptable for receivers with sensitivity in the microvolt range.

The timing control circuits are a key element of the receiver, and the standard method of generating a sequence of incremented pulses is by means of a dual-ramp circuit as shown in Figure 5.42. The fast ramp is at the same rate as the pulse repetition time, whereas the slow ramp is set to provide the desired number of samples, that is, 256, 512, or 1024. A generic example of the timing sequence for this is shown in Figure 5.43 in the upper panel where the slow ramp intersects the fast ramps. This provides an output from the comparator, which then generates the signal to open the sampling gate.

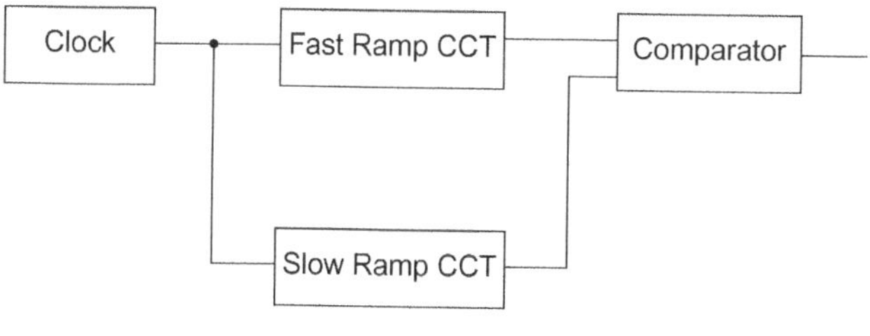

FIGURE 5.42 Dual-ramp timing circuit.

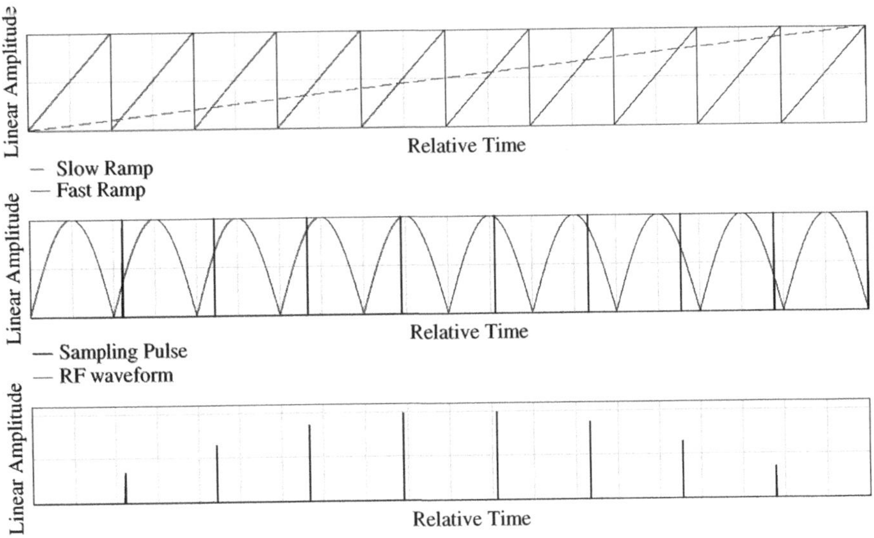

FIGURE 5.43 Principle of time-domain down sampling.

As can be seen in Figure 5.43 the sampling pulses are gradually incremented in time, and in the example shown a half sine wave is used to show how the values of the sampled waveform in the lowest panel are obtained. The sampling gate is opened at incremental intervals, and the amplitude of the sampled pulse is a copy of the RF waveform but at a slower frequency. The amplitude of the sampled pulse is normally held constant until the next sample is acquired, but this has been omitted for clarity.

The ramp circuits can be designed using analog or digital circuits. In the case of analog circuits the main building block is an integrator circuit, whereas in the case of a digital design suitable integrated circuits are available in the form of analog devices, for example, AD9500 digital delay integrated circuit (IC).

Evidently, time stability of these circuits is vitally important and, for example, if 256 samples in a 51.2-ns time window were required, the sample increment would be 200 ps. The sampling increment for a 1-MHz repetition frequency would occur at 1 μs + $n \times 200$ ps, where n is the nth sample number. To achieve accuracy of sampling, the timing accuracy of the sample should be ± 20 ps, which then requires the overall timing stability to be 1 μs + $n \times 200$ ps ± 20 ps or $\pm 0.002\%$. Great care is therefore needed in circuit design to achieve adequate stability.

A graph of received signal strength versus time is shown in Figure 5.44. From this it can be seen that the operating range of the radar system lies between the boundaries defined by the functions defined by the clutter profile, target reflection loss, and the limit of sensitivity due to the noise figure of the receiver. It can be readily appreciated that the limited dynamic range of the sampling receiver limits the performance of the radar.

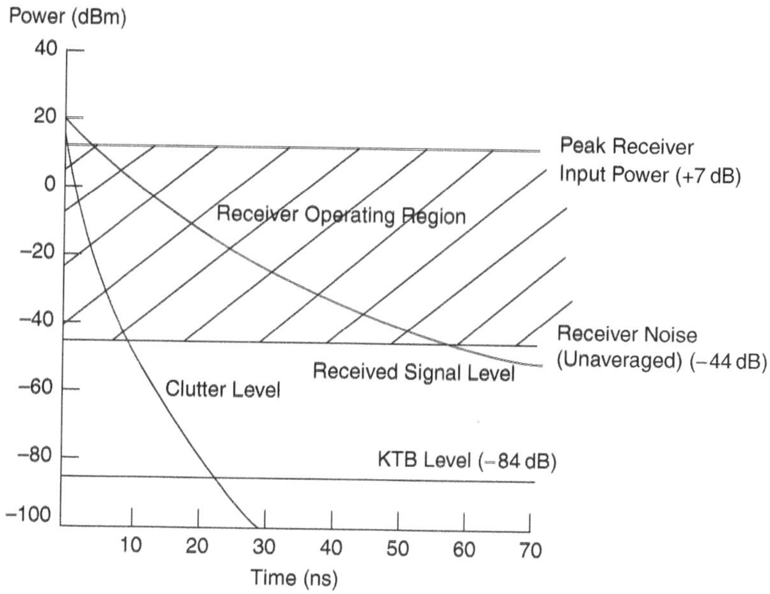

FIGURE 5.44 Received signal against time for a sequential sampling receiver.

The poor noise figure of the sampling gate can be improved by using a wideband low-noise RF amplifier prior to the gate. The typical noise figure of a 1-GHz amplifier is 2.4 dB, hence an immediate improvement in system noise figure is achieved. However, the sampling gate may now be vulnerable to saturation by high-level signals caused by targets at very short ranges.

The solution to this problem is to incorporate an additional RF amplifier whose gain can be varied as a function of time. In practice, this is most easily achieved in synchronism with the pulse repetition rate. This avoids undesirable intermodulation effects, which can occur if the gain is changed in real time. This technique enables the receiver to be operated at maximum sensitivity without encountering overload problems. Ideally, the gain/time characteristic should be related to the attenuation and reflection characteristics of the material under investigation. Hence, an adaptive calibration method is advisable.

Generally, a compression range of up to 40 dB can be expected, and this is adequate to compress most of the high-level close-range signals. The optimum technique is to use an adaptive signal-level compression whereby the peak received signal as a function of time is adaptively set to predetermined value by means of LNA gain adjustment.

It is also possible to improve the dynamic range by averaging the received signal, and this improvement is given in decibels by

$$A = 20 \log \sqrt{N} \tag{5.49}$$

where N is the number of signals averaged.

However, the rate of improvement quickly reduces as N is increased and practically 16 averages provides a reasonable improvement without excessive time penalties.

The frequency range of the output signal usually occupies a bandwidth up to 20 kHz. In the case of a radar operating at a repetition rate of 1 μs with 256 samples each averaged 16 times, the ensemble down-converted signal is repeated at the time given by

$$\tau' = \tau_r N_a N_s \tag{5.50}$$

which, for the values given above, is equal to $\tau' = 1 \times 10^{-6}$ (16) (256) = 4 ms.

Hence the bandwidth of the down-converted signal is given by

$$B = 1/5\tau' \tag{5.51}$$

where τ' is the equivalent down-converted time per sample:

$$B = 1/5\tau_r N_a = 12,500\,\text{Hz} \tag{5.52}$$

Note that the true noise bandwidth of the receiver is defined by the RF bandwidth and the noise energy is converted together with the signal.

The dynamic range of the ADC, which follows the sampling head, should be matched to the dynamic range of the latter, and, typically, a 12-bit or 16-bit converter is used. When constructing impulse radar systems, it is necessary to ensure that adequate decoupling of the internal power supplies is achieved as effect of impulsive noise from switched mode power supplies on the sampling circuits can result in serious degradation of overall system performance. Hence, good engineering practice must be maintained in the design and layout of the RF circuits.

It is also important to consider the physical layout of the sampling receiver, pulse generator, and antennas. Two options are available. The antenna can be directly connected to the transmitter and receiver circuits, or it can be interconnected via a length of high-quality RF cable. In the latter case the physical length serves to electrically separate the reflected signals caused by the antenna and the transmitter/receiver circuits. However, the cable acts as a low-pass filter, which degrades the system resolution. When the antennas are directly connected, multiple echoes can prove difficult to reduce to low levels, and some design skill is needed to achieve acceptable results. Additionally, the physical proximity of electronic components to the antennas may disturb their radiation characteristics.

5.8 SUMMARY

There are a considerable variety of radar systems for the detection of concealed targets. Doppler radar and harmonic radar systems offer significant improvement in signal-to-clutter performance over ranging radars that inherently detect all scatterers that appear in the main, side, and back lobes of the radar antenna. However, they have particular design issues that must be taken into account. Doppler radar systems only detect the radial velocity of moving targets, and noise, self-generated by the source and local oscillators, limits the inherent sensitivity that can be obtained. When Doppler radar is used to detect buried targets such as humans buried in earthquake rubble, the amplitude of the signal from the target may be of the same order as unwanted signals appearing in the side and back lobes unless action is taken to reduce these. Harmonic radar has been successfully exploited for the detection at close range of nonlinear junctions and for tags on insects, but the range dependence of the harmonics and the practical difficulties of minimizing self-generated harmonics make successful implementation of standoff designs a challenge. Spatial modulated radar relies on accurate spatial registration and recording of data prior to processing, and this limits the applicability to well-defined physical situations. The most successful modulation schemes have been amplitude and frequency modulation. Given that the inherent performance of any radar system is related to the mean power scattered from the target and detected in the receiver, the frequency modulation has an inherent advantage in performance over amplitude modulation. However, noise self-generated by the source and local oscillators limits the inherent sensitivity that can be obtained, and sweep linearity, in the case of the FMCW radar, is a key parameter in terms of achievable system range resolution. Amplitude-modulated radar, therefore, still has a major role for the application. Very short ranges have led to the successful sampling receiver design, and longer range systems use the conventional local oscillator and down-conversion schemes. Noise radar offers some benefits in terms of immunity to interference and difficulty of detection but has particular limitations in terms of the trade-offs between signal recovery and integration time, particularly where multiple targets are concerned.

CHAPTER 6

Passive Systems

6.1 INTRODUCTION

Passive millimeter-wave detection and imaging has much in common with infrared (IR) detection and imaging. William Herschel is credited with the discovery of infrared in 1800, and it is reported that even as early as World War I, British scientists developed the first IR search and track system, which could detect aircraft at a range of one mile. By 1937, Jones [281] in the United Kingdom had designed and built a basic IR airborne system that could detect other aircraft and proposed to build a light detection and ranging (LIDAR) system. Subsequent developments have resulted in extremely powerful and sensitive IR detection and imaging systems using large arrays of detectors to provide images of the radiometric scene.

Passive millimeter- and submillimeter-wave detection uses the same principles of operation but has remained comparatively underdeveloped until recently because of the difficulties associated with the technology. The performance, compactness, and cost-effectiveness of IR systems have always been a main reason for choice compared with millimeter-wave systems. However, advances in millimeter-wave technology since the late 1990s have improved the potential of millimeter- and submillimeter-wave systems.

The main advantage of millimeter- and submillimeter-wave systems compared with IR systems is their ability to penetrate through fog and dust in the transmission path and through clothing, as discussed in the chapter on the physics of propagation. Applications for millimeter- and submillimeter-wave systems include scanning people for concealed weapons and explosives, helicopter landing systems for poor visibility conditions, and the detection of overhead wires and cables.

EM Detection of Concealed Targets, by David J. Daniels

Current millimeter- and submillimeter-wave systems have not yet developed to the image resolution of IR systems, which range from 100,000 to 500,000 individual detectors. The size of individual elements and their cost are still high, and linear arrays, which are mechanically scanned, are common. Scanning speeds are on the order of 5 to 10 Hz, and work is in hand to increase the speed to video frame rates.

The following sections discuss the basic principles behind passive millimeter-wave imaging, or radiometry, and its application to detection or imaging systems. The detection of a concealed target against the background of the human body as a function of the incoming energy is the main situation that will be considered. This is followed by a discussion on the current state-of-the-art in practical imaging systems, considering the scanner and receiver hardware and the subsequent processing of the received signals to form the final image seen by the operator.

6.2 PRINCIPLES OF RADIOMETRY

A radiometer measures the incident energy, in defined frequency bands, within the submillimeter-to-centimeter wavelength range. The incident energy may be caused by man-made or natural sources reflecting from the target onto the receiver or receivers of the radiometer. Usually, the incident energy is natural and is incoherent or noise like. The radiometer has a very sensitive low noise receiver and measures the energy falling on the receiving aperture.

The total radiation received from a target contains two parts:

1. The radiation from the object due to it having a physical temperature greater that absolute zero and an emissivity. The emissivity of a material is the ratio of the energy radiated to the energy radiated by a blackbody at the same temperature and is a measure of a material's ability to radiate absorbed energy and is dimensionless. Emissivity depends on factors such as temperature, radiation emission angle, and wavelength.
2. The radiation due to the target reflecting the radiation from other surrounding sources. The emission spectrum of a perfect blackbody is dependent on its absolute physical temperature, and Planck showed that the power spectrum is given by

$$P_R = \frac{2hf^3}{c^2} \times \left(\frac{1}{e^{hf/kT} - 1}\right) \Delta v \, d\Omega \qquad (6.1)$$

where P_R = power radiated into solid angle $d\Omega$ in bandwidth Δv
h = Planck's constant = 6.63×10^{-34} J-sec

f = frequency (Hz)
c = velocity of light = 3.0×10^8 m/s
k = Boltzmann's constant = 1.38×10^{-23} J/K
T = temperature (K)

At microwave and millimeter-wave frequencies, we can approximate this expression by using the first two terms of the exponential expansion to obtain

$$P_R = \frac{1}{2} \times \frac{2kT\nu^2}{c^2} \Delta\nu \, d\Omega \tag{6.2}$$

This is known as the Rayleigh–Jeans approximation.

From this equation, the total power radiated by the blackbody medium P_R into the antenna can be shown to be

$$P_R = \frac{1}{2} \times \frac{2kT\nu^2}{c^2} \Delta\nu \, A(\theta, \phi) \, d\Omega \tag{6.3}$$

where $A(\theta,\phi)$ is equal to the absorption cross section of the antenna for spherical coordinates θ and ϕ. The factor of one-half arises because a practical antenna only accepts one polarization. From the reciprocity theorem for antennas, it can be shown that

$$A(\theta, \phi) d\Omega = \frac{c^2}{\nu^2} \tag{6.4}$$

Hence Eq. (6.3) becomes

$$P_R = kT \, \Delta\nu \tag{6.5}$$

Since the bandwidth $\Delta\nu$ is usually written as B in electrical circuits, Eq. (6.5) becomes

$$P_R = kTB \tag{6.6}$$

Therefore, it can be seen that the total power intercepted by the antenna from the absorbing medium is identical to the power radiated from a resistive load given the same bandwidth B and physical temperature T. This correspondence between power and temperature has led to widespread use of the term antenna temperature, which is defined by

$$T_A = \frac{P_R}{k_B} \tag{6.7}$$

This temperature T_A corresponds to the temperature that a blackbody must have at the antenna terminals in order to produce a signal of the observed power P_R.

In practice, a perfect blackbody does not exist, but the same concept of antenna temperature is still valid if the radiometrically observed medium is characterized by an equivalent blackbody temperature. The antenna temperature T_A can be described by an equivalent blackbody temperature T_{EB}, which is related either to an emissive medium or to a reflective target or to a combination of the two. The antenna temperature is then given by

$$T_A = T_{\mathrm{EB}} = \varepsilon_T T_T + (1 - \varepsilon_T) T_{\mathrm{AS}} \tag{6.8}$$

As

$$\varepsilon_T = (1 - R_T) \tag{6.9}$$

where R_T = reflectivity of the target (0 to 1)
ε_T = emissivity of the target (0 to 1)
T_T = physical temperature of the target (K)
T_{AS} = apparent temperature of the surroundings (K)

At millimeter waves, metal targets are very good reflectors, whatever their physical temperature. Therefore, the effective temperature of such a target is the reflected temperature of the surroundings. The human body is generally a poor reflector compared with metal, that is, it has a high emissivity. Hence the effective temperature of the human appears close to the actual body temperature. Hence, providing the temperature of the surroundings is at a different temperature from the human body, a contrast or temperature difference can be observed for a target on a human body. In practice, this temperature contrast is degraded due to the attenuation of the subject's clothing.

In the open and away from buildings the sky temperature, which is significantly colder than the human body, will illuminate the subject; hence a metal object appears as a cold spot against a warm background. Inside a building, the temperature contrast may be reduced or even reversed depending on the temperature of the surroundings. In order for a passive system to have the maximum sensitivity, the effective temperature difference between that of a human body and the surroundings must be as large as possible.

The sky temperature is fundamentally defined by the cosmic microwave radiation and is considered to be 2.73 K. However, gases, dust, and particles forming the Earth's atmosphere affect this value, so that at zenith and in clear air the sky temperature rises to 14.29 K at 32 GHz as noted by Klein [282]. Very heavy clouds can increase the sky temperature to 171 K.

The sky temperature, observed from the ground, varies as a function of angle, with the horizon being higher and the zenith being the lowest. The effect of the nonuniform illumination is to cause variation of the illumination of the subject as a function of angle, with the coldest parts being upper surfaces and the warmest being shadowed surfaces.

6.3 TOTAL POWER RADIOMETER

The total power radiometer (TPR) shown in Figure 6.1 is the simplest radiometer design, and despite its limitations it is most often encountered as a basic system design. The millimeter or microwave energy incident on the antenna is amplified with a low-noise amplifier (LNA) and then down-converted via a mixer to an intermediate frequency. The RF bandwidth of the TPR is dependent on the bandwidths of the antenna and LNA, and its sensitivity is dependent on the system noise figure.

The down-converted energy is amplified, filtered, and rectified to produce a low-frequency signal that is related to the incident energy. The main limitation of the TPR is that changes in the gain of the amplifiers and conversion loss of the mixer, often as a function of temperature, cannot be separated from the incident radiometric energy, and this can limit its application where stable performance over a long time is required. When the integration time is short, the use of the TPR is feasible and compensation for the changes in system gain

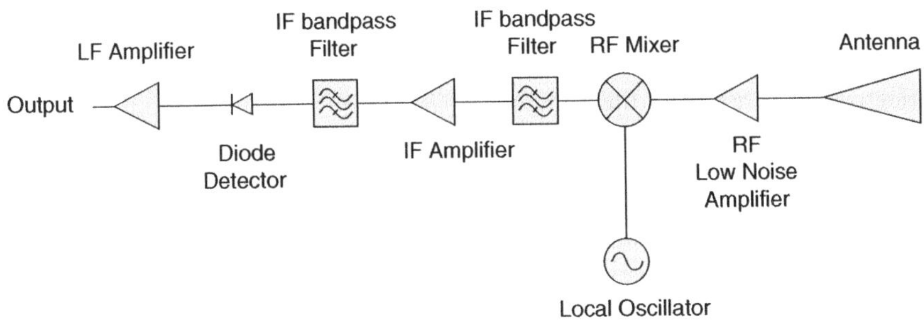

FIGURE 6.1 Total power radiometer.

can be built into the system, albeit creating a more complex system. The output of the TPR is a small signal corresponding to the antenna temperature superimposed on a large signal corresponding to the rest of the system temperature. In practice subtracting this large dc offset is difficult due to its varying with small gain changes in the receiver chain. Therefore, it is necessary to periodically recalibrate the system by injecting a calibration signal into the antenna or use an alternative design such as the Dicke radiometer.

6.4 DICKE RADIOMETER

The Dicke radiometer is a design that overcomes the inherent instability of the TPR by means of self-calibration.The basic principle of the Dicke radiometer is a comparison of the signal from a known source of energy, which can take the form of a matched load maintained at a known temperature, with that from the antenna. This is shown in Figure 6.2, and it can be seen that the Dicke radiometer is essentially a TPR with the addition of a switching system to compare the signal from the antenna with the signal from the calibration temperature and produce a synchronously demodulated output. In the diagram the antenna is connected to the plus channel and the calibrated temperature to the minus channel.

The switching rate of the system is selected to ensure a constant system gain over the switching interval, and it should be noted that this defines the bandwidth and hence dynamic response of the Dicke radiometer to fast changing changes of signal temperature. As the Dicke radiometer only samples the signal from the antenna for half of the time that a TPR would spend, it has half of the sensitivity of a TPR. The optimum sensitivity is achieved when the reference temperature is equal to the expected antenna temperature and its losses because the changes in signal level due to system gain changes cancel, thus optimizing the system sensitivity to the wanted signal fluctuations.

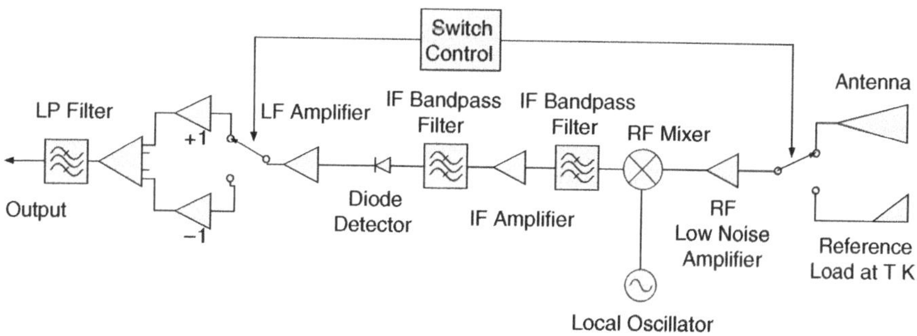

FIGURE 6.2 Dicke radiometer.

6.5 MINIMUM DETECTABLE TEMPERATURE

The sensitivity of an ideal radiometer with zero normalized system gain variation is given by

$$\Delta T_{\min} = cT_{\text{sys}}\sqrt{\frac{A}{B\tau}} \tag{6.10}$$

where $T_{\text{sys}} = T_A + T_{\text{RF}}$
$\Delta T_{\min}$ = minimum resoluble temperature difference (MRTD) of receiver (K)
c = radiometer constant, e.g., 1 for a total power system and 2 for Dicke radiometer
T_{sys} = system temperature (K)
A = signal-to-noise ratio for detection
B = noise bandwidth (Hz)
τ = postdetection integration time (s)
T_A = antenna temperature (K)
T_{RF} = receiver temperature (K)

In a TPR, the mixer and antenna losses are largely constant over a large bandwidth. If we assume that all parts of the receiver are at the same ambient temperature, then

$$T_{\text{sys}} = \alpha F_m F_{\text{rf}} F_{\text{IF}} T_0 \tag{6.11}$$

where T_{sys} = system temperature (K)
α = antenna feed loss
F_{rf} = RF amplifier noise figure
F_m = mixer double sideband noise figure
F_{IF} = intermediate frequency (IF) amplifier noise figure
T_0 = ambient temperature (K)

Many radiometers now have multiple receiver channels either in the form of a linear array, which is mechanically scanned, or a staring array of i columns and j rows. The integration time for a scanning multiple-channel radiometer is defined by the composite dwell time on each portion of the target.

Hence

$$\tau = \frac{A_S}{A_{\mathrm{FOV}}} \times \frac{N}{R_F} \tag{6.12}$$

where A_S = antenna spot size, or area, on the target (m^2)
A_{FOV} = field of view or target area under surveillance (m^2)
R_F = frame rate (frames/s)
N = number of identical receive channels

Combining Eqs. (6.10) to (6.12) gives

$$\Delta T_{\min} = c\alpha T_0 \sqrt{A} F_{\mathrm{rf}} F_m \frac{F_{\mathrm{IF}}}{\sqrt{\mathrm{B}}} \sqrt{\frac{A_{\mathrm{FOV}} R_F}{A_S N}} \tag{6.13}$$

As can be seen from the above, the sensitivity of a radiometer is directly proportional to the noise figure of the receiver including the LNA. As the mixer folds the spectra of the incident energy, the mixer must have as low a double sideband noise figure as possible, given that the IF bandwidth and hence the RF bandwidths on either side of the local oscillator (LO) are defined by the IF amplifier. The IF amplifier should have as low noise figure per root bandwidth. When the radiometer is scanned over the scene of interest, the sensitivity is related to the dwell time of the scanning spot on each portion of the target and is also increased in proportion to the square root of the number of channels.

The minimum detectable temperature is related to the radiometer bandwidth, and a high minimum detectable temperature reduces the bandwidth by a squared proportionality. Obviously, the radiometer factor needs to be as low as possible.

6.6 TEMPERATURE RESOLUTION

The ability of a radiometer to detect concealed targets is defined by its minimum detectable temperature and the temperature contrast of the target. As many radiometers for the detection of concealed target are operated at comparatively short ranges, it is reasonable to assume that the absorption and scattering in the air can be neglected. The human body has a high emissivity and emits microwave/millimeter-wave energy at around 300 K. Hence outside buildings it will appear warmer than the general background, which is on the order of 20 K to 100 K, depending on wavelength and also cloud cover. In contrast, a concealed weapon has a low emissivity and a high reflectivity, reflects the background energy, and appears as cold on the radiometric image. The temperature differential with respect to the background enables the weapon or object to be detected. In contrast, indoors there is little free illuminating energy, and the temperature difference between the target and the human body is much reduced.

In the case of a target with a particular emissivity and temperature in a background with an emissivity and temperature, the temperature difference can be determined. The antenna temperature T_A can be described by an equivalent blackbody temperature T_{EB}, which is related either to an emissive medium or to a reflective target or to a combination of the two as shown in Eq. (6.8).

When the objective is to detect a target against the background of the human body, both will be at the same thermal temperature and both will reflect the surrounding millimeter-wave temperature.

Hence:

$$T_{AO} = \varepsilon_O T_B + (1 - \varepsilon_O) T_S \tag{6.14}$$

$$T_{AB} = \varepsilon_B T_B + (1 - \varepsilon_B) T_S \tag{6.15}$$

where T_{AO} = equivalent blackbody temperature (K) of object with emissivity $\varepsilon_1 = 1$

T_{AB} = equivalent blackbody temperature (K) of human body with emissivity ε_B

T_B = physical temperature of human body (K)

T_S = temperature of surroundings or illumination (K)

ε_O = emissivity of object

ε_B = emissivity of human body

From Eqs. (6.14) and (6.15) we can determine the measurable temperature difference ΔT.

$$\Delta T = (\varepsilon_O - \varepsilon_B)(T_B - T_S) \tag{6.16}$$

This equation shows that the temperature difference is the product of the difference between the emissivities and the difference between the body and surrounding or illuminating temperature.

At first sight, it appears that the illuminating temperature could be higher or lower than the body temperature so long as the temperature difference is as large as practicable. However, the use of a high illuminating temperature increases the overall system temperature of the radiometer more than when a lower illuminating temperature is used. This in turn decreases the minimum temperature difference that can be measured. Practical measurements have also confirmed that it is preferable to use a cold illuminating source rather than a hot one, and there are practical difficulties in finding a suitable blackbody material for a hot illuminating source, although noise sources can be used to create suitable illumination.

From the previous sections and also with reference to Klein [282], it can be shown that Eq. (6.16) has to account for the antenna main-lobe beam efficiency η_b as well as its radiation efficiency, η_r, hence

$$\Delta T_{\min} = \eta_b \eta_r (e_O - e_B)(T_B - T_S) \tag{6.17}$$

The graph shown in Figure 6.3 illustrates the measurable temperature difference in kelvins as a function of background temperature in kelvins, for the case of a background emissivity of 0.95 and values of target emissivity of 0.1, 0.3, 0.5, 0.7, and 0.9, respectively, assuming a beam efficiency of 1. The effect of the gradually increasing emissivity causes the measurable temperature difference to gradually decrease.

When the target is concealed, then the covering material has to be taken into account, and its temperature, emissivity, and attenuation must be considered. In determining the measurable temperature, a number of factors need to be considered. These include the attenuation, thickness, and the emissivity of the clothing, as well as the temperature of the target and its emissivity.

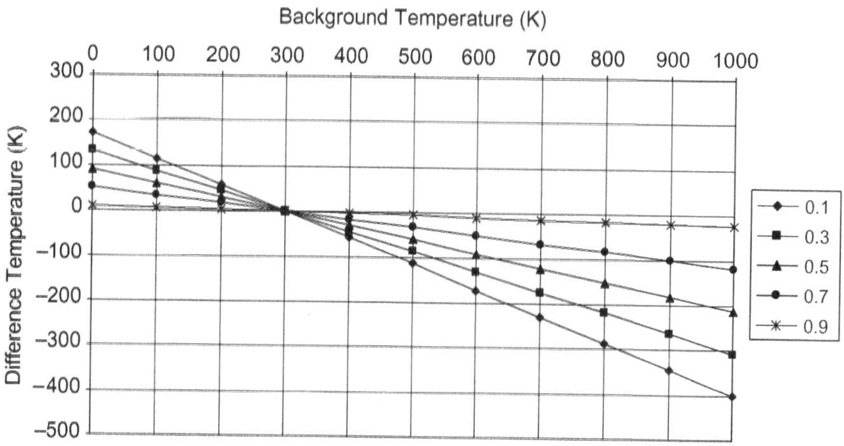

FIGURE 6.3 Temperature difference in kelvins versus background temperature in kelvins for various values of target emissivity.

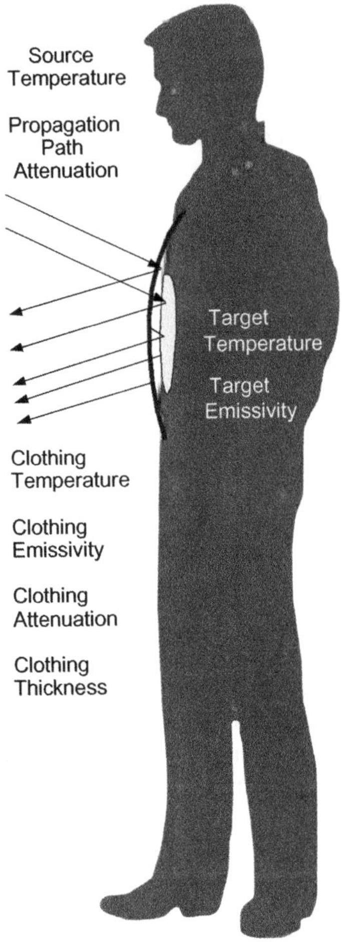

FIGURE 6.4 Schematic representation of the contributions to measurable target temperature.

A schematic representation of the various terms contributing to the measurable temperature of a target is shown in Figure 6.4. The main contributions comprise the source energy reflected from the outer clothing, the energy passing through the clothing and reflecting off the target, energy radiating from the target due its temperature, energy radiating from the human due to its temperature, and energy radiating from the target due to the temperature of the clothing.

Timms et al. [283] considers this case and proposed a six-term representation for the target temperature. However, the effects of view angle and propagation paths also need to be considered and an alternative approximate representation is that the measured target temperature is given by

$$T_{\text{meas}} = k_a[\Gamma_{\text{mat}} T_{\text{amb}} + k_{\text{mat}}^2(1 - \varepsilon_{\text{tar}})T_{\text{amb}} + k_{\text{mat}}(1 - \varepsilon_{\text{tar}})\,\varepsilon_{\text{mat}} T_{\text{mat}} + k_{\text{mat}}\varepsilon_{\text{tar}} T_{\text{tar}} + \varepsilon_{\text{mat}} T_{\text{mat}}] + T_s(1 - k_b)(1 - \beta \cos \varphi) \tag{6.18}$$

where k_a = fraction of energy lost from subject to receiver
k_b = fraction of energy lost from source to subject
k_{mat} = fraction of energy lost through material
Γ_{mat} = reflectivity of material
T_s = source temperature
T_{amb} = ambient microwave/millimeter-wave temperature
T_{mat} = material microwave/millimeter-wave temperature
ε_{tar} = emissivity of target
ε_{mat} = emissivity of material
φ = angle above ground
β = ground to zenith multiplying factor

Outdoors, the temperature of the illuminating source increases with angle from illumination, from above to illumination, from the horizon, and is lower in temperature than either ambient or that of the target. Indoors, the illuminating source may be higher in temperature, but this has disadvantages as explained earlier. An alternative option is that indoors a refrigerated panel covered with RAM or similar material could provide the illuminating source. This in turn would be covered by a layer of expanded polystyrene- or a double-or triple-glazed glass panel to prevent frosting up and ice forming on the surface exposed to ambient temperature and humidity. If the target is illuminated by a series of artificial sources, the contrast between metal objects and the human body can be enhanced. An increase in ΔT_{min} of 10 would reduce the receive bandwidth requirements by 100, assuming all other components in the system remained unchanged. Modulating the artificial source and synchronously detecting the signal using a lock-in amplifier in a similar way to the Dicke radiometer could gain an increase in sensitivity.

A major issue with passive millimeter-wave imaging is that the specular nature of reflection from human skin is such that the upper surfaces of the human body will tend to appear brighter than shadowed regions with outdoor imagers as shown in Figure 6.5. This can make automatic detection of suspect areas more difficult than in targets concealed in shadow zones, and targets placed in cold zones may be more difficult to isolate with automatic image recognition techniques. One way of overcoming this in indoor situations is to ensure that the body is illuminated from all angles.

FIGURE 6.5 Representation of the millimeter-wave illumination of a human in an outdoor situation with cold temperature shown bright and warm dark.

6.7 IMAGING SYSTEMS

This section provides a brief introduction to the fundamental principles of imaging system. An imaging system needs to provide the spatial resolution needed to recognise or identify targets by their shape. This resolution is a function of the field of view as well as the range to the target and for recognition and identification a minimum number of resolution cells are needed in the image. In the case of thermal imaging cameras, a typical staring array has 480 vertical and 640 horizontal has pixels.

While such an array density is easy to achieve at infrared wavelengths of 10.6 μm, it becomes physically difficult to achieve at millimetre or sub millimetre wavelengths. Hence the spatial resolution of millimetre or sub millimetre wave staring array imagers is more limited and many existing design use a linear array that is mechanically scanned to achieve the desired image resolution.

In addition to the spatial resolution requirements the speed of image formation is fundamentally defined by the temperature sensitivity of the individual receiver channels.

The temperature sensitivity ΔT of a radiometer is given by equation 6.19

$$\Delta T = \frac{T_{sys}}{\sqrt{B\tau}} \tag{6.19}$$

where T_{sys} = system noise temperature
B = RF bandwidth
τ = integration time

The number of resolution cells (i.e. pixels) can be determined from equation 6.20

$$N = \frac{X \cdot Y}{\delta^2} \tag{6.20}$$

where X = image width
Y = image height
δ^2 = resolution cell size

For a staring array imager the number of receivers that generate simultaneous beams (receivers) is given by n, hence the image acquisition time T is given by equation 6.21 and then 6.22

$$T = \frac{N}{n} \cdot \tau \tag{6.21}$$

$$T = \frac{N}{n} \cdot \frac{1}{B} \left(\frac{T_{sys}}{\Delta T} \right)^2 \tag{6.22}$$

It can be readily seen that a large number of channels are needed to reduce the image acquisition time for reasonable spatial resolution. Reducing the unit cost of these is a major challenge for the designers of passive millimetre wave imagers.

6.8 SUMMARY

Passive systems or radiometers have been extensively developed over the last decade for passenger screening at airports and access points. There are a number of companies with products in this market working at frequencies from tens of gigahertz up to submillimeter-wave frequencies, and some of these are described in the following chapter. There are technical limits to such systems, and they relate to the radiometric sensitivity that can be obtained from the receiver as well as the stability of calibration of arrays of receivers. Sensitivity is predominantly defined by the inherent noise of the receiver or its noise figure, and, whereas at 10 GHz noise figures of less than 1 dB are common, for LNAs at 90 GHz it is hard to achieve much better than 4 to5 dB with current uncooled technology. The effect of the higher noise at millimeter- and submillimeter-wave frequencies is that the poor radiometric sensitivity limits the system detection range to 30 to 50 m for many products. The spatial resolution of passive systems is generally much inferior to thermal imagers because the effective number of sensing elements is low. A typical thermal imager or IR camera might have 640 by 480 detecting elements, that is, 307,000 elements in the array, and the cost per detecting element of a complete camera system is typically less than half a dollar.

In contrast, a millimeter-wave imager, many of which are quasi-mechanical scanned linear arrays of 50 to 200 elements, has an order of magnitude lower number of effective receiving channels and hence a commensurately lower spatial resolution. With some of the systems that have been developed, the field of view is limited, and this means that unless correctly focused, the image loses resolution because the image is outside the field of view. Efforts have been made to develop staring arrays, and this is the approach that was adopted by TRW in the United States. An array of 640 by 480 detecting elements would be prohibitively expensive using current technology because the cost of each individual receiving element for millimeter or submillimeter-wave systems is still high and is several orders of magnitude higher than that of the equivalent in the thermal imager.

Thus, the cost of current mechanically scanned millimeter-wave radiometer is an order of magnitude greater than that of a thermal imaging camera. However, the enormous advantage of being able to "see" through clothing is such that research and development is being actively pursued. The goal of many researchers is to develop staring array radiometers that are self-calibrating at a low cost and with low noise. Such arrays may then be able to be electronically scanned using beam–forming techniques and could then further develop the market for passive systems.

CHAPTER 7

Applications and Technology

7.1 INTRODUCTION

The situations in which targets are concealed are extremely varied and can range from a target concealed on a person to people concealed within buildings, buried under earthquake rubble, or in avalanches. This chapter reviews the majority of the applications as well as some of the current technology that is used to detect targets of many types. The author is aware that those systems described in this chapter are a small sample of the available technology and many designs and systems have not been included. This is a deliberate omission for two reasons. In some cases organizations have not responded to requests to publish information, and in other cases the technology has not reached full production status. As technology is developed, more systems become commercially available and any material that is in print becomes out of date. The aim of this chapter is to provide a snapshot at the time of writing of some of the technology that is available. There is a wealth of literature describing research projects and products, and the reader can easily carry out his or her own search on IEEE Explore or other convenient search engines such as scirus, FreePatentsOnline, INSPEC, IET digital library etc.

7.2 PHYSIOLOGICAL MONITORING

An important detection technique for humans is based on the measurement of heartbeat and respiration and this has received much attention. Detection of heartbeat, respiration, and movement can be achieved by means of measurement of the micro-Doppler generated by the physical movement of either organs or limbs as described in Chapter 4.

EM Detection of Concealed Targets, by David J. Daniels

Researchers from Georgia Tech Research Institute developed a radar vital signs monitor (RVSM). The most recent version of the system was developed to measure the heart rate of Olympic rifle and bow-and-arrow competitors to determine if their training allowed them to detect their heartbeats and, if so, whether they were capable of using that training to avoid an approximate 5 mrad movement of the bow or rifle that occurs each time the heart beats. The RVSM that was developed was tested to detect the shooter's heartbeat at a distance of 10 m without the requirement of a physical connection to the subject. It was found that a second channel could be added to the RVSM to detect the shooter's respiration rate from a distance of 20 m without physical connection between the RVSM and the shooter. The radar signature of a heartbeat from a 10-GHz Doppler radar as reported by Greneker [284, 285] is shown Figure 7.1.

It is thought that the signature detected by the RVSM is the shock wave propagating from the beating heart as it spreads across the thorax region of the chest wall, rather than the detection of the movement of the beating heart. Studies have shown that there is little penetration of the chest wall by radio frequency (RF) energy at 24.1 GHz at the low power densities of 0.1 mW cm^{-2}, which is typical of those produced by the RVSM at a range of 3 m. It is thought that this shock wave is the same phenomenon that is heard by a health care provider using a stethoscope.

Considerable research effort has been made to improve the performance of such systems, and the work reported by Lin and Li [286] exploited double-sideband transmission and detection to replace quadrature detection and the

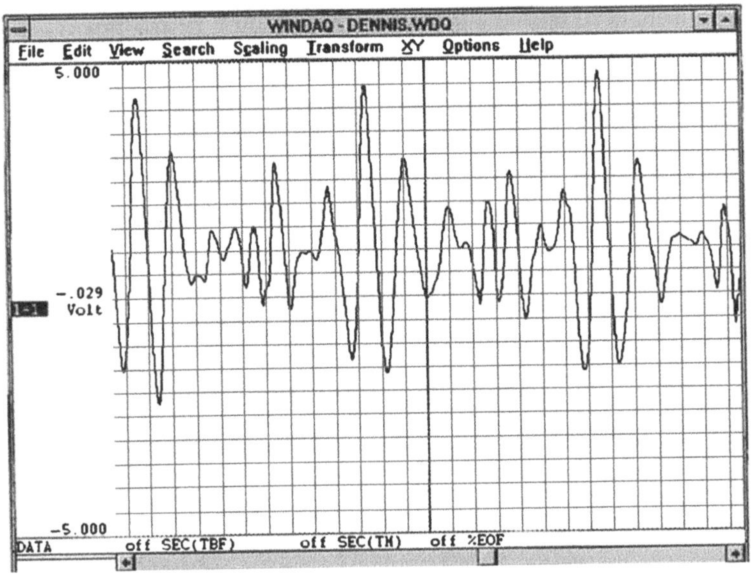

FIGURE 7.1 Doppler signature of human heartbeat (with kind permission ©1997 IET).

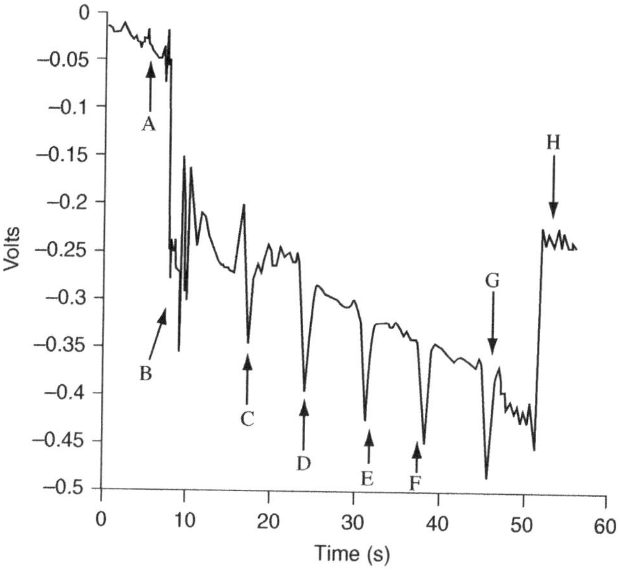

FIGURE 7.2 Radar signature at 10 GHz of human breathing (with kind permission ©1997 IET).

use of higher frequencies to improve sensitivity. Xiao et al. [287] reported the development of a portable noncontact heartbeat and respiration monitoring system operating in 5-GHz band with a claimed detection accuracy of better than 80% at a distance of 2.8 m radiating 20 μW. Ivashov et al. [288] discusses the reduction of clutter with radar for detection of breathing and heartbeat, and Nezirovic et al. [289] carried out sufficient experiments to show that breathing responses are strongest when the subject is facing to or away from the radar, both in terms of expected value and variance with the Doppler RCS response being some 30 to 40 dB lower when the subject is side on. Nezirovic and co-workers found there is no significant difference between bistatic radar cross section (BRCS) values for HH and VV co-polarization, as they are comparable in terms of mean and variations of breathing cross-section values. Cross-polarized antenna pairs (HV) generally produced very weak responses and thus should be avoided, as there is no expected polarization diversity gain. More detailed studies have been carried out by Yarovoy et al. [290] who investigated a UWB radar with an 11.7-GHz bandwidth and measured respiratory movement of a person in laboratory conditions and derived the wideband radar cross section in the frequency band from 1 to 12 GHz.

7.3 EARTHQUAKE AND AVALANCHE RADAR SYSTEMS

Chen et al. [291] from the Electromagnetics Laboratory of Michigan State University describes several radar systems that were developed and tested in realistic conditions. The earthquake rubble test jig was constructed with advice from the Federal Emergency Management Agency (FEMA) Task Force with bricks, cinder blocks, and steel rebars. The dimensions of the rubble test jig was about 5 ft wide, 6 ft long, and 6 ft high. Two layers of steel rebars separated by 8 in. were placed perpendicularly through bricks. The human subject to be tested was able to lie down in the cavity at the bottom of the rubble.

Two systems were constructed: one at 450 MHz and the other at 1150 MHz. Typical experimental results of the breathing and heartbeat signals of a human subject lying in the rubble cavity obtained with the 450-MHz system are shown in Figure 7.3. It was found that the EM wave of 450 MHz has difficulty in penetrating layers of reinforced concrete slabs with embedded metallic wire of 4-in. spacing.

However, the 1150-MHz radar was found to be able to penetrate a rubble with layers of reinforced concrete slabs with metallic wire mesh easier than that of 450 MHz. Chen [291] concluded that an EM wave of 450 MHz may penetrate deeper into rubble without metallic wire mesh than that of 1150 MHz.

Radar can also be used to detect people buried in avalanches, and trials with a 1-GHz radar system by ERA Technology showed that detection of volunteers buried up to 1 to 1.5 m was quite practical. A typical radar reflection is shown

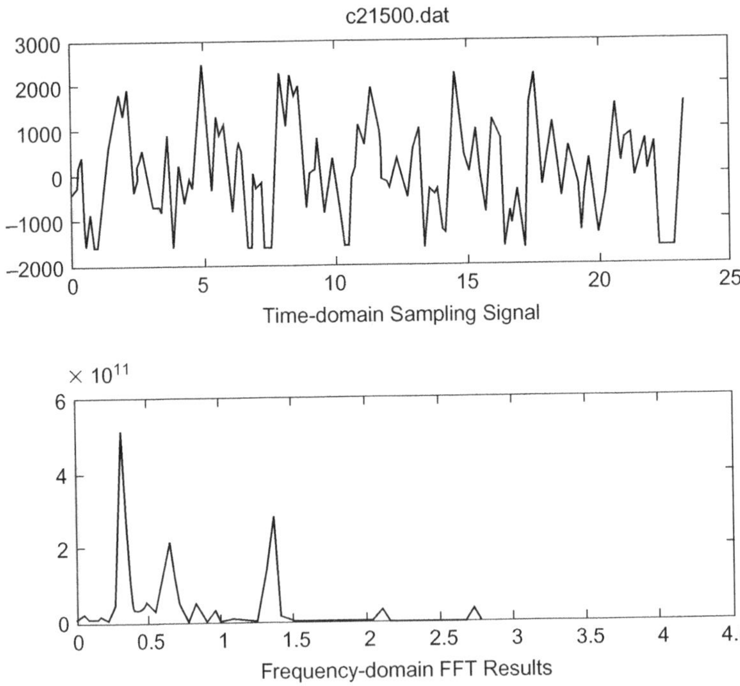

FIGURE 7.3 Breathing and heartbeat signals of a female human subject recorded at MSU simulated rubble using a 450-MHz radar (with kind permission ©IEEE).

FIGURE 7.4 Detection of humans buried in avalanches (with kind permission ERA Technology).

in Figure 7.4 for a depth of cover of 0.75 m and a distance to the bedrock of 1.5 m for trials in the French Pyrenees.

Further trials in the Italian Alps showed that operators, without previous experience of radar, easily detected targets, buried at 50 cm depth of cover of consolidated natural snow, associated with skiers such as the backpack of mountaineering equipment, frozen meat of a mass of about 5 kg, skis, and the like. Live human volunteers buried under 1.50 m of snow cover were also detected. A demanding blind search with two objects buried at 75 cm depth of cover in soft snow and hidden in an area 10 m by 20 m was carried out in 10 min.

7.4 FORENSIC APPLICATIONS

Forensic investigation using GPR has become a recognized method of forensic archaeology by means of some high-profile case histories. GPR techniques are one among a number of methods that can be used in forensic investigations and can greatly assist police investigations by pinpointing suspicious areas and thus saving unnecessary excavation. GPR provides the means to conduct rapid, nondestructive investigations, which can considerably reduce the need to carry out extensive and expensive excavations. The interpretation of GPR images is not easily carried out by the novice because of the variability of the targets and sites. Only those operators who have adequate experience of both forensic work and GPR radar should carry out forensic surveys. In skilled hands, GPR can be employed to search for and locate the types of target sought by the forensic investigators. Unlike searching for utilities and services, where the radar signatures tend to be well defined and typically hyperbolic in form, radar signatures of forensic targets are varied and due to the wide variety of targets and situations. Experienced operators develop an expertise in the understanding and interpretation of the radar images.

For good GPR survey results it is also vital that the ground is undisturbed by either the passage of heavy vehicles (which tend to compress the soil and give rise to unwanted reflections) or by digging. Once digging has been carried out, the soil structure is disturbed and effective GPR surveying becomes extremely difficult. The key to a successful forensic search is, as with all surveying techniques, a systematic procedure. The site to be examined should be scanned in two orthogonal directions because the GPR antennas often exhibit polarization dependence. The basis of the survey is to locate anomalous features in the area. By a process of elimination, anything that cannot be readily interpreted can be categorized as anomalous. The principal targets of most forensic searches are human remains, or more specifically graves that are a backfilled excavation in the ground, with buried human remains in some state.

7.5 THROUGH-WALL RADAR (TWR) FOR SURVEILLANCE

The safety of victims, hostages, or security forces placed at risk often depends on the ability to assess situations quickly and accurately. For example, hostage and terrorism situations require knowledge of building interiors and the location and identification of individuals. It is vital that this assessment be carried out remotely so as to avoid endangering more people. Many dangerous situations also require that assessment be done discreetly so as not to alert criminals or the enemy to the intentions of security force personnel. This information must then be presented to the operator in a clear and simple way. In addition, the system must be easily deployable and ultimately be capable of producing a simple representation of both moving and fixed targets within the room. Through-wall radar systems capable of monitoring the location of people in a room, by looking through the walls, floor, and/or ceiling of the room have been developed by a number of organizations.

In considering the designs that can be fielded, the key decision is whether an image of the building is required or whether movement alone will suffice. A single-channel Doppler radar is probably the simplest and lowest cost design option but would be limited to detecting target movement. However, Doppler radar is only fundamentally capable of providing information on the radial velocity of the target, and multiple moving targets are more difficult to resolve unless their differential velocities are large. Zhou et al. discusses a signal-processing technique that can be used to distinguish between the presence of 2, 1, or 0 subjects, even with a single antenna. However, the simplicity of the single-channel radar leads to operational limitations. An alternative arrangement is by means of a dual-channel Doppler radar system with both antennas arranged to form a stereo pair. The output from the radar could be a low-frequency audio signal monitored by an operator using headphones. Movement by the target across the field of view of the antennas would cause an audio sound field to be created that would indicate the bearing of the target. One of the challenges of Doppler through-wall radar is that range and bearing to the target are not easily determined and signals detected from unwanted friendly sources in the side and back lobes of the antenna can be confused with those from the wanted target.

In order to resolve the structure of a building, the radar needs to possess the capability of range resolution. This means that a bandwidth of at least 1 GHz is required and possibly more depending on the resolution requirement. The type of modulation technique could be time-domain impulse radar (TDIR), stepped frequency continuous wave (SFCW), noise or pseudo-random coded (PRC). The PRC offers the most efficient use of the spectrum and results in an efficient power budget and the greatest immunity to interference or electromagnetic interference. The issues of refraction of the radar beam, strong echoes caused by corner reflectors, and the generally high levels of clutter would be real difficulties, and the system would have to have a reasonable power output to overcome path loss, wall attenuation, and reflection losses.

This has led to the use of UWB radar systems, which have been used for a number of years to monitor bulk body movement behind walls. A single-channel radar can be used to detect the whole body movement of targets behind walls. An example of the detection of a person walking toward and away from the radar over a distance of several meters behind a 12-cm brick wall is shown in Figure 7.5.

The information supplied by a single channel does not enable nonambiguous, positional information to be determined. To achieve this, several antennas and receiver channels would be required, ideally arranged in a multistatic configuration. The plan view of an area is shown in Figure 7.6 in which four radar system antennas have been placed around the perimeter of an area in different, but known, locations. The targets are denoted by the black ellipses.

In such a system each radar transmits a pulse from a low-gain antenna, which propagates into the area. When it reflects from any impedance mismatch, energy will be reflected back to the antenna. Depending on the location of targets within the building, each antenna will receive reflections from the target at different times. Data from each channel is processed so that these reflected

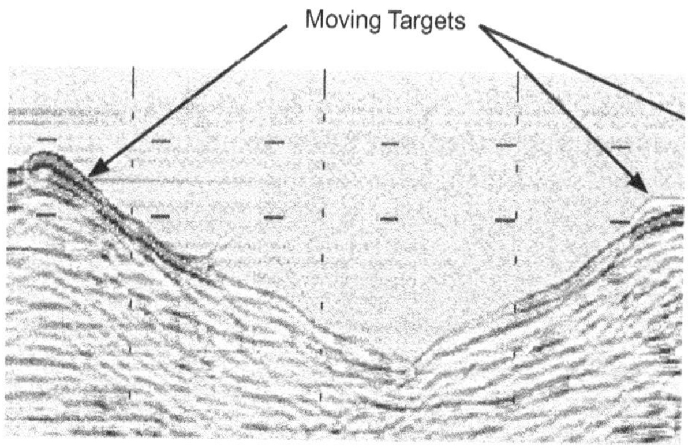

FIGURE 7.5 Radar image of target moving toward and away from a UWB radar working through a brick wall (with kind permission ©ERA Technology).

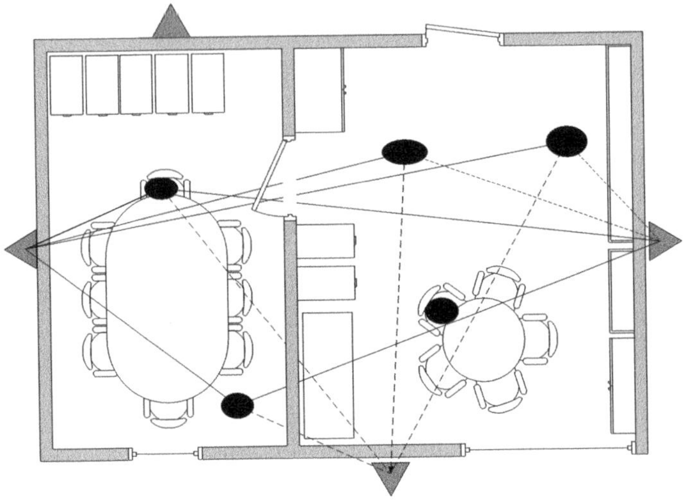

FIGURE 7.6 Multistatic through-wall radar.

events are detected and extracted. As the wave is traveling through air with a known velocity, and the travel time to the target is known, the distance from the antenna to the target can be calculated. Each radar gathers data on the set of targets, which are at known distances away. The positions of all unique targets are calculated by comparing all the targets from all the radars. If the locations of the antennas are known accurately, the locations of each target can be obtained by multilateration and can then be displayed on a plan view, showing the relative locations of the targets. It can be seen in the illustration that the radar system placed against the upper boundary cannot illuminate targets because there is a row of filing cabinets, which block the view of the radar. Much research work has been carried out to image the interior of buildings, but the challenge of producing realistic images with a sparse set of radars is significant. There are a number of commercially available TWR, which are described below. The selection is arbitrary and is intended to provide a snapshot of typical systems.

The author is grateful to the respective organizations for permission to use their material and its inclusion does not imply any view on the suitability of the equipment. The Camero Corporation, based in Vienna, Virginia, has developed the Xaver™ 800 through-wall radar system for special operation teams working in an urban environment that need accurate and reliable real-time information for successful execution of their missions. Camero believes that information about the number of people and their location inside a room is critical in determining situation awareness for urban operations. The Xaver™ 800 shown in Figure 7.7 is a portable, high-performance through-wall radar system, optimized for gathering instant and accurate real-time information from behind solid walls and is claimed to provide true 3D imaging of objects behind walls. Camero states that the

FIGURE 7.7 The Xaver™ 800 through-wall radar system (with kind permission ©2008 Camero).

Xaver™ 800's superior object resolution allows quick location of people hidden by walls and barriers and tracking of their movement patterns to support differentiation between victims and hostiles. The result is a high-end system that primarily suits urban intelligence, surveillance, and reconnaissance (ISR) missions, providing exceptional object resolution and image clarity. Also available is the Xaver™ 400 system.

Camero states that the Xaver™ 800 has detection ranges of 4, 8, and 20 m with a 128° field of view in both azimuth and elevation and can penetrate through cement, plaster, brick, concrete, reinforced concrete, adobe, stucco, and drywall. The Xaver™ 800 operates over the frequency range of 3 to 10 GHz with a range resolution of less than 0.2 m at the end of normal operational range and features 3D view, 2D plain view; 1.5 D (range with time history) display modes. Weighing in at 15 kg, it measures 84 cm by 84 cm by 15 cm when in operation and 47 cm by 47 cm by 15 cm for transportation and gives 2.5 hr battery operation.

Cambridge Consultants Limited (CCL) in the United Kingdom has developed the Prism 200 ultra-wideband radar, which is a lightweight, durable, and highly sophisticated through-wall radar. It is designed to provide police, special forces, and the emergency services with accurate covert intelligence of the location and movement of people in situations where it would otherwise be impossible to gain such insight. The system aims to identifies an individual person as a cluster of targets of the same color to give a clear representation of the movement of each person in the space being monitored and has the ability to focus on living/moving targets to give a tracking history of individuals and to build up a picture of the static objects in the room.

The Prism 200 can see through brick, block, and concrete walls and doors, and over a range of up to 20 m, providing comprehensive coverage of internal spaces. It can be used for the rapid evaluation of room occupancy, or it can be used for more detailed intelligence gathering. Cambridge Consultants states that the Prism 200 operates over the frequency range of 1.7 to 2.2 GHz, with a field of view of 120° horizontally and 90° vertically and will detect personnel up to 20 m.

Time Domain Corp., based in the United States, offers two handheld TWR systems. One is called SoldierVision A1, while the other is a slightly less powerful version for police called RadarVision 2. SoldierVision A1 can be used at a standoff distance of 10 m, while detecting motion 10 m beyond the other side of a wall. Placed flat against a wall, SoldierVision A1 can sense movement within a 180° range. That motion is displayed as a moving image viewed from a top-down vantage point on a small screen built into the unit (Figures 7.9 and 7.10).

FIGURE 7.8 Prism 200 through-wall radar (with kind permission © CCL).

FIGURE 7.9 Soldier vision (with kind permission © Time Domain Systems).

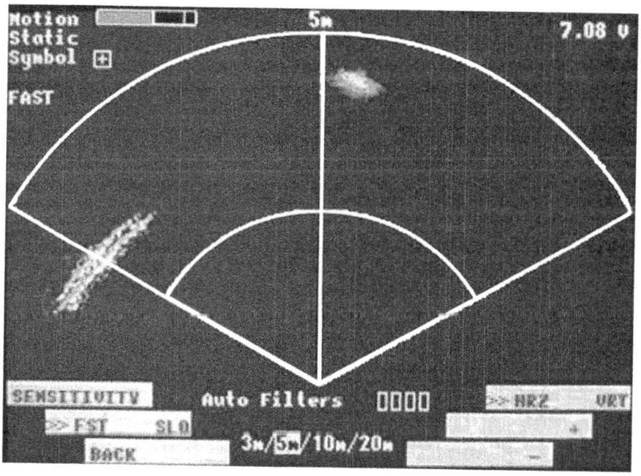

FIGURE 7.10 Soldier vision display (with kind permission © Time Domain Systems).

Defense contractor Raytheon is developing a low-cost Enhanced Motion and Ranging System (EMARS) that weighs less than 3 lb and can be held in one hand. Three EMARS units can be used to triangulate received radar signals for a better picture of how many people are behind a wall. This data can be wirelessly transmitted to a laptop up to 300 ft away. Details of other systems are given by Borek [293] who describes the U.S. Air Force Research Laboratory Information Directorate (AFRL/IF) research and development program for advanced through the wall surveillance (TWS) technologies and five prototype TWR systems delivered to AFRL/IF for test and evaluation. TWR is capable of detecting gross body movements, but it is not generally designed for the small variations in chest movement typical of humans that are concealed or buried and who may be lying very still.

7.6 HARMONIC RADAR SYSTEMS

Harmonic radar systems have been developed for two main applications: detecting nonlinear junctions and tracking insects, which have cooperative tags fitted. The nonlinear junction detectors have more relevance to the detection of concealed targets and have ranged from the powerful METTRA radar to more compact commercially available systems such as the REI Orion or Winkelmann Hawk 450XD Non-Linear Junction Detector. Typical systems transmit between 1 and 5 W at frequencies in the range of 850 to 1005 MHz and receive at the second (1700 to 2010 MHz) and third (2550 to 3015 MHz) harmonic bands. The typical receiver sensitivity is on the order of −120 to −140 dBm with an IF bandwidth of several kilohertz. The systems typically weigh a few kilograms and require a careful search procedure to ensure the targets are correctly detected. The useful range is often less than a meter but some manufacturers offer a remote operation capability.

7.7 LAND-MINE DETECTION RADAR SYSTEMS

7.7.1 Handheld Land-Mine Detection Radar Systems

The detection of buried ordnance, improvised explosive devices (IEDs), and land mines has proved to be a successful application of UWB radar, and commercially available handheld systems are now in production. Backscattering of electromagnetic energy from a land mine results from the impedance differences of the land mine compared with the host material. The mine may have a number of scattering centers, each with their own angular radiation pattern, and in the case of plastic land mines the internal structure of the mine may generate additional scatterers. Most minimum metal land mines may be considered as multiple layered dielectric cylinders, each interface causing a reflection, the impact of the small internal metallic fuse being minimal. The land mine is surrounded by soil, which is a lossy dielectric whose relative dielectric constant depends mainly on the water content. Typically, the relative dielectric constant of the soil varies from 3 in dry sand to greater than 16 in wet and waterlogged soils. The ground also contains significant levels of radar scatterers such as stones, animal burrows, as well as man-made detritus, and, consequently, radar encounters extremely high levels of clutter at short ranges.

Typical handheld radar systems use separate, man-portable, transmit-and-receive antennas, which are placed just above the surface of the ground and moved in a known pattern over the surface of the ground under investigation. This generates, in real time, data or an image. By systematically surveying the area in a regular pattern, a radar image of the ground can be built up. Alternatively, the equipment may be designed to provide an audible warning of target presence while the antenna is swept over the ground. Vehicle-based or airborne systems use much larger linear arrays of antennas to illuminate a swathe of the ground surface ahead of the platform and rely on the movement of the vehicle to create the data, which may be processed using SAR techniques.

Currently available handheld dual-sensor technology has enabled improved discrimination against metal fragments compared with the standard metal detector to be demonstrated in live mine fields, and reductions of up to 7 : 1 have been achieved in the field by handheld systems such as MINEHOUND™ [294] and AN/PSS-14 [295]. The AN/PSS-14 is produced by L-3 Communications Cyterra, and production for the U.S. Army commenced in 2006. The AN/PSS-14 system was developed for the U.S. Army especially for U.S. Army requirements. Further details can be obtained from L-3 Communications Cyterra. The MINEHOUND™ was designed for humanitarian demining operations and has also proved highly successful in a military role due its simplicity of operation and ability to detect IEDs as well as land mines. Vallon GmbH of Germany produces the MINEHOUND™ detector and ERA Technology supplies the radar subsystem to Vallon GmbH. The MINEHOUND™ radar module is comprised of two main sections: the RF and the digital sections, as shown in Figure 7.11.

The RF subsection is responsible for signal conditioning and down-conversion, while the digital subsection is responsible for signal processing, control functions, and the generation of the audio output. These two sections are placed side by side on an extended single Eurocard (220 mm by 100 mm)

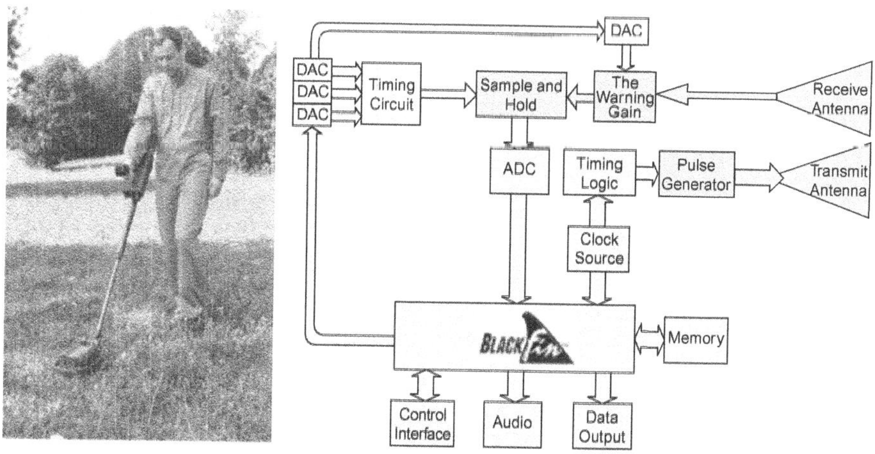

FIGURE 7.11 MINEHOUND detector and radar module system diagram (with kind permission ©ERA Technology).

six-layer printed circuit board (PCB). The UWB radar generates 240 ps duration, 12-V impulses every 1 μs using an integrated step recovery diode pulse generator and transmits these via an UWB antenna with a center frequency of 1 GHz, operating from 200 MHz to 2 GHz, thus radiating 1 ns duration wavelets. The UWB receiver applies a time-varying gain profile to the incoming signal before sampling every 50 ps. These samples are then down-converted to a slower time frame using interleaving and an averaging process. The waveform-sampling window is 13.2 ns, and this waveform is reconstructed over a period of 16.4 ms at a refresh rate of 61 Hz. Classical repetitive sampling down-conversion is used. The key RF component block is the wide bandwidth Schottky diode sampling system developed by ERA, and RF sampling is carried out by sequentially incrementing the sample time position each pulse repetition interval up to 512 samples and then repeating the process.

The radar software runs on an Analog Devices Blackfin™ DSP and is responsible for data acquisition at the full data rate of 16 Mbit s^{-1}, the updating of timing circuit digital to analog converters every 32 μS and general housekeeping tasks. In addition to this, the DSP runs the signal-processing algorithms, which convert the radar data to audio every 16 ms. Further tasks include communication with other devices that are connected to the radar board. These can include devices such as metal detectors in the case on MINEHOUND™ and PDA devices (Figure 7.12).

In parallel with this process, the Blackfin™ is responsible for dealing with control functions typically fulfilled by a microcontroller. The Blackfin™ core has been designed with control and DSP functions in mind and as such is ideally suited to this application. There are a number of configuration changes that need to take place at the 32-μ s averaging rate; this results in a 14-bit parallel, 14-Mbit per second data rate, which, again, is dealt with by the dynamic

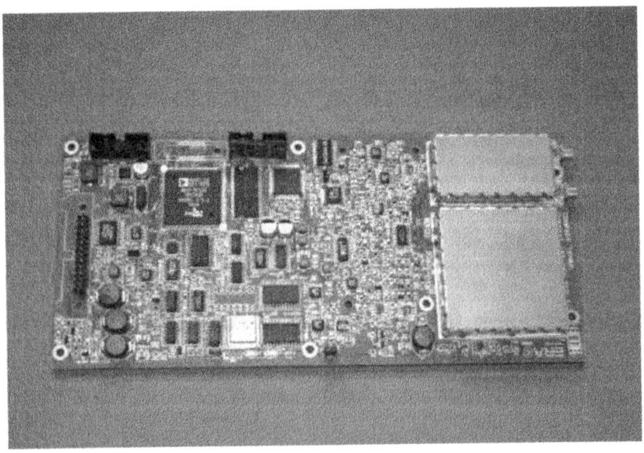

FIGURE 7.12 MINEHOUND radar transmitter receiver and processor (with kind permission ©ERA Technology).

memory allocation (DMA) controller. Another part of the control task is implementing the drift compensation technique. This ensures that drift in the start point of the sampling window, which is due to thermal changes affecting component tolerances, is removed before the data is passed to the signal-processing algorithm. Two feedback loops are used to ensure that drift in both amplitude and time position with respect to changes in temperature is minimized.

An excellent performance is achieved in controlling amplitude and timing variations. Timing variations are held to within 3 samples in 256 (1.2%) and standard deviation variations in amplitude to within +2.3 and −4.3%. Stability of amplitude and timing drift is critical for the signal subsequent processing. The radar power consumption is 2.3 W and contains all the processing needed to provide audible warning of a land-mine target.

It should be noted that the performance of such radar systems is defined by four main factors: path losses that are predominantly dominated by soil attenuation, target radar cross section, surface clutter, and buried clutter. Typically, in average loss soils a minimum metal antitank mine can be detected to a depth of cover of 40 cm, whereas a minimum metal antipersonnel mine can be detected to a depth of cover of 15 cm.

7.7.2 Vehicle-Mounted Land-Mine Detection Radar Systems

Another application of UWB radar is for vehicle-based land-mine detection, and an example of a 4-m-wide swathe radar system with a total of 32 antenna elements, which was developed for the UK Minder CAP program on behalf of the UK MoD, is shown in Figure 7.13.

FIGURE 7.13 Photograph of downward look 32-channel array system used on MINDER system (with kind permission ©ERA Technology).

The architecture of the system is based around 16 receivers each of which sequentially sample the signal incident on receive antenna elements. The transmitters are synchronized by adjacent receivers and a central master clock. The system is designed to be modular in that it can be either increased in width, or alternatively in density and up to 64 channels can be configured. The transmitter–receiver modules can be used either singly or up to $N/2$ where N is the number of antenna elements.

A recent development by Cobham Technical Services is a more compact 8-channel radar system as shown in Figure 7.14. Each radar transmit–receive channel operates as a self-contained module and is triggered by an interface board, which ensures that each pulse is transmitted in its own preassigned time slot. The data from each radar transmit–receive channel is concatenated and fed by a USB interface to a laptop computer. The pulse repetition interval of 1 μS and the sampling window of 25 ns of each module define the timing of the radar boards. This enables a theoretical maximum of 40 time slots but practically a maximum 32 time slots is available, allowing for a guard band around each.

The complete radar system includes the antenna array, radar system, laptop interconnecting cables, as well as a shaft encoder to provide position information. The rack electronics is contained within a standard format 19 in. 4U height rack-mount enclosure. The electronics requires approximately 25 W at an operating voltage of 10 to 30 Vdc. The rack comprises eight independent radar channels that are synchronized to minimize interference, drift compensated to remove the effects of environmental temperature, and controlled through an liquid-crystal display (LCD) interface on the front of the panel. The radar sampling is triggered through the use of a shaft encoder for vehicle applications or may be set to free run at a user-defined speed. Eight independent transmit-and-receive antenna pairs are mounted in a secure, ruggedized antenna enclosure. The lateral antenna spacing is approximately 18 cm, which allows

FIGURE 7.14 Complete 8-channel radar system (with kind permission ©ERA Technology).

for detection of AT and IED-sized targets. The antenna cabling length is 2.5 m, which is sufficient to allow for installation of the system on a typical four-wheel-drive vehicle. The radar has been tested on a number of platforms including autonomous vehicles and Land Rover as shown in Figure 7.15.

The radar software provides a rolling map display of the radar data. The software allows the user to change color palette and gain and also allows for recording and playback of radar data. The software will run on a 32-bit Windows platform, on an installation running 32-bit Windows XP. The display is real time and in the version shown in Figure 7.16, the operator can mark GPR detections, which then initiates a paint spray system to mark the detected target. The mark is first placed as a cross by the cursor, and when it passes the line denoting the paint array changes to a spray simulation.

In general, GPR systems for land-mine detection have a loop gain on the order of 120 dB, which sets their order of magnitude performance. Their radiated power is limited by licence restrictions and EMC considerations as well as the need to avoid detonation of certain types of fuse.

Most path losses are such that penetration is limited to 50 cm depth of cover for most GPR systems, particularly for the detection of antipersonnel land mines. At 1 GHz the total losses in typical soils mean that, in ideal conditions, detection ranges of 20 to 30 cm are feasible. It should be noted that the attenuation losses in materials rapidly increase with frequency, which means that most systems will receive frequencies in the range of 300 MHz to 1.5 GHz. The use of transmitted frequencies above 2 GHz is unlikely to provide useful performance in real-world conditions and will severely limit depth performance. The attenuation losses in materials will reduce the effectiveness of multi-look antenna arrays by effectively putting a window taper across the array. In dry soils the dielectric contrast between the soil and

FIGURE 7.15 Eight channel GPR system on trials on Land Rover (with kind permission ©ERA Technology).

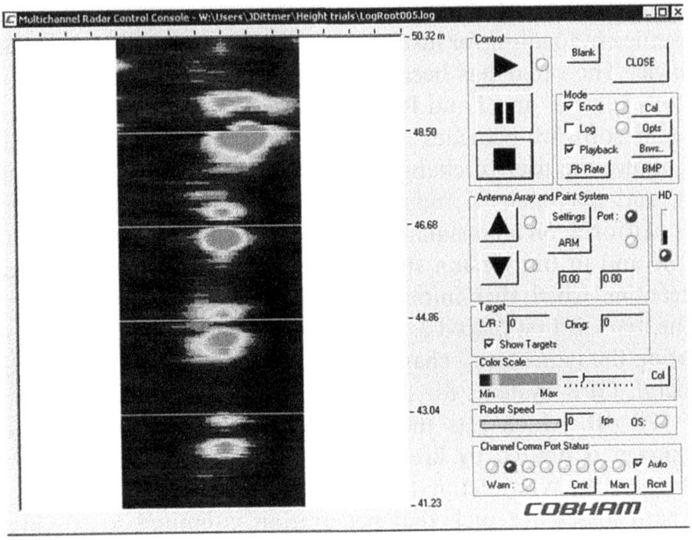

FIGURE 7.16 Laptop screen display of buried target (with kind permission ©ERA Technology).

mine reduces, and this can make the detection of mines with minimal air voids more difficult.

Most GPR systems will achieve optimum performance in terms of range when the antennas are operated in close proximity to the ground. As the antenna-to-ground spacing increases, the antenna radiation pattern results in reduction of the received signal from small targets and increased vulnerability to clutter from free space sources. Rough surfaces, ruts, potholes, and the like degrade the signal-to-clutter ratio and reduce the system performance. The angular response of mines that are tilted relative to the ground surface may not be coincident with their physical position, and this should be considered when neutralizing. Standoff SAR radar systems have limits to performance at shallow grazing angles, which constrains their forward look range to between 10 and 20 m.

7.8 RADAR FOR GENERAL SEARCH OPERATIONS

Ground-penetrating radar systems are available from a number of manufacturers, who have designs suitable for many applications. A typical example of a GPR system, the Seeker SPR is produced by U.S. Radar of New Jersey and is shown in Figure 7.17 and consists of a pair of antennas, one for transmit and one for receive, connected to the transmitter–receiver and processor and contained within a sealed enclosure, a battery and control, processor, and display

FIGURE 7.17 Seeker SPR radar (with kind permission ©US Radar).

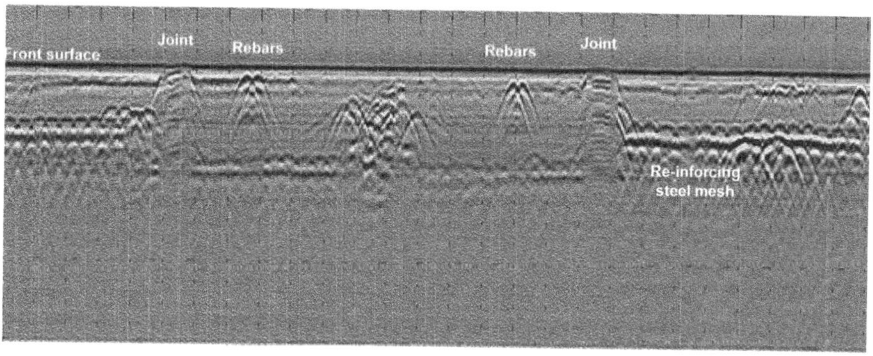

FIGURE 7.18 Typical GPR B-scan image.

unit. The wheels drive a shaft encoder that triggers data acquisition and hence the display, which is synchronized to the movement of the system. An example of the display, which takes the form of a cross section of the ground surveyed by the GPR, is shown in Figure 7.18. The horizontal scale is 10 cm per marker and the vertical scale is time in nanoseconds (51 ns). The key features are marked on the radar image.

Ground-penetrating radar has become an established technique for the detection of concealed targets, and some if not all of the applications are as follows: archaeological investigations, bridge deck analysis, forensic investigations (detecting buried bodies), pipe and cable detection, rail track and bed inspection, and road condition survey. The literature is well served with material on the research, design, and application of GPR techniques, including the signal processing, and the reader is referred to Daniels [2] for more information on the technique.

7.9 SPATIALLY MODULATED SYSTEMS

The main benefit of the spatially modulated detection system is that its spectral occupancy is lower, and it is able to detect anomalies near the surface of the material more easily than other types of modulation. Rascan Systems has developed a holographic system that can detect such anomalies and is described by Ivashov et al. [296]. The value of such a system is that with very shallow targets the range resolution of conventional impulse radars and the high levels of internal clutter caused by multiple reflections means such radars cannot detect anomalies. However, the holographic radar described by Ivashov and co-workers was able the detect anomalies in examples of the space shuttle tiles and could determine regions in which the tile bonding to the metal backing was defective. However, such systems have poorer range performance because unlike radars that use time-varying gain, holographic systems have a constant gain with range profile, and this results in weak signals at depth. In addition close-in targets shade the signal from deeper targets.

7.10 MILLIMETER-WAVE RADAR SYSTEMS

A variety of millimeter-wave imaging systems have been developed by a number of companies such as Brijot Imaging Systems who produce a passive millimeter-wave imager (Gen-2) and the Scan port, Teraview in the United Kingdom who offer spectroscopic systems, Trex Enterprises in San Diego who are developing and supplying millimeter-wave camera systems, and L3 Cyterra who produce the ProVision™ Body Scanning System, which claims to screen people to reveal and pinpoint hidden weapons, explosives, drugs, and, other contraband at a rate of 600 people per hour. These systems have the capability of detecting weapons, drugs, and explosives concealed on people.

The UK-based company QinetiQ has developed an advanced imaging system that can detect stowaways hidden inside vehicles and even weapons concealed under a person's clothing or in their baggage. The QinetiQ Millimetre Wave Camera system provides a real-time moving-image scanning capability and works by passively detecting naturally occurring radiation as it reflects off different objects.

The QinetiQ iSPO-30 Passive Imaging System (Figure 7.19) is a standoff system based on a Schmidt Cassegrain Mirror and offers a field of view of 20 in. elevation and 10 in. azimuth, with a frame time of 100 ms. It has an angular resolution of 5 mrad and a focus range of 6 to 30 m. The millimeter-wave temperature sensitivity is 4 K and operates in the 90 to 100 GHz frequency band.

FIGURE 7.19 iSPO-30 Passive Imaging System (with kind permission QinetiQ).

A typical image from a 94-GHz imager is shown in Figure 7.20 with the visual image for comparison. QinetiQ also produces the SPO-20 IED defeat millimetre wave detector as shown in Figure 7.21 to counter suicide bombers. This provides instant IED detection at ranges up to 20 m using a robust, all-weather system with an integrated visible sensor.

Smiths Detection has developed an electronic, real-time imager in a standard checkpoint layout to detect concealed weapons or explosives as shown in Figure 7.22. The patented technology is based on millimeter-wave imaging and uses an array of antenna elements covering the 2-m high panel to create a three-dimensional image of the person standing in front of it. Variations in spatial reflection coefficient from the person generate an image that reveals any kind of hidden material threat—metal, liquid, ceramic, or explosive. The system provides a clear, moving picture of the person passing in front of a flat panel, and a remote operator can check for any threat items hidden beneath clothing. The Smith "EQO" system uses active radar techniques and incorporates privacy filters such as face blurring built into the system (Figure 7.23). The equipment is designed to be incapable of storing, transmitting, or printing any images. Travelers pause to face the panel and then slowly turn around before continuing their journey. Smiths Detection indicates that future configurations should allow the traveler to pass through a series of panels without any pausing or turning. A typical image from a 94-GHz system is shown in Figure 7.20.

FIGURE 7.20 A 94-GHz system millimeter-wave image taken outdoors with visual image for comparison (with kind permission QinetiQ).

FIGURE 7.21 QinetiQ SPO-20 IED defeat sensor (with kind permission QinetiQ).

FIGURE 7.22 EQO scanning system (with kind permission Smiths Detection, Ireland).

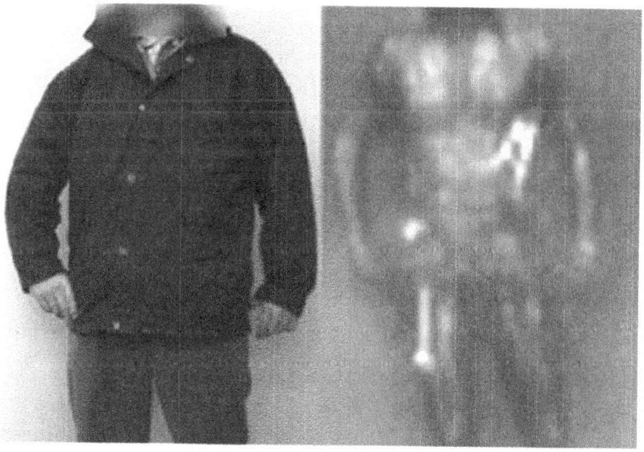

FIGURE 7.23 Images from a 94-GHz passive scanning system (with kind permission Smiths Detection, Ireland).

A novel and low-cost approach is described by Daniels and Hunt [297] in which a patented microwave search detector using noise radar technology operating in the 27 to 40-GHz range of frequencies is described. The basic system concept is based on wideband differential noise radar with a transmitted power of a few nanowatts. The total transmitted power is some 30 dB above

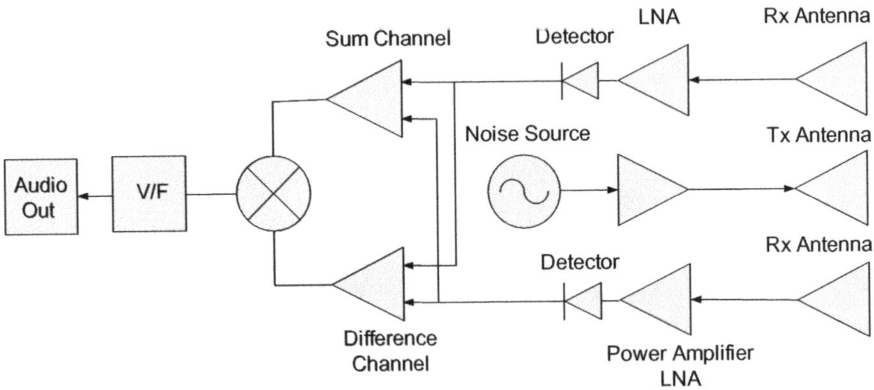

FIGURE 7.24 Noise radar for the detection of concealed targets (with kind permission ©ERA Technology).

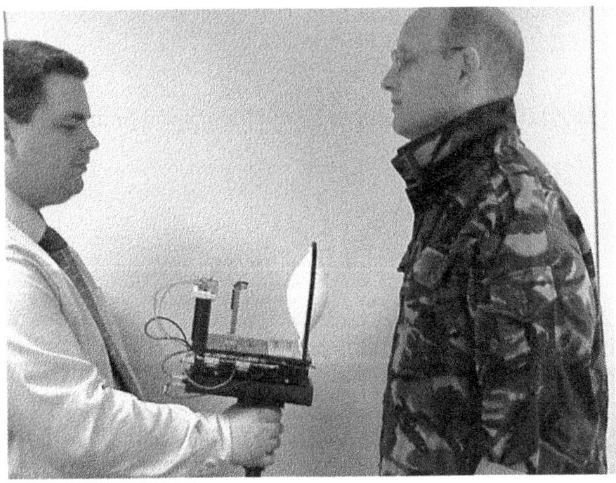

FIGURE 7.25 Prototype handheld scanner under test (with kind permission ©ERA Technology).

thermal noise (<0.05 μW). The receiver is based on a pair of low-noise amplifiers of about 40 dB gain whose outputs are rectified and then subtracted.

The noise source generates energy that illuminates the target and raises the reflected temperature difference from the subject. The easiest way to determine temperature difference is to illuminate a small area of the subject, and measure the microwave temperature of an adjacent area with two antennas. The detector is scanned over the subject, and where there is a change in reflected energy caused by a concealed object this is measured and can be heard by the operator. A prototype system under test showing the lens antenna and feeds is shown in Figure 7.25.

7.11 SUMMARY

There is a wide range of equipment for the detection of concealed targets. The equipment has been developed and optimized to meet particular targets and environments, and this has resulted in specific designs for radar systems and radiometer systems. Research and development continues to explore novel methods of improving detection capability, and the impetus given to RF technology by the telecommunications market and the availability of high-performance and low-cost digital RF components has given the designer many more options to choose from in terms of subsystem component design. Passive millimeter-wave systems have public acceptance in terms of radiation, but recently the privacy issue has made the use of these systems less acceptable, although in reality media reporting has exaggerated the quality of the imaging, which is often on a par with TV images of the 1930s. Active systems must meet stringent radiological and licensing requirements, and it is interesting to speculate whether objections to RF illumination would be so strongly held if the subject was being saved from an avalanche or recovered from earthquake debris. GPR techniques are well established for forensic and other search applications, and the developments in through-wall radar, which exploit similar technology, are providing much needed help to security and military operations.

CHAPTER 8

Summary

The market for electromagnetic sensors for the detection of concealed targets has attracted considerable interest by potential end users from the military, security, and industrial bases. Considerable investment has been made in developing products for niche markets, and there is a growing interest in the various application areas. The threat from terrorist activities has galvanized research and development in this area, but the challenge of the basic physics in what is often an uncontrolled environment with high levels of clutter still remains.

The most successful developments operate in a more controlled situation, and this may be one of the precursors to success. The opportunities are considerable, but issues of licensing, public acceptance of electromagnetic radiation, albeit at very low levels, as well as the reliability of detection and reduction of false alarms may be just as much a challenge to overcome as those related to the basic physics.

The main physical phenomena that influence the detection of concealed targets have been described. Some parameters are within the control of the designer and operator, and these relate to frequency of operation, antenna radiation patterns, polarization, and whether the system operates as radar or as a passive system. The factors that can be exploited as means of detection, recognition, and identification relate to specific indicators of the target, whether this is associated with explosives (absorption or reflection spectra) or with humans (Doppler or micro-Doppler signatures).

The difficulty of detection is increased by the intervening path losses, whether they are due to atmospheric absorbers for standoff systems or clothes or earth materials for close-in detection systems. The effect of clutter is to make

EM Detection of Concealed Targets, by David J. Daniels

detection of the target more difficult and is particularly noticeable for radar systems working through the rubble of earthquakes.

The antennas used for systems for the detection of concealed targets are defined by the frequency of operation as well as the requirements for portability and design architecture. When the radar system is operated in standoff mode, then conventional antenna designs can be employed, provided the antenna is working in its far-field mode.

For close-in detection the dielectric has a significant influence on the propagation characteristics of the antenna and must be fully considered in the design process. When the dielectric is lossy, the frequency of operation tends to be a few gigahertz, and hence the antennas are usually electrically small and consequently exhibit low gain.

All the classes of antenna discussed can be used in synthesized, FMCW, or noise-modulated radars. For time-domain systems the resistively loaded dipole, bow-tie, and TEM traveling-wave antenna are used because of the need for a linear phase response.

The NQR method is noninvasive, highly specific, and signals are generally seen in solids; thus, suspensions of materials and mixtures with other substances are often unsuitable for detection. Unlike the closely related technique of nuclear magnetic resonance (NMR), no static magnetic field is necessary, so remote materials and large volumes (the largest value known is 8000 liters) can be examined. The physics of detection is complex, and RFI as well as signals caused by magnetoacoustic and piezoelectric signals or the ringing signals from the NQR coil complicate system design.

In general, the signal-to-noise ratio increases as a function of $\omega^{3/2}$ so that materials whose resonance is below 1 MHz is more difficult to detect than materials whose resonance is above 1 MHz.

Times for detection depend on the substance and the need to average pulse sequences as well as the requirement to retune the system for different substances. All these factors have limited the widespread acceptance of NQR as a detection technique but some commercialization has taken place.

A major challenge in the design of coils for NQR applications is ensuring that the very weak signal from the sample is not masked by the decaying energy from the transmit pulse or by interference from radio transmitters. The material under investigation may be placed inside a coil (internal field detection) or outside a coil (external field detection). In the case of external fields and small samples of material, the re-radiated signal is on the order of 0.1 pT. Internal fields and large samples cause fields in the range greater than 100 pT. The design of the exciting coil and receiver is critical to achieving as high a signal-to-noise ratio as possible. These challenges still remain to be fully solved. NQR techniques have found most success where the sample can be placed inside a coil and the SNR is much higher than when the sample is external to the coil probe.

There are a considerable variety of radar systems for the detection of concealed targets. Doppler radar and harmonic radar systems offer significant improvement in signal-to-clutter performance over ranging radars that inherently detect all scatterers that appear in the main, side, and back lobes of the radar antenna. When Doppler radar is used to detect buried targets such as humans buried in earthquake rubble, the amplitude of the signal from the target may be of the same order as unwanted signals appearing in the side and back lobes, unless action is taken to reduce these. Harmonic radar has been successfully exploited for the detection at close range of nonlinear junctions and for tags on insects, but the range dependence of the harmonics and the practical difficulties of minimizing self-generated harmonics make successful implementation of standoff designs a challenge.

Spatial modulated radar relies on accurate spatial registration and recording of data prior to processing, and this limits the applicability to well-defined physical situations. The most successful modulation schemes have been amplitude and frequency modulation. Given that the inherent performance of any radar system is related to the mean power scattered from the target and detected in the receiver, the frequency modulation has an inherent advantage in performance over amplitude modulation. However, noise self-generated by the source and local oscillators, limits the inherent sensitivity that can be obtained and sweep linearity, in the case of the FMCW radar is a key parameter in terms of achievable system range resolution. Amplitude-modulated radar, therefore, still has a major role for the application. Very short ranges have led to the successful sampling receiver design, and longer-range systems use the conventional local oscillator and down-conversion schemes. Noise radar offers some benefits in terms of immunity to interference and difficulty of detection but has particular limitations in terms of the trade-offs between signal recovery and integration time, particularly where multiple targets are concerned.

Passive systems or radiometers have been extensively developed over the last decade for passenger screening at airports and access points. There are a number of companies with products in this market working at frequencies from tens of gigahertz up to submillimeter-wave frequencies. There are technical limits for such systems, and they relate to the radiometric sensitivity that can be obtained from the receiver as well as the stability of calibration of arrays of receivers. Sensitivity is predominantly defined by the inherent noise of the receiver or its noise figure, and whereas at 10 GHz noise figures of less than 1 dB are common for LNAs at 90 GHz, it is hard to achieve much better than 4 to 5 dB with current uncooled technology. The effect of the higher noise at millimeter- and submillimeter-wave frequencies is that the poor radiometric sensitivity limits the system detection range to 30 to 50 m for many products. The spatial resolution of passive systems is generally much inferior to thermal imagers because the effective number of sensing elements is low.

The enormous advantage of being able to "see" through clothing is such that research and development is being actively pursued. The goal of many researchers is to develop staring array radiometers that are self-calibrating at a low cost and with low noise. Such arrays may then be able to be electronically scanned using beam-forming techniques and could then further develop the market for passive systems.

There is a wide range of equipment for the detection of concealed targets. The equipment has been developed and optimized to meet particular targets and environments, and this has resulted in specific designs for radar systems and radiometer systems. Research and development continues to explore novel methods of improving detection capability, and the impetus given to RF technology by the telecommunications market and the availability of high-performance and low-cost digital RF components has given the designer many more options to choose from in terms of subsystem component design. Passive millimeter-wave systems have public acceptance in terms of radiation. However, recently, the privacy issue has made the use of these systems less acceptable, although in reality media reporting has exaggerated the quality of the imaging, which is often on a par with TV images of the 1930s. Active systems must meet stringent radiological and licensing requirements, and it is interesting to speculate whether objections to RF illumination would be so strongly held if the subject was being saved from an avalanche or recovered from earthquake debris. GPR techniques are well established for forensic and other search applications, and the developments in through-wall radar, which exploit similar technology, are providing much needed help to security and military operations.

The previous chapters have considered the physics and the system design considerations relevant to a variety of electromagnetic sensors and signal or image processing techniques have not been addressed. This is deliberate, as not only are these subjects in their own right, but they are well covered in the literature.

Many of the system block diagrams shown in previous chapters have shown functionality rather than hardware realization. This does not imply that such systems should be realized in component-level form and could equally well be designed using digital signal processors. Digital signal processors (DSPs) are microprocessors designed to perform digital signal processing, which is defined as the mathematical manipulation of digitally represented signals.

A variety of DSP algorithms are used in sensor applications. Traditional algorithms such as FFTs, convolutions, filtering, and the like, remain the core elements of sensor signal-processing design and implementation, and, while gradual improvements have been made in implementation and efficiency, the availability of greater performance has allowed techniques such as neural networking, time frequency analysis, and the like to be exploited in portable systems. The main signal-processing tasks comprise the following: pre-processing, detection, recognition, identification, compression, and data fusion.

Pre-processing is used in order to remove all object-independent information from the data. After basic signal conditioning and filtering, filters to reduce noise and optimize the signal-to-noise ratio (SNR) are often implemented, such as least squares or Wiener filtering. Detection algorithms detect whether a target is present in the data. Suitable algorithms often use a statistical measure such as deviation from a statistically based threshold and a likelihood ratio test to generate the decision. Constant false alarm rate (CFAR) detection methods are traditionally used in radar processing. When the data or image contains information that is specific to the wanted targets, it is analyzed to extract image-specific parameters and a range of image-processing techniques such as geometric transforms, convolution, morphological processing, wavelet transform filtering, edge finders, feature extraction, connected component shape features, image segmentation, and region growing are available for image feature processing. The parameters extracted in feature processing are input into a classifier, and a decision is made as to whether the target is likely to be a specific type. The accuracy is currently limited by processing power and the ability to transfer data at the rates required.

Other DSP techniques that are commonly used are compression techniques so that when transferring. data between systems with bandwidth limitations the intrinsic information content can be transferred. Data-intensive activities (e.g., image processing, voice) are normally compressed prior to transmission.

When multiple sensors are used for the detection or classification of an object, the sensor outputs from each can be combined to provide a better aggregate value. Methods based on Bayesian classification or statistical combination (e.g., Kalman filtering) are often used.

The hardware for digital signal and image processing is based on a variety of semiconductor integrated circuits such as general-purpose digital signal processors (DSP), application-specific integrated circuits (ASIC) to provide the main processing power, or low-cost field-programmable gate array (FPGA).

The dynamic range of DSPs and the range of numbers that must be processed in the course of an application define and classify DSPs. Most DSPs use fixed-point arithmetic because, in real-world signal processing, the additional range provided by floating point is not needed, and there is a large speed and cost benefit due to reduced hardware complexity. However, floating-point DSPs are useful in applications where a wide dynamic range is required. Floating-point DSPs can be used to reduce the cost and complexity of software development in exchange for a more expensive hardware since it is generally easier to implement algorithms in floating-point format. Each type of processor is suited to a particular range of applications. Sixteen-bit fixed-point DSPs are used for voice-grade systems, whereas 24-bit fixed-point DSPs are used for more demanding sensor applications. Image processing, 3D graphics, and scientific simulations have a generally much wider dynamic range and require 32bit floating-point processors. Several DSPs can be combined into a single system. This is called multiprocessing or parallel processing. Generally, DSPs are dedicated integrated circuits; however, DSP functionality can also be realized using field-programmable gate array chips.

An FPGA is a semiconductor device containing programmable logic components called "logic blocks" and programmable interconnects. Logic blocks can be programmed to perform the function of basic logic gates such as AND and XOR or more complex combinational functions such as decoders or simple mathematical functions. In most FPGAs, the logic blocks also include memory elements, which may be simple flip-flops or more complete blocks of memories. A hierarchy of programmable interconnects allows logic blocks to be interconnected as needed by the system designer, somewhat like a one-chip programmable breadboard. Logic blocks and interconnects can be programmed by the customer or designer, after the FPGA is manufactured, to implement any logical function—hence the name field programmable. FPGAs are usually slower than their application-specific integrated circuit (ASIC) counterparts, as they cannot handle as complex a design and draw more power. But their advantages include a shorter time to market, the ability to reprogram in the field, to fix bugs, and lower nonrecurring engineering costs. Rewriting to an FPGA is not as simple as recompiling a program to a new platform as essentially the process involves designing circuits. The potential gains in system efficiency have to be balanced against the investment needed in developing FPGA programs. For smaller designs and/or lower production volumes, FPGAs may be more cost effective than an ASIC design as the nonrecurring engineering costs to produce an ASIC can be significant.

An ASIC is an integrated circuit customized for a particular use, rather than intended for general-purpose use. As feature sizes have shrunk and design tools have improved over the years, the maximum complexity (and hence functionality) possible in an ASIC has grown from 5000 gates to over 100 million. Modern ASICs often include entire 32-bit processors, memory blocks including Flash, EEPROM, and ROM and RAM, as well as other large building blocks. Such an ASIC is often termed a system on a chip (SoC). Designers of digital ASICs use a hardware description language (HDL) to describe the functionality of ASICs.

Recent developments in the area of multifunction chips, such as the Analog Devices BlackFin™ and Texas Instruments (TI) OMAP™, which have both microprocessor and DSP functionality, have opened powerful and low-power capability for portable equipment.

REFERENCES

1. Baum, C. *Detection and Identification of Visually Obscured Targets*. UK HPA: Taylor and Francis, Inc., 1999.
2. Daniels, D. J. *Ground Penetrating Radar*. London: IET, 2004.
3. Astanin, L.Y., and Kostylev, A.A. *Ultra-Wideband Radar Mesurements Systems. IEE Radar Sonar Navigation and Avionics Series*, Vol. 7.
4. Beckner, C., and Shaheen, M. *Sensors and Homeland Security: A Market Assessment* Washington: Civitas Group, 2005.
5. Shaheen, M. *The Homeland Security Market Essential Dynamics and Trends*. Washington: Civitas Group, 2006.
6. Barami, B. Market Trends in Homeland Security Technologies. IEEE Conference on Technologies for Homeland Security. Volpe Center, Cambridge, MA, April 21–22, 2004.
7. http://www.hpa.org.uk/radiation/publications/w_series_reports/2004/nrpb_w59.htm.
8. NRPB. Electromagnetic Fields and the Risk of Cancer. Report of an Advisory Group on Non-ionizing Radiation. Doc. NRPB, 3, No. 1 (1992).
9. NRPB. Electromagnetic Fields and the Risk of Cancer. Summary of the views of the Advisory Group on Non-ionizing Radiation on epidemiological studies published since its 1992 report (23 March 1993). Doc NRPB, **4**(5), 65–69 (1993).
10. NRPB. Electromagnetic Fields and the Risk of Cancer. Supplementary report by the Advisory Group on Non-ionizing Radiation. Radiological Protection Bulletin, No. 142 (1993).
11. NRPB. Electromagnetic Fields and the Risk of Cancer. Supplementary report by the Advisory Group on Non-ionizing Radiation (12 April 1994). Doc NRPB, **5**(2), 77–81 (1994).
12. NRPB. ELF Electromagnetic Fields and Neurodegenerative Disease. Report of an Advisory Group on Non-ionizing Radiation. Doc NRPB, **12**(4), 5–24 (2001).

EM Detection of Concealed Targets, by David J. Daniels

13. NRPB. ELF Electromagnetic Fields and the Risk of Cancer. Report of an Advisory Group on Non-ionizing Radiation. Doc NRPB, **12**(1), 1–179 (2001).
14. NRPB. ELF Electromagnetic Fields and the Risk of Cancer. Report of an Advisory Group on Non-ionizing Radiation. Doc NRPB, **12**(1), 1–179 (2001).
15. NRPB. Health Effects from Radio Frequency Electromagnetic Fields. Report of an independent Advisory Group on Non-ionizing Radiation. Doc NRPB, **14**(2), 5–177 (2003).
16. HPA. Power Frequency Electromagnetic Fields, Melatonin and the Risk of Breast Cancer. Report of an independent Advisory Group on Non-ionizing Radiation. Doc HPA, RCE-1 (2006).
17. Floderus, B., Persson, T., Stenlund, C., Wennberg, A., and Knave, B. Occupational exposure to electromagnetic fields in relation to leukaemia and brain tumours: A case-control study in Sweden. *Cancer Cause Control*, **4**:465 (1993).
18. Guenél, P., Raskmark, P., Andersen, J. B., and Lynge, E. Incidence of cancer in persons with occupational exposure to electromagnetic fields in Denmark. *Br. J. Ind. Med.*, **50**:758 (1993).
19. Tynes, T., Andersen, A., and Langmark, F. Incidence of cancer in Norwegian workers potentially exposed to electromagnetic fields. *Am. J. Epidemiol.*, **136**:81 (1992).
20. Feychting, M., and Ahlbom A. Magnetic fields and cancer in people residing near Swedish high voltage power lines. Stockholm, IMM-rapport 6/92 (1992).
21. Olsen, J. H., Nielsen, A., and Schulgen, G. Residence near high-voltage facilities and risk of cancer in children. *Br. Med. J.*, **367**:891 (1993).
22. NRPB. Restrictions on Human Exposure to Static and Time Varying Electromagnetic Fields and Radiation: Scientific Basis and Recommendations for the Implementation of the Board's Statement. Doc. NRPB, 4, No. 5, 7 (1993).
23. ACGIH Annual. Report of the Committees on Threshold Limit Values and Biological Exposure Indices. Cincinnati, OH: American Conference of Governmental Industrial Hygienists, 1993.
24. ANSI. *American National Standard Safety Levels with Respect to Human Exposure to Radio Frequency Electromagnetic Fields, 300 kHz to 100 GHz* ANSI C95.1-1982. New York: American National Standards Institute, ANSI/IEEE C95.1, 1992.
25. Elder, J. A., and Cahill, D. F. (Ed.). *Biological Effects of Radio Frequency Radiation*, 1984. EPA Report No. EPA-600/8-83-026F. U.S. Environmental Protection Agency, Research Triangle Park, N.C. 27711.
26. http://www.osha.gov/SLTC/radiationionizing/standards.html.
27. http://ewh.ieee.org/soc/embs/comar/rf_mw.htm#ANSI-82.
28. http://www.emfs.info/expert_ICNIRP.asp.
29. http://www.eurogpr.org/licensing.htm.
30. http://www.g-p-r.com/.
31. Provost, F., and Kohavi, R. On applied research in machine learning. *Machine Learning*, **30**:127–132 (1998).
32. Simonson, K. M. Statistical Considerations in Designing Tests of Mine Detection Systems. Albuquerque New Mexico, 1998, SANDIA report SAND98-1769/1.

33. Voles, R. Confidence in trials of landmine detection systems. *Math. Today* (April, 2000).

34. Yarovoy, A. G., Qiu, W., Yang, B., and Aubry, P. J. Reconstruction of the Field Radiated by GPR Antenna into Ground, The Second European Conference on Antennas and Propagation ELICAP 2007, 11–16 November 2007, The IECC, Edinburgh, UK, pp. 1–6.

35. Lenler-Eriksen, H.-R., Planar Near-Field Measurements of GPR Antennas and Applications to Imaging. PhD Thesis. Technical University of Denmark, Denmark, August 2005.

36. IEEE. chap 2.4 ruck ref 11 IRE standards on radio propagation.

37. Born, M., and Wolf, E. Principles of Optics. Cambridge: Cambridge University Press, 1999.

38. Skolnik, M., *Radar Handbook*, 3rd ed. Europe: McGraw-Hill Education.

39. Ruck, G. T., Barrick, E., Stuart, W. D., and Krichbaum, C. K., Radar Cross Section Handbook, Vol. 1, New York: Plenum Press, 1970.

40. Blore, W. E., The radar cross section of ogives, double backed cones, double rounded cones and cone spheres, *IEEE Trans.*, **AP-12**:582–590 (Sept. 1964).

41. Borden, B., Radar Imaging of Airborne Targets. California: Taylor and Francis, pp. 90–94, 1999.

42. Debye, P. Polar Molecules, Chemical Catalog. New York 1929, reprinted by Dover Publications, New York, 1954.

43. Cole, K. S., and Cole, R. H., Dispersion and absorption in dielectrics. I. Alternative current characteristics. *J. Chem. Phys.*, **9**:341–351 (1941).

44. Currie, G. R. *Radar System Performance Modeling*, 2nd ed. New York: Artech House, 2005.

45. De Loor, G. P. The dielectric properties of wet materials. *IEEE Trans.*, **GE-21**: 364–369 (1983).

46. Hoekstra, P., and Delaney, A. Dielectric properties of soils at UHF and microwave frequencies. *J. Geophys. Res.*, **79**:1699–1708 (1974).

47. Hipp, J. E. Soil electromagnetic parameters as functions of frequency, soil density and soil moisture. *Proc. IEEE*, **62**(1):98–103 (1974).

48. Wang, J. R., and Schmugge, T. J. An empirical model for the complex dielectric permittivity of soil as a function of water content. *IEEE Trans.*, **GE-18**:288–295 (1980).

49. Hallikainen, M. T., Ulaby, F. T., Dobson, M. C., Elrayes, M. A., and Wu, L. K. Microwave dielectric behaviour of wet soil. Parts I and II. *IEEE Trans.*, **GE-23**(1):25–34 (1985).

50. Wobschall, D. A theory of the complex dielectric permittivity of soil containing water: The semidisperse model. *IEEE Trans.*, **GE-15**:49–58 (1977).

51. Parkhomenko, E. I. *Electrical Properties of Rocks*. New York: Plenum Press, 1967.

52. Keller, G. V. Electrical characteristics of the earth's crust. In: Wait, J. R. (Ed.). Electromagnetic Probing in Geophysics. Boulder: Golem Press, 1971.

53. Fuller, B. D., and Ward, S. K. Linear system description of the electrical parameters of rocks. *IEEE Trans.*, **GE-8**:7–17(1970).

54. Campbell, M. J., and Ulrichs, J. Electrical properties of rocks and their significance for lunar radar observations. *J. Geophys. Res.*, **74**:5867–5881 (1969).

55. Endres, A. L., and Knight, R. A theoretical treatment of the effect of microscopic fluid distribution on the dielectric properties of partially saturated rocks. Geophys. Prospecting, **40**:307–324 (1992).

56. Parkhomenko, E. I. *Electrical Properties of Rocks*. New York: Plenum Press, 1967, p. 314.

57. Peplinski N. R., Ulaby, F. T., and Dobson. M. C. Dielectric properties of soils in the 0.3–1.3 GHz range. *IEEE Trans. Geosci. Remote Sensing*. **33**(3):(1995).

58. Hallikainen, M. T., Ulaby, F. T., Dobson, M. C., Elrayes, M. A., and Wu, L. K. Microwave dielectric behaviour of wet soil. Parts I and II. *IEEE Trans.*, **GE-23**, (1):25–34 (1985).

59. Hipp, J. E. Soil electromagnetic parameters as functions of frequency, soil density and soil moisture. *Proc. IEEE*, **62**:(1):98–103 (1974).

60. Hoekstra, P., and Delaney, A. Dielectric properties of soils at UHF and microwave frequencies. *J. Geophys. Res.*, **79**:1699–1708 (1974).

61. Bhagat, P. K., and Kadaba, P. K. Relaxation models for moist soils suitable at microwave frequencies. *Mater. Sci. Eng.*, **28**:47–51 (1977).

62. De Loor, G. P. The dielectric properties of wet materials. *IEEE Trans.*, **GE-21**: 364–369 (1983).

63. King, R. W. P., and Smith, G. S. Antennas in Matter. Cambridge, MA: MIT Press, 1981.

64. Pottel, R. Dielectric properties In: Franks, F. (E.d). *Water: A Comprehensive Treatise, Vol 3: Aqueous Properties of Simple Electrolytes*. New York: Plenum Press, 1973, Chapter 8.

65. Cole, K. S., and Cole, R. H. Dispersion and absorption in dielectrics. I.Alternative current characteristics. *J.Chem. Phys.*, **9**:341–351 (1941).

66. Brussaard, G., and Watson, P. A. *Atmospheric Modelling and Millimetre Wave Propagation*. London: Chapman and Hall, 1995, Chapter 2, pp. 21–33.

67. Liebe, H. MPM—An atmospheric millimetre wave propagation model. *Int. J. Infrared Millimetre Waves*, **10**:(6):1989.

68. Crane, R. K. Prediction of atmospheric attenuation by rain. *IEEE Trans. Commun.*, **28**(9, pt1):1717–1733 (1980).

69. Appleby, R., and Wallace, B. *Standoff detection of weapons and contraband in the 100GHz to 1THz region. IEEE Trans. Antennas propagation*, **55**(11):2944–2956 (2007).

70. Goldstein, H. *Attenuation by condensed water*. In: Kerr, D.E (Ed.). *Propagation of Short Radio Waves*. New York: McGraw-Hill, 1951.

71. Knox, J. E. Millimetre wave propagation in smoke. IEEE Eascon-79 Conference Record, Washington, Vol. 2, pp. 357–361.

72. Ghobrial, S. I., Hemeidi, M. A., and Tawfig, M. E. Observations on 2 and 7.5 GHz microwave links during dust storms. *Electron. Lett.*, **23**:(1):44–45 (1987).

73. Ali, A. A., and Alhaider, M. A. Effect of Multipath Fading on Millimetre Wave Propagation. Microwaves, Antennas and Propagation. *IEE Proc.* **140**(5):343–346 (1993).

74. NRC. *Assessment of Millimeter-Wave and Terahertz Technology for Detection and Identification of Concealed Explosives and Weapons Committee on Assessment of Security Technologies for Transportation.* National Research Council, 2007; http://www.nap.edu/catalog/11826.html.

75. Goldhirsh, J. Attenuation and backscatter from a derived two-dimensional dust-storm model. *IEEE Trans. Antennas Propagation,* **49**(12):1703–1711 (2001).

76. Gabriel, C., and Gabriel, S. Compilation of the Dielectric Properties of Body Tissues at RF and Microwave Frequencies. Final Report Prepared for Afosr/Nl Bolling Afb Dc 20332-0001, June 1996, Physics Department, King's College, London.

77. Daniels, D. J., and Hunt, N. Hand-held microwave search detector. *SPIE,* April, 12–16 (2004).

78. Woodward, R. M. An analytical model of skin: Comparison with experimental results *in vivo. Proc. SPIE,* **5692-43**:241–254 (2005).

79. Ivashov, S. I., Razevig, V. V., Sheyko, A. P., and Vasilyev, I. A., Detection of Human Breathing and Heartbeat by Remote Radar, Progress in Electromagnetic Research Symposium, Pisa, Italy, March 28–31, 2004.

80. Staderini, E. M. UWB radars in medicine. *Aerospace Electronic Syst. Mag.,* **17**(1):13–18 (2002).

81. Committee on Assessment of Security Technologies for Transportation. *Assessment of Millimeter-Wave and Terahertz Technology for Detection and Identification of Concealed Explosives and Weapons,* National Materials Advisory Board Division on Engineering and Physical Sciences, National Academies Press, Washington, 2007.

82. Dickinson, J. C., Goyette, T. M., Gatesman, A. J., Joseph, C. S., Root, Z. G., Giles, R. H., Waldman, J., and Nixon, W. E. Terahertz imaging of subjects with concealed weapons. In: *Terahertz for Military and Security Applications IV,* Woolard, D. L., Hwu, R. J., Rosker, M. J., Jensen, J. O. (Ed.). Proceedings of the SPIE, Vol. 6212. Bellingham, WA: SPIE, 2006.

83. Gatesmana, A. J., Danylova, A., Goyettea, T. M., Dickinsona, J. C., Gilesa, R. H., Goodhuea, W., Waldmana, J., Nixonb, W. E., and Hoenc, W. Terahertz Behavior of Optical Components and Common Materials Technical Report. Massachusetts Univ., Lowell Submillimeter-Wave Tech Lab, Lowell, May 2006.

84. Bjarnason, J. E., Chan, T. L. J., Lee, A. W. M., Celis, M. A., and Brown, E. R. Mm wave, terahertz and mid-infrared transmission through common clothing. *App. Phys. Lett.* **85**(4):519–521 (2004).

85. AvaLAN Wireless Systems Application Note: Indoor Path Loss AvaLAN-AN001, Madison, August 2006.

86. Gatesmana, A. J., Danylova, A., Goyettea, T. M., Dickinsona, J. C., Gilesa, R. H., Goodhuea, W., Waldmana, J., Nixonb, W. E., and Hoenc, W. Terahertz Behavior of Optical Components and Common Materials Technical Report. Massachusetts Univ. Lowell Submillimeter-Wave Tech Lab, Lowell, May 2006.

87. Kemp, M. C. Millimetre wave and terahertz technology for the detection of concealed threats—a review. *Proc. SPIE,* **6402** 64020D-1 (2006).

88. Yamamoto, K., Yamaguchi, M., Miyamaru, F., Tani, M., and Hangyo, M. Non-invasive inspection of C-4 explosive in mails by terahertz time-domain spectroscopy. *Jpn. J. Appl. Phys.,* **42**:L414–L417 (2004).

89. Cook, D. J., Decker, B. K., Maislin, G., and Allen, M. G. Through container THz sensing: Applications for explosive screening. *SPIE*, **5354**:55 (2004).

90. Shen, Y. C., Taday, P. F., and Kemp, M. C. Terahertz spectroscopy of explosive materials. *SPIE*, **5619**:82–89 (2004).

91. Baker, C., Tribe, W. R., Lo, T., Cole, B. E., Chandler, S., and Kemp, M. C. People screening using terahertztechnology. *SPIE*, **5790**:1–10 (2005).

92. Liu, H.-B., Chen, Y., Bastiaans, G. J., and Zhang, X.-C. Detection and identification of explosive RDX by THz diffuse reflection spectroscopy. *Optics Express*, **14**(1):415–423 (2006).

93. Cook, D. J., Allen, M. G., Decker, B. K., Wainner, R. T., Hensley, J. M., and Kindle, H. S. Detection of High Explosives with THz Radiation, presented at Joint 30th International Conference on Infrared and Millimeter Waves and 13th International Conference on Terahertz Electronics, Williamsburg, VA, 19–23 September 2005.

94. Federici, J. F., Schulkin, B., Huang, F., and Dale, G. THz imaging and sensing for security applications—explosives, weapons and drugs. *Semicond. Sci. Technol.*, **20**:S266–S280 (2005).

95. Kemp, M. C., Taday, P. F., Cole, B. E., Cluff, J. A., Fitzgerald, A. J., and Tribe, W. R. Security applications of terahertz technology. *Proc. SPIE*, **5070**:44 (2003).

96. Balanis, C. A. Antenna Theory and Design, 2nd ed. New York: Wiley, 1997.

97. Cloude, S. *An Introduction to Electromagnetic Wave Propagation and Antennas*. London: UCL Press, 1995.

98. Collin, R. E. *Antennas and Radio Wave Propagation*. New York: McGraw-Hill, 1985.

99. Elliott, R. S. *Antenna Theory and Design*. Englewood Cliffs, NJ: Prentice Hall, 1981.

100. Jasik, H. *Antenna Engineering Handbook*. New York: McGraw-Hill, 1961.

101. Johnson, R. C. *Antenna Engineering Handbook*. New York: McGraw-Hill, 1993.

102. Kraus, J. D. *Antennas*. New York: McGraw-Hill, 1988.

103. King, R. W. P., and Smith, G. S. *Antennas in Matter*. Cambridge, MA: MIT Press, 1981.

104. Schelkunoff. S. A. *Advanced Antenna Theory*. New York: Wiley, 1952.

105. Rudge, A. W., Milne, K., Olver, A. D., and Knight, P. (Eds.). *The Handbook of Antenna Design*. Peter Peregrinus. London: Stevenage, 1982.

106. Rumsey, V. H. *Frequency Independent Antennas*. Electrical Science Series. New York: Academic Press, 1966.

107. Silver, S. (Ed.). *Microwave Antenna Theory and Design*. MIT Radiation Lab. Series, Vol. 12, New York: McGraw-Hill, 1949; reprinted by Peter Peregrinus, Stevenage, 1984.

108. Ramsey, J. F. *Fourier Transforms in Aerial Theory*. London: Marconi Company, Marconi Review, 1948.

109. Valle, S., Zanzi, L., Sgheiz, M., Lenzi, G., and Friborg, J. Ground penetrating radar antennas: Theoretical and experimental directivity functions. *IEEE Trans.* Geosci. Remote Sensing, **39**(4):749–759 (2001).

110. Junkin, G., and Anderson, A. P. Limitations in microwave holographic imaging over a lossy half space. *IEE Proc.*, **135**(4):321–329 (1988).

111. Brewitt-Taylor, C. R., Gunton, D. J., and I∼Ees, H. D. Planar antennas on a dielectric surface. *Electron. Lett.*, **17**(20):729–731 (1981).

112. Burke, G. J., Johnson, I. A., and Miller, E. K. Modelling of simple antennas near to and penetrating an interface. *Proc. IEEE*, **71**(1):174–175 (1983).

113. Rutledge, D. B., and Muha, M. S. Imaging antenna arrays. *IEEE Trans.*, **AP-30**(4):535–540 (1982).

114. Wait, J. R. Propagation of electromagnetic pulses in a homogeneous conducting earth. *Appl. Sci. Res. B*, **8**:213–253 (1960).

115. King, R. W. P., and Nu, T. T. The propagation of a radar pulse in sea water. *J. Appl. Phys.*, **73**(4):1581–1589 (1993).

116. Kappen, F. W., and Monich, C. Single Pulse Radiation from a Resistive Coated Dipole Antenna. Proceedings of Fifth International Conference on Antennas and Propagation, ICAP '87, York, UK, IEE Conf., Publ. 274, Vol. 1, 1987, pp. 90–93.

117. Randa, J., Kanda, M., and Orr, K. D. Resistively-Tapered-Dipole Electric-Held Probes up to 40 GHz. Proceedings of IEEE International Symposium on Electromagnetic Compatibility, Cherry Hill, NC, 1991, pp. 265–266.

118. Esselle, K. P., and Stuchly, S. S. A New Broadband Antenna for Transient Electromagnetic Field Measurements. Proceedings of IEEE Symposium on Antennas and Propagation—Merging Technologies for 90s, Dallas, TX, 1990, pp. 1584–1587.

119. Altshuler, E. E. The travelling-wave linear antenna. *IRE Trans.*, **AP-9**(4):324–329 (1961).

120. Wu, T. T., and King, R. W. P. The cylindrical antenna with non-reflecting resistive loading. *IEEE Trans.*, **AP-13**(5):369–373 (1961).

121. Rao, B. R. Optimised tapered resistivity profiles for wideband HF monopole—Antenna. Proceedings of the IEEE Antennas & Propagation Society Symposium, Vol. 2, 1991, pp. 711–713.

122. Nishioka, Y., Maeshima, O., Uno, T., and Adachi, S. FDTD analysis of resistor-loaded bow-tie antennas covered with ferrite-coated conducting cavity for subsurface radar. Antennas and Propagation Society. *IEEE Trans.*, **47**(6):970–977 (1999).

123. Maloney, J. G., and Smith, C. S. The Role of Resistance in Broadband, Pulse-Distortionless Antennas. Proceedings of IEEE Antennas & Propagation Society Symposium, 24–28 June 1991, Vol. 2, pp. 707–710.

124. Randa, J., Kanda, M., and Orr, K. D. Resistively-Tapered-Dipole Electric-Held Probes up to 40 GHz. Proceedings of IEEE International Symposium on Electromagnetic Compatibility, Cherry Hill, NC, 1991, pp. 265–266.

125. Morrow, I. L., Persijn, J., and van Genderen, P. Rolled edge ultra-wideband dipole antenna for GPR application. Antennas and Propagation Society International Symposium, IEEE, Vol. 3, 2002, pp. 484–487.

126. Brown, G. H., and Woodward, O. M. Experimentally determined radiation characteristics of conical and triangular antennas. *RCA Rev.*, **13**:425–452 (1952).

127. Young, M., Jonathan, D., Caldecott, R. Underground Pipe Detector. U.S. Pat. 4, 062, 010, 1977.

128. Chen, C. C. Higgins, M. B., O'Neill, K. and Detsch, R. UWB full-polarimetric horn-fed bow-tie GPR antenna for buried unexploded ordnance (UXO)

discrimination. Geoscience and Remote Sensing Symposium, 2000. Proceedings. IGARSS 2000. IEEE 2000 International, Vol. 4, 2000, pp. 1430–1432.

129. Lestari, A. A., Yarovoy, A. G., and Ligthart, L. P. Improvement of bow-tie antennas for pulse radiation. Antennas and Propagation Society International Symposium, 2002. IEEE, Vol. 4, 2002, pp. 566–569.

130. Mueller, G. E., and Tyrrell, W. A. Polyrod antennas. *Bell Syst. Tech. J.*, **26**:837–851 (1947).

131. Chen, C.-C., Rama Rao, K., and Lee, R. A tapered-permittivity rod antenna for ground penetrating radar applications. *J. Appl. Geophys.*, **47**(3–4):309–316 (2001).

132. Chen, C.-C., Rama Rao, K., and Lee R. A new ultra-wide bandwidth dielectric rod antenna for ground penetrating radar applications. *IEEE Trans. Antennas Propagation*, **51**(1):371–377, March (2003).

133. Chen, C.-C. Broadband Dielectric Probe Prototype Development for Near-Field Measurements. ElectroScience Laboratory Technical Report 743119, The Ohio State University, January 2003.

134. Diez, P. A., Chen, C.-C., and Burnside, W. D. Broadband, near-field dielectric probe. Proceeding of 2001 Antenna Measurement Techniques Association (AMTA) Symposium, Boulder, Colorado, 2001.

135. Yarovoy, A. G., Schukin, A. D., Kaploun, I. V., Ligthart, L. P. The dielectric wedge antenna. Antennas and Propagation Society. *IEEE Trans.*, **50**(10):1460–1472 (2002).

136. Venkatarayalu, N. V., Chen, C.-C., Teixeira, F. L., and Lee, R. Impedance characterization of dielectric horn antennas using FDTD. Antennas and Propagation Society International Symposium, 2002. IEEE, Vol. 4, 2002, pp. 482–485.

137. Adamiuk, G., Sturm, C., Zwick, T., and Wiesbeck, W. Dual polarised travelling wave antenna for UWB applications. ProcIRS 2008, 21–23 May 2008, Wroclaw, Poland.

138. Ilzuka, K. The travelling-wave V-antenna and related antennas. *IEEE Trans.*, **AP-15**(2):23–A43 (1967).

139. Theodorou, E. A., Gorman, M. R., Rig, R. R., and Kong, F. N. Broadband pulse optimised antenna. *IEE Proc H*, **128**(3)1:124–130.

140. Shen, H. M., King, R. W. P., and Wu, T. V-Conical antenna. *IEEE Trans.*, **AP-36**(11):1519–1525 (1988).

141. Wohlers, R. J. The CWIA, An Extremely Wide Bandwidth Low Dispersion Antenna. Abstracts of 20th Symposium, U.S. Air Force R&D Program, 1970.

142. Daniels, D. J. The use of radar in geophysical prospecting. IEE International Conference Radar 77, 1977, 25–28 October, London.

143. Evans, S., and Kong, F. N. TEM horn antenna: Input reflection characteristics in transmission. *IEE Proc. H*, **130**(6):403–409 (1993).

144. Reader, H. C., Evans, S., and Young, W. K. Illumination of a Rectangular Slot Radiator over a 3 Octave Bandwidth. Proceedings of the Fourth International Conference on Antennas and Propagation, Cap '85, Coventry, UK. IEE Conf. Publ. 248, 1985, pp. 223–226.

145. Foster, P. R., and Tun, S. M. Design and Test of Two Tem Horns for Ultra Wideband Use. IEE Colloquium on Antennas and Propagation Problems of Ultra Wideband Radar. *IEE Colloquium Digest*, 11/1–11/8, 1993.

146. Martel, C., Philippakis, M., and Daniels, D. J. Time Domain Design of a TEM Horn Antenna for Ground Penetrating Radar. AP2000 Millennium Conference on Antennas & Propagation, Davos, Switzerland, April 9–14, 2000.

147. Montoya, T. P., and Smith, G. S. Resistively-loaded Vee antennas for short-pulse ground penetrating radar. Antennas and Propagation Society International Symposium, July 21–26, 1996. *AP-S. Digest*, **3**:2068–2071 (1996).

148. Baum, C. E., and Farr, E. G. Impulse radiating antennas. In: Bertoni, H. L. et al. (Eds.). Ultra Wideband/Short-Pulse Electromagnetics, New York: Plenum Press, 1993, pp. 139–147.

149. Baum, C. E. Intermediate field of an impulse-radiating antenna. Ultra-Wideband Short-Pulse Electromagnetics, 14–19 June 1998, pp. 77–89.

150. Baum, C. E. Aperture efficiencies for IRAs. Antennas and Propagation Society International Symposium. Held in conjunction with URSI Radio Science Meeting and Nuclear EMP Meeting, IEEE, 18–25 July 1992. *Digest*, **AP-S3**:1228–1231 (1992).

151. Farr, E. G., and Baum, C. E. The radiation pattern of reflector impulse radiating antennas: Early-time response, *Sensor and Simulation*, Note 358, June, 1993.

152. Farr, E. G., and Frost, C. A. Development of a reflector IRA and a solid dielectric lens IRA, Part I: Design, predictions and construction. *Sensor and Simulation*, Note 396, April, 1996.

153. Farr, E. G., Baum, C. E., and Buchenauer, C. J. Impulse radiating antennas, Part II. In: Carin, L. et al. (Eds.). Ultra Wideband/Short-Pulse Electromagnetics, Vol. 2, New York: Plenum Press, 1995, pp. 159–178.

154. Farr, E. G. Analysis of the impulse radiating antenna. Antennas and Propagation Society International Symposium. Held in conjunction with URSI Radio Science Meeting and Nuclear EMP Meeting, IEEE, 18–25 July 1992. *Digest*, **AP-S3**: 1232–1235 (1992).

155. Farr, E. G., and Baum, C. E., Prather, W. D., and Bowen, L. H. Multifunction impulse radiating antennas: Theory and experiment. In: *Ultra-Wideband Short-Pulse Electromagnetics*, 14–19 June 1998, pp. 131–144.

156. Farr, E. G., and Bowen, L. H. Recent progress in impulse radiating antennas. Ultra Wideband Systems and Technologies, 2002. Digest of Papers 2002 IEEE Conference on 21–23 May 2002, pp. 337–340.

157. Buchenauer, C. J., Tyo, J. S., and Schoenberg, J. S. H. Aperture efficiencies of impulse radiating antennas. *Ultra-Wideband Short-Pulse Electromagnetics*, 14–19 June 1998, pp. 91–108.

158. Tyo, J. S. Optimisation of the TEM feed structure for four-arm reflector impulse radiating antennas. Antennas and Propagation, *IEEE Trans.* **49**(4):607–614 (2001).

159. Gibson, P. J. *The Vivaldi Aerial. Proceedings of the Seventh European Microwave Conference*, London: Microwave Exhibitions & Publishers, 1979, pp. 101–105.

160. Miller, E. K., and Landt, J. A. Short-pulse characteristics of the conical spiral antenna. *IEEE Trans.*, **AP-25**(3):621–626 (1977).

161. Pastol, Y., Arsavalingam, G., and Halbout, J.-M. Transient Radiation Properties of an Integrated Equiangular Spiral Antenna. Proceedings of IEEE Symposium on Antennas & Propagation—Merging Technologies for 90s, Dallas, TX, 1990, pp. 1934–1937.

162. Dyson, J. D. The equiangular spiral antenna. *IRE Trans.*, **AP-7**:181–187 (1959).

163. Morgan, T. E. Spiral antennas for ESM. *IEE Proc. F*, **132**(4):245–251 (1985).

164. Kooy, C. Impulse response of a planar sheath Equi-angular spiral antenna. *Arch. Elektron. Uebertrag. Tech.* **38**(2):89–92 (1984).

165. Deschamps, G. Impedance properties of complementary multiterminal planar structures. *IRE Trans.*, **AP-7**:5371–5378 (1959).

166. Bower, R., and Wolfe, J. J. The spiral antenna. *IRE National Convention Record*, Part J, 84–95 (1960).

167. Goldstone, L. L. Termination of a spiral antenna. *IBM Tech Disclosure Bull.* **25**(11a):5714–5715 (1983).

168. Phelan, M., Su, H, and LoVetri, J. Near field analysis of a wideband log-spiral antenna for 1–2 GHz. Canadian Conference on GPR Electrical and Computer Engineering, 2002. *IEEE CCECE 2002*. Vol. 1, 2002, pp. 336–341.

169. Kerr, J. L. Short axial length broad-band horns. *IEEE Trans.* **AP-21**:710–714 (1973).

170. Walton, K. L., and Sunberg, V. C. Broadband ridged horn design. *Microwaves J.*, **3**:96–101 (1964).

171. Daniels, D. J. Radar for Non-destructive Testing of Materials. IEE Colloquium Measurements, Modelling and Imaging for Non-Destructive Testing, London, 27 March 1991.

172. Sun, E. Y., and Rusch, W. V. T. Transient Analysis of Large Reflector Antennas under Pulse-Type Excitation. Proceedings of IEEE Antennas & Propagation Society Symposium, 1991, pp. 678–677.

173. Schelkunoff, S. A., and Friis, H. T. Antennas Theory and Practice. New York: Wiley, 1952, p. 150.

174. Tyo, J. S., Buchenauer, C. J., and Schoenberg, J. S. H. Beamforming in time-domain arrays. Antennas and Propagation Society. IEEE International Symposium, Vol. 3, 11–16 July 1999, pp. 2014–2017.

175. Rutledge, D. B., and Muha, M. S. Imaging antenna arrays. *IEEE Trans.*, **AP-30**(4):535–540 (1982).

176. Anders, R. On the Complete Calculation of the Equiangular Spiral Antenna. Proceedings of the 20th International Symposium on Antennas and Propagation, Tokyo, Japan, March 1971, pp. 19–20.

177. Smith, J. A. S. Chem. Soc. Rev., **15**:225–260 (1986).

178. Smith J. A. S. Nuclear quadrupole resonance spectroscopy, general principles. *J. Chem. Edu.*, **48**:39–45 (1971).

179. Hoult D. I., and Richards, R. E. The signal-to-noise ratio of the nuclear magnetic resonance experiment. *J. Mag. Reson.*, **24**:71–85 (1976).

180. Marino, R. A., and Klainer, S. M. Multiple Spin Echoes in Pure Quadrupole Resonance. *J. Chem. Phys.*, **67**:3388–3389 (1977).

181. Grechishkin, V. S., and Sinyavskii, H. Ya. New technologies: Nuclear quadrupole resonance as an explosive and narcotic detection technique. *Phys. UspeAhi*, **40**(4):393–406 (1977).

182. Garroway, A. N., Buess, M. L., Yesinowski, J. P., and Miller, J. B. Narcotics and explosives detection by 14N pure nuclear quadrupole resonance. *Proc. SPIE*, **2092**:318–327 (1994).

183. Garroway, A. N. *Nuclear quadrupole resonance, Appendix K.* In: MacDonald, J., and Lockwood, J. R. (Eds.). Alternatives for Landmine Detection. Rand Corporation.

184. Cantor, R. S., and Waugh, J. S. Pulsed spin locking in pure nuclear quadrupole resonance. *J. Chem. Phys.*, **73**:1054–1063 (1980).

185. Rudakov, T. N. Modification of SSFP technique for the effective detection of NQR signals. *Phys. Lett. A*, **358**(4):322–326 (2006).

186. Garroway, A. N., Miller, J. B., Zax, D. B., and Liao, M-Y. A Means for Detecting Explosives and Narcotics by Stochastic Nuclear Quadrupole Resonance (NQR). U.S. Pat. 5,608,321, March 4, 1997.

187. Garroway, A. N., Miller, J. B., Zax, D. B., and Liao, M.-Y. Method and apparatus for detecting target species have quadrupolar nuclei by stochastic nuclear quadrupole resonance. U.S. Pat. 5,608,321 March 4, 1997.

188. Buess, M. L., Garroway, A. N., and Yesinowski, J. P. Removing the effects of acoustic ringing and reducing temperature effects in the detection of explosives by NQR. U.S. Pat. 5,365,171, 1994.

189. Ostafin, M., and Nogaj, B. Temperature dependence of 14N-NQR spectra in commercial 2,4,6-trinitrotoluene (TNT). 32rd Polish Seminar on Nuclear Magnetic Resonance and Its Applications, Krakow, Poland, 1–2 Dec. 1999, *Mol. Phys. Rep.*, **29**:64–67 (1999).

190. Jakobsson, A., Mossberg, M., Rowe, M. D., and Sinith, J. A. S. Exploiting temperature dependency in the detection of NQR signals. IEEE International Conference on Acoustics, Speech, and Signal Processing, Philadelphia, PA, 18–23 March 2005, Part No.: Vol. 4

191. Grechishkin, V. S., and Grechishkina, R. V. Optimal conditions for observations of signals in an integrated NQR detector. *Russ. Phys. J.*, **44**(8):892–893(2001).

192. Hirschfield, T., Klainer, S. M. Short range remote NQR measurements. *J. Mol. Struc.* **58**:63–77 (1980).

193. Fukushima, E., and Roeder, S. *Experimental Pulse NMR: A Nuts and Bolts Approach.* Reading MA: Addison-Wesley, 1981.

194. Suits, B. H., Garroway, A. N., and Miller, J. B. Surface and gradiometer coils near a conducting body: The lift-off effect. *J. Mag. Reson.*, **135**:373–379 (1998).

195. Suits, B. H., and Garroway, A. N. Optimizing surface coils and the self-shielded gradiometer. *J. Appl. Phys.*, **94**:4170–4178 (2003).

196. Barrall G. A., Burnett, L. J., and Sheldon, A. G. Method and system for cancellation of extraneous signals in nuclear quadrupole resonance Spectroscopy: U.S. Pat. No. 6,392,408, 2002.

197. Garroway, A. N., Michael, L., Buess, J. B., Miller, B. H., Suits, A. D., Hibbs, G. A., Barrall, R. M., and Burnett, L. J. Remote sensing by nuclear quadrupole resonance. *IEEE Trans. Geosci. Remote Sensing*, **39**(6):1108–1118 (June 2001).

198. Wilker, C., McCambridge, J. D., Laubacher, D. B., Alvarez, R. L., Guo, J. S., Carter, C. F., III, Pusaten, M. A., and Schiano, J. L. HTS sensors for NQR spectroscopy. *IEEE MTT-S Digest*, Vol. 1, 143–146 (6–11 June 2004).

199. Hoult, D. I., and Richards, R. E. The signal-to-noise ratio of the nuclear magnetic resonance experiment. *J. Mag. Reson.*, **24**:71–85 (1976).

200. Peshkovsky, A. S., Forguez, J., Cerioni, L., and Pusiolab, D. J. RF probe recovery time reduction with a novel active ringing suppression circuit. *J. Mag. Reson.*, **177**(1):67–73(2005).

201. Miller, J. B., Suits, B. H., Garroway, A. N., and Hepp, M. A. Interplay among recovery time, signal, and noise. *Concepts Mag. Reson.*, **12**(3):125–136 (2000).

202. Klainer, S. M., Hirschfeld, T. B., and Marino, R. A. Fourier transform nuclear quadrupole resonance spectroscopy. In: Marshall A. G., (Ed.). Fourier, Hadamard, and Hilbert Transforms in Chemistry. New York: Plenum, 1982, pp. 147–182.

203. Tantum, S., Collins, L., Carin, L., Gorodnitsky, I., Hibbs, A., Walsh, D., Barrall, G., Gregory, D., Matthews, R., and Vierko?tter, S. Signal processing for NQR discrimination of buried land mines. *Proc. SPIE*, **3710**:474–482 (1999).

204. Tantum, S., Collins, L. et al. Signal processing for NQR discrimination of buried landmines. *Proc. SPIE*, **3710**:474–482 (1999).

205. Kercel, S. W., Dress, W. B., Hibbs, A. D., and Barrall, G. A. *Investigation of wavelet-based enhancements to nuclear quadrupole resonance explosives detectors. Proc. SPIE*, **3391**:424–434 (1998).

206. Somasundaram, S. D., Jakobsson, A., Smith, J. A. S., and Althoefer, K. A. Frequency selective detection of nuclear quadrupole resonance (NQR) spin echoes, detection and remediation. Technologies for Mines and Minelike Targets XI, Orlando, FL, 17 April 2006. *Proc. SPIE*, **6217**(1):62172D-1–12 (2006).

207. Tan, Y., Tantum, S. L., and Collins, L. M. Kalman filtering for enhanced landmine detection using quadrupole resonance. *IEEE Trans. Geosci. Remote Sensing*, **43**(7):1507–1516 (2005).

208. Miller, G. R., and Garroway, A. N. A Review of the Crystal Structures of Common Explosives. Part I: RDX, HMX, TNT, PETN, and Tetryl. NRL Memorandum Report NRL/MR/6120 01 8585, October 15, 2001.

209. Garroway, A. N., Buess, M. L., Yesinowski, J. P., Miller, J. B., and Krauss, R. A. Explosives detection by nuclear quadrupole resonance (NQR). *Proc. SPIE*, **2276**:139–149 (1994).

210. Garroway, A. N., Buess, M. L., Yesinowski, J. P., and Miller, J. B. Narcotics and explosives detection by ^{14}N pure NQR. *Proc. SPIE*, **2092**:318–327 (1993).

211. Hibbs, A. Advanced detection technology nuclear quadrupole resonance. *Quant. Mag.*,

212. Barras, J., Gaskell, M. J., Hunt, N., Jenkinson, R. I., Mann, K. R., Pedder, D. A. G., Shilstone, G. N., and Smith, J. A. S. Detection of ammonium nitrate inside vehicles by nuclear quadrupole resonance. *J. Appl. Mag. Reson.*, **25**:411–438 (2004).

213. Crowley, C., Petrov, T., Mitchell, O., Shelby, R., Ficke, L., Kumar, S., and Prado, P. A Novel Shoe Scanner Using an Open-Access Quadrupole Resonance and Metal Sensor, Technologies for Homeland Security and Homeland Defense VI, Carapezza, E. M. (Ed.). Proc. of SPIE, Vol. **6538**, 65380J (2007).

214. Blauch, A. J., Schiano, J. L., Ginsberg, M. D. Landmine detection using feedback NQR. *Proc. SPIE-Int. Soc. Opt. Eng.,*, **3710**:464–473 (1999).

215. Barrall, G. A., Burnett, L. J., Derby, K. A., Drew, A. J., Ermolaev, K. V., Huo, S., Lathrop, D. K., Petrov, T. R., Steiger, M. J., Stewart S. H., and Turner, P. J. Nuclear quadrupole resonance for landmine detection. Proceedings of the Sixth Joint International Military Sensing Symposium, 2005, Washington, 05-Jul-2005.

216. Hibbs, A. D., Barrall, G. A., Czipott, P. V., Lathrop, D. K., Lee, Y. K., Magnuson, E. E., Matthews, R., and Vierkotter, S. A. Landmine detection by nuclear quadrupole resonance. *Proc. SPIE,* **3392**:522–532 (1998).

217. Hibbs, A. D., Barrall, G. A., Czipott, P. V., Drew, A. J., Gregory, D. M., Lathrop, D. K., Lee, Y. K., Magnuson, E. E., Matthews, R., Skvoretz, D. C., Vierkotter, S. A., and Walsh, D. O. Detection of TNT and RDX landmines by stand-off nuclear quadrupole resonance. *Proc. SPIE,* **3710**:454–463 (1999).

218. Nolte M., Privalov, A., Altmann, J., Anferov, V., and Fujara, F. 1H–14N cross-relaxation in trinitrotoluene: A step toward improved landmine detection. *J. Phys. D: Appl. Phys.*, **35**:939–942 (2002), City.

219. Schubert, H., and Kuznetsov, A. Detection of Explosives and Landmines: Methods and Field Experience. St. Petersburg, Russia, 2005: Springer, 2002, **920** p. 217.

220. Miller, G. R., and Garroway, A. N. A Review of the Crystal Structures of Common Explosives, Part I: RDX, HMX,TNT, PETN, and Tetryl. NRL Memorandum Report NRL/MR/6120–01–8585, October 15, 2001.

221. Hibbs, A. D. In-situ Real Time Detection of Explosive/Chemical Compounds in Mines Using NQR. Technical Report NATICK/TR-01/007 March 2001.

222. Arakawa, M., Barabash, L. S., Barrall, G. A., Bobroff, S., Chepin, J. F., Derby, K. A., Drew, A. J., Ermolaev, K. V., Huo, S., Lathrop, D. K., Steiger, M. J., Stewart, S. H., and Turner, P. J. Advances in the engineering of quadrupole resonance landmine detection systems. *Proc. SPIE,* Vol. **5794**, 10 June 2005.

223. Haq, M. M. I. Ph.D. thesis, University of London, 1986.

224. Yesinowski, J. P., Buess, M. L., Garroway, A. N., Ziegeweid, M., and Pines, A. Detection of 14N and 35Cl in cocaine base and hydrochloride using NQR, NMR, and SQUID techniques. *Anal. Chem.*, **67**(13):2256–2263 (July 1995).

225. King, R. J. Microwave Homodyne Systems, IEE Electromagnetic Series, London, 1978.

226. Nguyen, D. P., Petrochilos, N., Host-Madsen, A., Lubecke, V., and Boric-Lubecke, O. A versatile through-the-wall Doppler radar using bss algorithms. Proc. of APS, Honolulu, Hawaili, 2007.

227. Nguyen, D., Yamada, S., Park, B.-K., Lubecke, V., Boric-Lubecke, O., and Host-Madsen, A. Noise considerations for remote detection of life signs with microwave doppler radar. *Conf. Proc. IEEE Eng. Med. Biol. Soc.*, **1**:1667–1670 (2007).

228. Bilik, I., and Tabrikian, J. Radar target classification using Doppler signatures of human locomotion models. *IEEE Trans. Aerospace Electron. Syst.*, **43**:(4)1510–1522 (2007).

229. Massagram, W., Hafner, N. M., Park, B.-K., Lubecke, V. M., Host-Madsen, A., and Boric-Lubecke, O. Feasibility of heart rate variability measurement from

quadrature Doppler radar using arctangent demodulation with DC offset compensation. *Conf. Proc. IEEE Eng. Med. Biol. Soc.*, **1**:1643–1646 (2007).

230. Aggelopoulos, E. G., Karabetsos, E., Constan tinou, and Uzunoglu, N. Mobile microwave sensor for detection of trapped human beings. *Measurement: J. Int. Measure. Conf.*, **18**(3):177–183 (1996).

231. Boric-Lubecke, O., Ong, P. W., and Lubecke, V. M. 10 GHz Doppler radar sensing of respiration and heart movement. In: Proc. of the IEEE 28th Ann. Northeast Bioeng. Conference, 2002, pp. 55–56.

232. Chen, K. M., Mirsa, D., Wang, H., Chuang, H. R., and Postow, E. An X-band microwave life detection system. *IEEE Trans. Biomed. Eng.*, **33**:697–70 (1986).

233. Droitcour, A. D., Lubecke, V. M., Lin, J., and Boric-Lubecke, O. A microwave radio for Doppler radar sensing of vital signs. *IEEE MTT-S Digest*, **1**:175–178 (2001).

234. Droitcour, A., Boric-Lubecke, O., Lubecke, V. M., and Lin, J. 0.25 μm CMOS and biCMOS single chip direct conversion Doppler radars for remote sensing of vital signs. *IEEE ISSCC Digest Tech. Papers*, **2**:348–349 (2002).

235. Lohman, B. B., Boric-Lubecke, O., Lubecke, V. M., Ong, P. W., and Sondhi, M. M. A digital signal processor for Doppler radar sensing of vital signs. 23rd Annual International Conference of the IEEE Engineering in Medicine and Biology Society, 2001, Engineering in Medicine and Biology Magazine, IEEE **21**(5):161–164, (Sept.–Oct. 2002).

236. Lubecke, V., Boric-Lubecke, O., and Beck, E. A compact low-cost add-on module for Doppler radar sensing of vital signs using a wireless communications terminal. IEEE MTT-S International Microwave Symposium, Microwave Symposium Digest, **3**:1767–1770 (2–7 June 2002).

237. Lin, J. C. Non-invasive microwave measurement of respiration. *Proc. IEEE*, **63**:1530 (1975)

238. Lin, J. C. Microwave apexcardiography. *IEEE Trans. MTT*, **27**:618–620 (1979).

239. Lin, J. C. Microwave sensing of physiological movement and volume change: A review. *Bioelectromagnetics*, **13**:557–565 (1992).

240. Park, B. K., Vergara, A., Boric-Lubecke, O., Lubecke, V., and H½st-Madsen, A. Center tracking quadrature demodulation for a Doppler radar motion detector. *IEEE Trans. Microwave Theory Tech.* pp. 1323–1326, 3–8 June 2007.

241. Xiao, Y., Jenshan, L., Boric-Lubecke, O., and Lubecke, V. M. A Ka-Band Low Power Doppler Radar System for Remote Detection of Cardiopulmonary Motion Engineering in Medicine and Biology Society, 2005. 27th Annual International Conference, 2005, pp. 7151–7154.

242. Zhou, Q., Petrochilos, N., H½st-Madsen, A., Boric-Lubecke, O., and Lubecke, V. Detection and monitoring of heartbeat using Doppler radar. *IEEE Trans. Signal Process* (2007).

243. Khlopov, G., Lutsenko, V., and Popov, I. Informative Characters in Spectral-Polarization Images of Backscattering from Surface Objects. International Radar Symposium, 2006. 24–26 May 2006, Krakow, Poland, pp. 1–4, 24–26 May 2006.

244. Araki, T., Tomoyuki, N., Araki, M., and Ono, H. A two-frequency CW radar system for short-range distance measurements. *IEE Mag.*, Japan (2008), **129**(1), January 2009.

245. Langman, A. The Design of Hardware and Signal Processing for a Stepped Frequency Ground Penetrating Radar. PhD Thesis, University of Capetown, 2002.

246. Langman, A., and Inggs, M. R. Pulse versus Stepped Frequency Continuous Wave Modulation for Ground Penetrating Radar. Proceedings of IGARSS 2001, page CDROM.Institute of Electrical and Electronic Engineers, July, 2001.

247. Inggs, M. R., and Garvin, A. D. M. A Stepped Frequency CW Ground Penetration Radar. Proceedings of the Microwave Signatures-92 Conference, Igls, Austria, July, 1992. Inst. for Radio Frequency Technology, Oberpfaffenhofen, July 1992. German Aerospace Research Establishment (DLR), pp. 1c13–1c16.

248. Iizuka, K., and Freundorfer, A. P. Detection of nonmetallic buried objects by a step frequency radar. *IEEE Proc.*, **71**(2):276–279 (1983).

249. Stickley, G. F., Noon, D. A., Cherniakov, M., and Longstaff, I. D. Gated stepped-frequency ground penetrating radar. 16 March 1999, *J. Applied Geophysics* **43**(2–4):259–269 (March 2000).

250. Ilzuka, Z., and Freundorfer, A., Step frequency radar. *J. Appl. Phys.*, **56**:2572–2583 (1984).

251. Ji-Chang, F. A new method of signal processing with high range resolution and reduced wide-bandwidth for linear-FMCW radar. *AMSE Rev.*, **14**(1):59–63 (1990).

252. Hua, L., Cooper, D. C., and Shearman, E. D. R. A prototype FMCW radar using an analogue linear frequency sweep. Proceedings of the IEE Colloquium on High Time-Bandwidth Product Waveforms in Radar and Sonar, London, 1 May 1991. *IEE Digest*, **93**:11/1–3 (1991).

253. Peebles, P. Z., and Stevens, G. H. A technique for the generation of highly linear FM pulse radar signals. *IEEE Trans. Military Electr.*, **9**:32–38 (1965).

254. Johnson, D. G., and Brooker, G. M. Wide Band, Linearization of a Millimetre-Wave, Linear Frequency Modulated Radar Employing a Surface Acoustic Wave, Delay Line Discriminator. The 2nd International Conference on Sensing Technology, November 26–28, 2007, Palmerston North, New Zealand.

255. Stove, A. G. Linear FMCW radar techniques. *IEE Proc. F*, **139**(5):343–351 (1992).

256. Dennis, P., and Gibbs, S. E. Solid-State Linear FM/CW Radar Systems—Their Promise and Their Problems, IEEE Proc. Int. Microwave Symposium, Atlanta, GA, June 1974, pp. 340–342.

257. Brooker, G. M. Understanding Millimetre Wave FMCW Radars, 1st International Conference on Sensing Technology, November 21–23, 2005, Palmerston North, New Zealand.

258. Johnson, D. G., and Brooker, G. M. Wide Band, Linearization of a Millimetre-Wave, Linear Frequency Modulated Radar Employing a Surface Acoustic Wave, Delay Line Discriminator, The 2nd International Conference on Sensing Technology, November 26–28, 2007, Palmerston North, New Zealand.

259. Morgan, D., Beasley, P. D. L., Ball, K. E., and Jones, W. P. Exploitation of direct digital synthesis for sweep generation in FMCW radar, *ARMMS Conference*, 7–8th November 2005, Newport Pagnell UK.

260. Kay, S., and Marple, S. L. Spectrum analysis: A modern perspective. *Proc. IEEE*, **69**(11):1380–1419 (1981).

261. Al-Attar, A., Daniels, D. J., and Scott, H. F. A Novel Method of Suppressing Clutter in Very Short-Range Radars RADAR-82. Institution of Electrical

Engineers, Proceedings of the International Radar Conference, London, 18–20 October 1982, pp. 419–423.

262. Flemming, M. A., Mulllins, F. H., and Watson, A. Harmonic Radar Detection Systems, Radar-77. IEE Conference, London, pp. 552–555, 25–28 October, 1977.

263. Shefer, J., Klensch, R., Kaplan, G., and Johnson, H. *Wireless World* **80**:117–121, 199–202 (May, June), 1974.

264. O'Neal, M. E., Landis, D. A., Rothwell, E., Kempel, L., and Reinhard, D. Tracking insects with harmonic radar: A case study. *Am. Entomol.*, **50**:212–218, Winter 2004.

265. Colpitts, B. G., and Boiteau, G. Harmonic radar transceiver design: miniature tags for insect tracking. *IEEE Trans. Antennas Propagation*, **52**(11):2825–2832 (2004).

266. Elsner, R. F. Vehicular variable parameter METRRA system. Report E6224, U.S. Army Mobility Equipment Research and Development Center, May 1974.

267. Bucci, O. M., Bonitatibus, A., and Pinto, I. Harmonic cross section of bistatic nonlinear responder, *Alta Frequenza*, **53**(3):172–176 (1984).

268. Harger, R. O. Harmonic radar systems for near-ground in foliage nonlinear scatterers. *IEEE Trans.*, **AES-12**:230–245 (1976).

269. Guosui, L., Hong, G., and Weimin, S. Development of random signal radars. *IEEE Trans. Aerospace Electronic Syst.*, **35**(3):770–777 (1999).

270. Dawood, M., Narayanan, R. M. Ambiguity Function of an Ultrawideband Random Noise Radar. IEEE Antennas and Propagation Society International Symposium, 2000, Salt Lake City, UT, 2000, vol. 4, pp. 2142–2145.

271. Xu, X., and Narayanan, R. M., Range sidelobe suppression technique for coherent ultra wide-band random noise radar imaging, *IEEE Trans. Antennas Propagation*, **49**(12):1836–1842 (2001).

272. Axelsson, S. R. J. Noise radar using random phase and frequency modulation, *IEEE Trans. Geosci. Remote Sensing*, **42**(11):2370–2384 (2004).

273. Axelsson, S. R. J. Suppression of Noise Floor and Dominant Reflectors in Random Noise Radar. Radar Symposium, 2006, Krakow, Poland, 24–26 May 2006, pp. 1–4.

274. Gray, D. A. Multi-channel Noise Radar. Radar Symposium, 2006, Krakow, Poland, 24–26 May 2006.

275. Daniels, D. J. *Ground Penetrating Radar*. London: IET, 2004.

276. Sachs, J., Kmec, M., Herrmann, R., Peyerl, P., and Rauschenbach, P. An Ultra-Wideband Pseudo-Noise Radar Family Integrated in SiGe. Radar Symposium, 2006. Krakow, Poland, 24–26 May 2006.

277. Pasmurov, A. J., and Zimoviev, J. S. *Radar Imaging and Holography*. London: IET, 2005.

278. Osumi, N., and Ueno, K. Microwave holographic imaging of underground objects. *IEEE Trans. Antennas Propagation*, **AP-33**:152–159 (1985).

279. Ivashov, S. I., Razevig, V. V., Sheyko, A. P., and Vasilyev, I. A. Application of Passive and Active Radars for Detection and Classification of Land Mines. *Proceedings of the Progress in Electromagnetics Research Symposium*, PIERS 2004, Pisa, Italy, 28–31 March 2004, pp. 93–96.

280. Popov, A. V., Kopeikyn, V. V., Vinogradov, V. A., and Zapunidi, S. A. Reconstruction Algorithms and Experiments with a Prototype of Holographic Subsurface Radar. International Conference on Antenna Theory and Techniques, 9–12 September, 2003, Sevastopol, Ukraine, pp. 561–563.

281. Jones, R. V. *Most Secret War*. Hodder and Stoughton: Coronet Books, 1978.

282. Klein L. A. *Millimeter Wave and Infra-red Multi-sensor Design and Signal Processing*. Artech House, 19xx, Table 6–2, p. 202.

283. Timms, G. P., Bunton, J. D., Brothers, M. L., and Archer, J. W. 190 GHz Millimetre-wave Imaging using MMIC-based Heterodyne Receivers. International Conference on Wireless Broadband and Ultra Wideband Communications, 27–30 Aug. 2007, pp. 32–34.

284. Greneker, E. F. Radar Sensing of Heartbeat and Respiration at a Distance with Applications of the Technology. IEE RADAR 97 Edinburgh, UK, 14–16 Oct. 1997, pp. 150–154.

285. Greneker, E. F. Radar Sensing of Heartbeat and Respiration at a Distance with Security Applications, Proceedings of SPIE, Radar Sensor Technology II, Vol. 3066, Orlando, FL, April, 1997, pp. 22–27.

286. Lin, J., and Li, C. Wireless Non-Contact Detection of Heartbeat and Respiration Using Low-Power Microwave Radar Sensor. Microwave Conference, 2007. 11–14 Dec. 2007, pp. 1–4.

287. Xiao, Y., Li, C., and Lin, J. A portable noncontact heartbeat and respiration monitoring system using 5-GHz. *Radar Sensors J.*, **7**(7):1042–1043 (2007).

288. Ivashov, S. I., Razevig, V. V., Sheyko, A. P., and Vasilyev, I. A. Detection of Human Breathing and Heartbeat by Remote Radar. Progress in Electromagnetics Research Symposium *(*PIERS 2004*)*, Pisa, Italy, March 28–31, 2004.

289. Nezirovic, A., Tesfay, S., Valavan, S. A., and Yarovoy, A. Experimental Study on Human Breathing Cross Section Using UWB Impulse Radar. Proceedings of the 5th European Radar Conference, 2008, Amsterdam, October.

290. Yarovoy, A. G., Ligthart, L. P., Matuzas, J., and Levitas, B. UWB Radar for Human Being Detection. *IEEE Aerospace Electr. Syst. Mag.*, **23**(5):36–40 (2008).

291. Chen, K.-M., Huang, Y., Zhang, J., and Norman, A. Microwave life-detection systems for searching human subjects under earthquake rubble or behind barrier. *IEEE Trans. Biomed. Eng.*, **47**(1):105–114 (2000).

292. Zhou, Q., Liu, J., Host-Madsen, A., Boric-Lubecke, O., and Lubecke, V. Detection of Multiple Heartbeats Using Doppler Radar. 2006 IEEE International Conference on Acoustics, Speech and Signal Processing, **2**(14–19):II-1160–II-1163 (2006).

293. Borek, S. E. An overview of through the wall surveillance for homeland security. Applied Imagery and Pattern Recognition Workshop, 2005. Proceedings. 34th, 19–21 Oct. 2005.

294. Daniels, D. J., and Curtis, P. MINEHOUND trials in Bosnia, Angola and Cambodia. Proceedings of the SPIE Defense and Security Conference 2006, 17–23 April, 2006, Orlando, FL.

295. Doheny, R. C., Burke, S., Cresci, R., Ngan, P., Walls, R., and Chernoff, J. Handheld Standoff Mine Detection System (HSTAMIDS) field evaluation in Namibia. Proceedings of the SPIE Defense and Security Conference, 16–21 April 2006, Orlando, FL.

296. Ivashov, S., Razevig, V., Sheyko, A., Vasilyev, I., Zhuravlev, A., and Bechtel, T. Holographic Subsurface Radar Technique and Its Applications. 12th International Conference on Ground Penetrating Radar, June 16–19, 2008, Birmingham, UK.

297. Daniels, D. J., and Hunt, N. Hand-held microwave search detector. Proceedings of the SPIE Defense and Security Conference SPIE 2004, 12–16 April, Orlando, FL.

298. Frost, and Sullivan. European Homeland Security. A Market Opportunity Analysis, July 2005.

299. Ground Penetrating Radar. 2nd Edition, IEE ISBN: 0-86341-360-9, July 2004, p. 440.

300. Hanai, T. Dielectric theory on the interfacial polarisation for two phase mixtures, Bulletin Inst. Chem. Res. Kyoto University, 39, pp. 314–368 (1961).

301. Smith, G. E., Woodbridge, K., Baker, C.J. Micro-Doppler Signature Classification. CIE International Conference on Radar, Shanghai, China (2006).

302. Griffiths, H. D. Phase and linearity errors in FM radar, Radar 89 Conference, Paris, pp. 103–106, 24–28 April 1989.

303. Griffiths, H. D. New ideas in FM radar, Electronics and Communication Engineering Journal, **2**(5):185–194 (October 1990).

Index

Printed and bound by CPI Group (UK) Ltd, Croydon, CR0 4YY
16/09/2026
14963459-0003